Numbers, Groups and Codes

Second Edition

Numbers, Groups and Codes

Second Edition

J. F. HUMPHREYS
Senior Fellow in Mathematics, University of Liverpool

M. Y. PREST
Professor of Mathematics, University of Manchester

CAMBRIDGE UNIVERSITY PRESS
Cambridge, New York, Melbourne, Madrid, Cape Town, Singapore, São Paulo, Delhi

Cambridge University Press
The Edinburgh Building, Cambridge CB2 8RU, UK

Published in the United States of America by Cambridge University Press, New York

http://www.cambridge.org
Information on this title: www.cambridge.org/9780521540506

© Cambridge University Press 2004

First published 2004
Fifth printing 2008

Printed in the United Kingdom at the University Press, Cambridge

A catalogue record for this publication is available from the British Library

ISBN 978-0-521-54050-6 paperback

To Sarah, Katherine and Christopher *J. F. Humphreys*
To the memory of my parents *M. Y. Prest*

Contents

Preface to first edition

This book arose out of a one-semester course taught over a number of years both at the University of Notre Dame, Indiana, and at the University of Liverpool. The aim of the course is to introduce the concepts of algebra, especially group theory, by many examples and to relate them to some applications, particularly in computer science. The books which we considered for the course seemed to fall into two categories. Some were too elementary, proceeded at too slow a pace and had far from adequate coverage of the topics we wished to include. Others were aimed at a higher level and were more comprehensive, but had correspondingly skimpy presentation of the material. Since we could find no text which presented the material at the right level and in a way we felt appropriate, we prepared our own course notes: this book is the result. We have added some topics which are not always treated in order to increase the flexibility of the book as the basis for a course. The material in the book could be covered at an unhurried pace in about 48 lectures; alternatively, a 36-hour unit could be taught, covering Chapters 1, 2 (not Section 2.4), 4 and 5.

Preface to second edition

We have prepared this second edition bearing in mind the fact that students studying mathematics at university, at least in the UK, are less well prepared than in the past. We have taken more time to explain some points and, in particular, we have not assumed that students are comfortable reading formal statements of theorems and making sense of their proofs. Especially in the first chapter, we have added many comments designed to help the reader make sense of theorems and proofs. The more 'mathematically sophisticated' reader may, of course, read quickly through these comments. We have also added a few more straightforward exercises at the ends of some sections.

Two major changes in content have been made. In Chapter 3 we have removed material (some propositional logic, Boolean algebras and Karnaugh maps) around Boolean algebras. We have retained most of the material on propositional logic, added a section designed to help students deal with quantifiers and added a further section on proof strategies. The emphasis of this chapter is now on the use of logic within mathematics rather than on the Boolean structure behind propositional logic.

The second major change has been the inclusion of a new chapter, Chapter 6, on the algebra of polynomials. We emphasise the similarity with the arithmetic of integers, including the usefulness of the notion of congruence class, and we show how polynomials are used in constructing cyclic codes.

Introduction

'A group is a set endowed with a specified binary operation which is associative and for which there exist an identity element and inverses.' This, in effect, is how many books on group theory begin. Yet this tells us little about groups or why we should study them. In fact, the concept of a group evolved from examples in number theory, algebra and geometry and it has applications in many contexts. Our presentation of group theory in this book reflects to some extent the historical development of the subject. Indeed, the formal definition of an abstract group does not occur until the fourth chapter. We believe that, apart from being more 'honest' than the usual presentation, this approach has definite pedagogic advantages. In particular, the student is not presented with a seemingly unmotivated abstract definition but, rather, sees the sense of the definition in terms of the previously introduced special cases. Moreover, the student will realise that these concepts, which may be so glibly presented, actually evolved slowly over a period of time.

The choice of topics in the book is motivated by the wish to provide a sound, rigorous and historically based introduction to group theory. In the sense that complete proofs are given of the results, we do not depart from tradition. We have, however, tried to avoid the dryness frequently associated with a rigorous approach. We believe that by the overall organisation, the style of presentation and our frequent reference to less traditional topics we have been able to overcome this problem. In pursuit of this aim we have included many examples and have emphasised the historical development of the ideas, both to motivate and to illustrate. The choice of applications is directed more towards 'finite mathematics' and computer science than towards applications arising out of the natural sciences.

Group theory is the central topic of the book but the formal definition of a group does not appear until the fourth chapter, by which time the reader will have had considerable practice in 'group theory'. Thus we are able to present the idea of a group as a concept that unifies many ideas and examples which the reader already will have met.

One of the objectives of the book is to enable the reader to relate disparate branches of mathematics through 'structure' (in this case group theory) and hence to recognise patterns in mathematical objects. Another objective of the book is to provide the reader with a large number of skills to acquire, such as solving linear congruences, calculating the sign of a permutation and correcting binary codes. The mastery of straightforward clearly defined tasks provides a motivation to understand theorems and also reveals patterns. The text has many worked examples and contains straightforward exercises (as well as more interesting ones) to help the student build this confidence and acquire these skills.

The first chapter of the book gives an account of elementary number theory, with emphasis on the additive and multiplicative properties of sets of congruence classes. In Chapter 2 we introduce the fundamental notions of sets, functions and relations, treating formally ideas that we have already used in an informal way. These fundamental concepts recur throughout the book. We also include a section on finite state machines. Chapter 3 is an introduction to the logic of mathematical reasoning, beginning with a detailed discussion of propositional logic. Then we discuss the use of quantifiers and we also give an overview of some proof strategies. The later chapters do not formally depend on this one. Chapter 4 is the central chapter of the book. We begin with a discussion of permutations as yet another motivation for group theory. The definition of a group is followed by many examples drawn from a variety of areas of mathematics. The elementary theory of groups is presented in Chapter 5, leading up to Lagrange's Theorem and the classification of groups of small order. At the end of Chapter 5, we describe applications to error-detecting and error-correcting codes. Chapter 6 introduces the arithmetic of polynomials, in particular the division algorithm and various results analogous to those in Chapter 1. These ideas are applied in the final section, which depends on Chapter 5, to the construction of cyclic codes.

Every section contains many worked examples and closes with a set of exercises. Some of these are routine, designed to allow the reader to test his or her understanding of the basic ideas and methods; others are more challenging and point the way to further developments.

The dependences between chapters are mostly in terms of examples drawn from earlier material and the development of certain ideas. The main dependences are that Chapter 5 requires Chapter 1 and the early part of Section 4.3

and, also, the examples in Section 4.3 draw on some of Chapter 1 (as well as Sections 4.1 and 4.2).

The material on group theory could be introduced at an early stage but this would not be in the spirit of the book, which emphasises the development of the concept. The formal material of the book could probably be presented in a book of considerably shorter length. We have adopted a more leisurely presentation in the interests of motivation and widening the potential readership.

We have tried to cater for a wide range in ability and degree of preparation in students. We hope that the less well prepared student will find that our exposition is sufficiently clear and detailed. A diligent reader will acquire a sound basic knowledge of a branch of mathematics which is fundamental to many later developments in mathematics. All students should find extra interest and motivation in our relatively historical approach. The better prepared student also should derive long-term benefit from the widening of the material, will discover many challenging exercises and will perhaps be tempted to develop a number of points that we just touch upon. To assist the student who wishes to learn more about a topic, we have made some recommendations for further reading.

Changes in teaching and examining mathematics in secondary schools in the UK have resulted in first-year students of mathematics having rather different skills than in the past. We believe that our approach is well suited to such students. We do not assume a great deal of background yet we do not expect the reader to be an uncritical and passive consumer of information.

One last word: in our examples and exercises we touch on a variety of further developments (for example, normal subgroups and homomorphisms) that could, with a little supplementary material, be introduced explicitly.

Advice to the reader

Mathematics cannot be learned well in a passive way. When you read this book, have paper and pen(cil) to hand: there are bound to be places where you cannot see all the details in your head, so be prepared to stop reading and start writing. Ideally, you should proceed as follows. When you come to the statement of a theorem, pause before reading the proof: do you find the statement of the result plausible? If not, why not? (try to disprove it). If so, then why is it true? How would you set about showing that it is true? Write down a sketch proof if you can: now try to turn that into a detailed proof. Then read the proof we give.

Exercises   The exercises at the end of each section are not arranged in order of difficulty, but loosely follow the order of presentation of the topics. It is essential that you should attempt a good portion of these.

Understanding the proofs of the results in this book is very important but so also is doing the exercises. The second-best way to check that you understand a topic is to attempt the exercises. (The best way is to try to explain it to someone else.) It may be quite easy to convince yourself that you understand the material: but attempting the relevant exercises may well expose weak points in your comprehension. You should find that wrestling with the exercises, particularly the more difficult ones, helps you to develop your understanding. You should also find that exercises and proofs illuminate each other.

Proofs   Although the emphasis of this text is on examples and applications, we have included proofs of almost all the results that we use. Since students often find difficulty with formal proofs, we will now discuss these at some length. Attitudes towards the need for proofs in mathematics have changed over the centuries.

The first mathematics was concerned with computations using particular numbers, and so the question of proof, as opposed to correctness of a computation, never arose. Later, however, in arithmetic and geometry, people saw patterns and relationships that appeared to hold irrespective of the particular numbers or dimensions involved, so they began to make general assertions about numbers and geometrical figures. But then a problem arose: how may one be certain of the truth of a general assertion? One may make a general statement, say about numbers, and check that it is true for various particular cases, but this does not imply that it is always true.

To illustrate. You may already have been told that every positive integer greater than 1 is a product of primes, for instance $12 = 2 \cdot 2 \cdot 3, 35 = 5 \cdot 7$, and so on. But since there are infinitely many positive integers it is impossible, by considering each number in turn, to check the truth of the assertion for every positive integer. So we have the assertion: 'every positive integer greater than 1 is a product of primes'. The evidence of particular examples backs up this assertion, but how can we be justifiably certain that it is true?

Well, we may give a proof of the assertion. A proof is a sequence of logically justified steps which takes us from what we already know to be true to what we suspect (and, after a proof has been found, know) to be true.

It is unreasonable to expect to conjure something from nothing, so we do need to make some assumptions to begin with (and we should also be clear about what we mean by a valid logical deduction). In the case of the assertion above, all we need to assume are the ordinary arithmetic properties of the integers, and the principle of induction (see Section 1.2 for the latter). It is also necessary to have defined precisely the terms that we use, so we need a clear definition of what is meant by 'prime'. We may then build on these foundations and construct a proof of the assertion. (We give one on p. 28.)

It should be understood that current mathematics employs a very rigorous standard of what constitutes a valid proof. Certainly what passed for a proof in earlier centuries would often not stand up to present-day criteria. There are many good reasons for employing such strict criteria but there are some drawbacks, particularly for the student.

A formal proof is something that is constructed 'after the event'. When a mathematician proves a result he or she will almost certainly have some 'picture' of what is going on. This 'picture' may have suggested the result in the first place and probably guided attempts to find a proof. In writing down a formal proof, however, it often is the case that the original insight is lost, or at least becomes embedded in an obscuring mass of detail.

Therefore one should not try to read proofs in a naive way. Some proofs are merely verifications in which one 'follows one's nose', but you will probably be able to recognise such a proof when you come across one and find no great trouble with it – provided that you have the relevant definitions clear in your mind and have understood what is being assumed and what is to be proved. But there are other proofs where you may find that, even if you can follow the individual steps, you have no overview of the structure or direction of the proof. You may feel rather discouraged to find yourself in this situation, but the first thing to bear in mind is that you probably will understand the proof sometime, if not now, then later. You should also bear in mind that there is some insight or idea behind the proof, even if it is obscured. You should therefore try to gain an overview of the proof: first of all, be clear in your mind about what is being assumed and what is to be proved. Then try to identify the key points in the proof – there are no recipes for this, indeed even experienced mathematicians may find difficulty in sorting out proofs that are not well presented, but with practice you will find the process easier.

If you still find that you cannot see what is 'going on' in the proof, you may find it helpful to go through the proof for particular cases (say replacing letters with numbers if that is appropriate). It is often useful to ignore the given proof (or even not to read it in the first place) but to think how *you* would try to prove the result – you may well find that your idea is essentially the same as that behind the proof given (or is even better!).

In any event, do not allow yourself to become 'stuck' at a proof. If you have made a serious attempt to understand it, but to little avail, then *go on*: read through what comes next, try the examples, and maybe when you come back to the proof (and you should make a point of coming back to it) you will wonder why you found any difficulty. Remember that if you can do the 'routine' examples then you are getting something out of this text: understanding (the ideas behind) the proofs will deepen your understanding and allow you to tackle less routine and more interesting problems.

Background assumed   We have tried to minimise the prerequisites for successfully using this book. In theory it would be enough to be familiar with just the basic arithmetic and order-related properties of the integers, but a reader with no more preparation than this would, no doubt, find the going rather tough to begin with. The reader that we had in mind when writing this book has also seen a bit about sets and functions, knows a little elementary algebra and geometry, and does know how to add and multiply matrices. A few examples and exercises refer to more advanced topics such as vectors, but these may safely be omitted.

1 Number theory

This chapter is concerned with the properties of the **set of integers** $\{\ldots, -2, -1, 0, 1, 2, \ldots\}$ under the arithmetic operations of addition and multiplication. We shall usually denote the set of integers by $\mathbb{Z}$. We shall assume that you are acquainted with the elementary arithmetical properties of the integers. By the end of this chapter you should be able to solve the following problems.

1. What are the last two digits of 3^{1000}?
2. Can every integer be written as an integral linear combination of 197 and 63?
3. Show that there are no integers x such that $x^5 - 3x^2 + 2x - 1 = 0$.
4. Find the smallest number which when divided by 3 leaves 2, by 5 leaves 3 and by 7 leaves 2. (This problem appears in *Sūn tzĭ suàn jīng* (*Master Sun's Arithmetical Manual*) which was written around the fourth century.)
5. How may a code be constructed which allows anyone to encode messages and send them over public channels, yet only the intended recipient is able to decode the messages?

1.1 The division algorithm and greatest common divisors

We will assume that the reader is acquainted with the elementary properties of the order relation '$\leq$' on the set $\mathbb{Z}$. This is the relation 'less than or equal to' which allows us to compare any two integers. Recall that, for example, $-100 \leq 2$ and $3 \leq 3$. The following property of the set $\mathbb{P} = \{1, 2, \ldots\}$ of **positive integers** is important enough to warrant a special name.

Well-ordering principle   Any non-empty set, X, of positive integers has a smallest element (meaning an element which is less than or equal to every member of the set X).

You are no doubt already aware of this principle. Indeed you may wonder why we feel it necessary to state the principle at all, since it is so 'obvious'. It is, however, as you will see, a key ingredient in many proofs in this chapter. An equivalent statement is that one cannot have an unending, strictly decreasing, sequence of positive integers.

Note that the principle remains valid if we replace the set of positive integers by the set $\mathbb{N} = \{0, 1, 2, \ldots\}$ of **natural numbers**. But the principle fails if we replace $\mathbb{P}$ by the set, $\mathbb{Z}$, of all integers or, for a different kind of reason, if we replace $\mathbb{P}$ by the set of positive rational numbers (you should stop to think why). We use $\mathbb{Q}$ to denote the set of all **rational numbers** (fractions).

A typical use of the well-ordering principle has the following shape. We have a set X of positive integers which, for some reason, we know is non-empty (that is, contains at least one element). The principle allows us to say 'Let k be the least element of X'. You will see the well-ordering principle in action in this section.

The well-ordering principle is essentially equivalent to the method of proof by mathematical induction. That method of proof may take some time to get used to if it is unfamiliar to you, so we postpone mathematical induction until the next section.

The proof of the first result, Theorem 1.1.1, in this section is a good example of an application of the well-ordering principle. Look at the statement of the result now. It may or may not be obvious to you what the theorem is 'really saying'. Mathematical statements, such as the statement of 1.1.1, are typically both general and concise. That makes for efficient communication but a statement which is concise needs thought and time to draw out its meaning and, when faced with a statement which is general, one should always make the effort (in this context, by plugging in particular values) to see what it means in particular cases.

In this instance we will lead you through this process but it is something that you should learn to do for yourself (you will find many opportunities for practice as you work through the book).

The first sentence, 'Let a and b be natural numbers with $a > 0$', invites you to choose two natural numbers, call one of them a and the other b, but make sure that the first is strictly positive. We might choose $a = 175, b = 11$.

The second sentence says that there are natural numbers, which we will write as q and r, such that $0 \leq r < a$ and $b = aq + r$. The first statement, $0 \leq r < a$, says that r is strictly smaller than a (the '$0 \leq r$' is redundant since any natural number has to be greater than or equal to 0, it is just there for emphasis).

The second statement says that b is an integer multiple of a, plus r.

With our choice of numbers the second statement becomes: 'There are natural numbers q, r such that $0 \le r < 175$ and $11 = 175q + r$'. In other words, we can write 11 as a non-negative multiple of 175, plus a non-negative number which is strictly smaller than 175. But that is obvious: take $q = 0$ and $r = 11$ to get $11 = 175 \cdot 0 + 11$.

You would be correct in thinking that there is more to 1.1.1 than is indicated by this example! You might notice that 1.1.1 says more if we take $b > a$. So let us try with the values reversed, $a = 11, b = 175$. Then 1.1.1 says that there are natural numbers q, r with $r < 11$ such that $175 = 11q + r$. How can we find such numbers q, r? Simply divide 175 by 11 to get a quotient (q) and remainder (r): $175 = 11 \cdot 15 + 10$, that is $q = 15, r = 10$.

So the statement of 1.1.1 is simply an expression of the fact that, given a pair of positive integers, one may divide the first into the second to get a quotient and a remainder (where we insist that the remainder is as small as possible, that is, strictly less than the first number).

Now you should read through the proof to see if it makes sense. As with the statement of the result we will discuss (after the proof) how you can approach such a proof in order to understand it: in order to see 'what is going on' in the proof.

Theorem 1.1.1 (Division Theorem) *Let a and b be natural numbers with $a > 0$. Then there are natural numbers q, r with $0 \le r < a$ such that:*

$$b = aq + r \qquad q = 0 \quad r = b$$
$$b = b$$

*(r is the **remainder**, q the **quotient** of b by a).*

Proof If $a > b$ then just take $q = 0$ and $r = b$. So we may as well suppose that $a \le b$.

Consider the set of non-negative differences between b and integer multiples of a: *reordering* $r = b - aq$

$D = \{b - ak : b - ak \ge 0 \text{ and } k \text{ is a natural number}\}$.

(If this set-theoretic notation is unfamiliar to you then look at the beginning of Section 2.1.)

This set, D, is non-empty since it contains $b = b - a \cdot 0$. So, by the well-ordering principle, D contains a least element $r = b - aq$ (say). If r were not strictly less than a then we would have $r - a \ge 0$, and therefore

$$r - a = (b - aq) - a = b - a(q + 1). \qquad b - qa - a$$
$$b - a(q+1)$$

So $r - a$ would be a member of D strictly less than r, contradicting the minimality of r.

Hence r does satisfy $0 \leq r < a$; and so r and q are as in the statement of the theorem. $\square$

For example, if $a = 3$ and $b = 7$ we obtain $q = 2$ and $r = 1$: we have $7 = 3 \cdot 2 + 1$. If $a = 4$ and $b = 12$ we have $q = 3$ and $r = 0$: that is $12 = 4 \cdot 3 + 0$.
  The symbol '$\square$' above marks the end of a proof.

Comments on the proof   Let us pull the above proof apart in order to see how it works.

You might recognise the content of the first sentence from the discussion before 1.1.1: it is saying that if $a > b$ then there is nothing (much) to do – we saw an example of that when we made the choice $a = 175$, $b = 11$. The next sentence says that we can concentrate on the main case where $a \leq b$.

The next stage, the introduction of the set D, certainly needs explanation. Before you read a proof of any statement you should (make sure you understand the statement! and) think how you might try to prove the statement yourself. In this case it is not so obvious how to proceed: you know how to divide any one number into another in order to get a quotient and a remainder, but trying to express this formally so that you can prove that it always works could be quite messy (though it is possible). The proof above is actually a very clever one: by focussing on a well chosen set it cuts through any messy complications and gives a short, elegant path to the end. So to understand the proof we need to understand what is in the set D.

Now, one way of finding q and r is to subtract integer multiples of a from b until we reach the smallest possible non-negative value. The definition of the set D is based on that idea. That definition says that the typical element of D is a number of the form $b - ak$, that is, b minus an integer multiple of a (well, in the definition k is supposed to be non-negative but that is not essential: we are after the *smallest* member of D and allowing k to be negative will not affect that). In other words, D is the set of non-negative integers which may be obtained by subtracting a non-negative multiple of a from b (so, in our example, D would contain numbers including 175 and $98 = 175 - 7 \cdot 11$).

What we then want to do is choose the least element of D, because that will be a number of the form $b - ak$ which is the smallest possible (without dropping to a negative number). The well-ordering principle guarantees that D, a set of natural numbers, has a smallest element, but only if we first check that D has at least one element. But that is obvious: b itself is in D.

So now we have our least element in D and, in anticipation of the last line of the proof, we write it as r. Of course, being a member of D it has the form $r = b - aq$ for some q (again, in anticipation of how the remainder of the proof will go, we write q for this particular value of what we wrote as 'k' in the definition of D).

Rearranging the equation $r = b - aq$ we certainly have $b = aq + r$ so all that is left is to show that $0 \le r < a$. We chose r to be in D and it is part of the definition of D that all its elements should be non-negative so we do have $0 \le r$. All that remains is to show $r < a$.

The last part of the proof is an example of what is called 'proof by contradiction' (we discuss this technique below). We want to prove $r < a$ so we say, suppose not – then $r \ge a$ – but in that case we could subtract a at least once more from r and still have a number of the form $b - ak$ which is non-negative. Such a number would be an element of D but strictly smaller than r and that contradicts our choice of r as the smallest element of D. The conclusion is that we do, indeed, have $r < a$ and, with that, the proof is finished.

Proof by contradiction   Suppose that we want to prove a statement. Either it is true or it is false. What we can do is suppose that it is false and then see where that leads us: if it leads us to something that is wrong then we must have started out by supposing something that is wrong. In other words, the supposition that the statement is false must be wrong. Therefore the original statement must be true.

For instance, suppose that we want to prove that there is no largest integer. Well, either that is correct or else there *is* a largest integer. So let us suppose for a moment that there is a largest integer n say. But then $n + 1$ is an integer which is larger than n, a contradiction (to n being the largest integer). So supposing that there is a largest integer leads to a contradiction and must, therefore, be false. In other words, there is no largest integer.

Definition   Given two integers a and b, we say that a **divides** b (written '$a \mid b$') if there is an integer k such that $ak = b$.

For example, $7 \mid 42$ but 7 does not divide 40, we write $7 \nmid 40$ (it is true that $40/7$ makes sense as a rational number but here we are working in the integers and insist that k in the definition should be an integer: positive, negative or 0).

Thus a divides b exactly if, with notation as in Theorem 1.1.1, $r = 0$.

Note that this definition has the consequence (take $k = 0$) that every integer divides 0.

Another idea with which you are probably familiar is that of the greatest common divisor (also called highest common factor) of two integers a and b. Usually this is described as being 'the largest integer which divides both a and b'. In fact, it is not only 'the largest' in the sense that every other common divisor of a and b is less than it: it is even the case that every common divisor of a and b *divides* it.

This is essentially what the next theorem says. The proof should be surprising: it proves an important property of greatest common divisor that you may not have come across before, a property which we extract in Corollary 1.1.3.

Theorem 1.1.2 *Given positive integers a and b, there is a positive integer d such that*

(i) *d divides a and d divides b, and*
(ii) *if c is a positive integer which divides both a and b then c divides d (that is, any common divisor of a and b must divide d).*

Proof Let D be the set of all positive integers of the form $as + bt$ where s and t vary over the set of *all* integers:

$$D = \{as + bt : s \text{ and } t \text{ are integers and } as + bt > 0\}.$$

Since $a(a = a \cdot 1 + b \cdot 0)$ is in D, we know that D is not empty and so, by the well-ordering principle, D has a least element d, say. Since d is in D there are integers s and t such that

$$d = as + bt.$$

We have to show that any common divisor c of a and b is a divisor of d. So suppose that c divides a, say $a = cg$, and that c divides b, say $b = ch$. Then c divides the right-hand side ($cgs + cht$) of the above equation and so c divides d. This checks condition (ii).

We also have to check that d does divide both a and b, that is we have to check condition (i). We will show that d divides a since the proof that d divides b is similar (a and b are interchangable throughout the statement and proof so 'by symmetry' it is enough to check this for one of them). Applying Theorem 1.1.1 to 'divide d into a', we may write

$$a = dq + r \text{ with } 0 \le r < d.$$

We must show that $r = 0$. We have

$$r = a - dq$$
$$= a - (as + bt)q$$
$$= a(1 - sq) + b(-tq).$$

Therefore, if r were positive it would be in D. But d was chosen to be minimal in D and r is strictly less than d. Hence r cannot be in D, and so r cannot be positive. Therefore r is zero, and hence d does, indeed, divide a. $\square$

Comment   Note the structure of the last part of the proof above. We chose d to be minimal in the set D and then essentially said, 'The remainder r is an integer combination of a and b so, if it is not zero, it must be in the set D. But d was supposed to be the *least* member of D and $r < d$. So the only possibility is that $r = 0$.' There is a definite similarity to the end of the proof of 1.1.1.

Given any a and b as in 1.1.2, we claim that there is just one positive integer d which satisfies the conditions (i) and (ii) of the theorem. For, suppose that a positive integer e also satisfies these conditions. Applying condition (i) to e we have that e divides both a and b; so, by condition (ii) applied to d and with e in place of 'c' there, we deduce that e divides d. Similarly (the situation is symmetric in d and e) we may deduce that d divides e. So we have two integers, d and e, and each divides the other: that can only happen if each is $\pm$ the other. But both d and e are positive, so the only possibility is that $e = d$, as claimed.

Note the strategy of the argument in the paragraph above. We want to show that there is just one thing satisfying certain conditions. What we do is to take two such things (but allowing the *possibility* that they are equal) and then show (using the conditions they satisfy) that they *must* be equal.

Definition   The integer d satisfying conditions (i) and (ii) of the theorem is called the **greatest common divisor** or **gcd** of a and b and is denoted (a, b) or $\gcd(a, b)$. Some prefer to call (a, b) the **highest common factor** or **hcf** of a and b. Note that, just from the definition, $(a, b) = (b, a)$.

For example, $(8, 12) = 4$, $(3, 21) = 3$, $(4, 15) = 1$, $(250, 486) = 2$.

Note   It follows easily from the definition that if a divides b then the gcd of a and b is a. For instance $\gcd(6, 30) = 6$.

The proof of 1.1.2 actually showed the following very important property (you should go back and check this).

Corollary 1.1.3   *Let a and b be positive integers. Then the greatest common divisor, d, of a and b is the smallest positive integral linear combination of a and b. (By an **integral linear combination** of a and b we mean an integer of the form $as + bt$ where s and t are integers.) That is, $d = as + bt$ for some integers s and t.*

For instance, the gcd of 12 and 30 is 6: we have $6 = 30 \cdot 1 - 12 \cdot 2$. In Section 1.5 we give a method for calculating the gcd of any two positive integers.

We make some comment on what might be unfamiliar terminology. A 'Corollary' is supposed to be a statement that follows from another. So often, after a Theorem or a Proposition (a statement which, for whatever reason, is judged by the authors to be not quite as noteworthy as a Theorem) there might be one or more Corollaries. In the case above it was really a corollary of the proof, rather than the statement, of 1.1.2. The term 'Lemma', used below, indicates a result which we prove on the way to establishing something more notable (a Proposition or even a Theorem).

Before stating the next main theorem we give a preliminary result.

Lemma 1.1.4 *Let a and b be natural numbers and suppose that a is non-zero. Suppose that*
   $b = aq + r$ with q and r positive integers.
Then the gcd of b and a is equal to the gcd of a and r.

Proof Let d be the gcd of a and b. Since d divides both a and b, d divides the (term on the) right-hand side of the equation $r = b - aq$: hence d divides the left-hand side, that is, d divides r. So d is a common divisor of a and r. Therefore, by definition of (a, r), d divides (a, r).

   Similarly, since the gcd (a, r) divides a and r and since $b = aq + r$, (a, r) must divide b. So (a, r) is a common divisor of a and b and hence, by definition of $d = (a, b)$, it must be that (a, r) divides d.

   It has been shown that d and (a, r) are positive integers which divide each other. Hence they are equal, as required. $\square$

Discussion of proof of 1.1.4 Sometimes, if the structure of a proof is not clear to you, it can help to go through it with some or all 'x's and 'y's (or in this case, a and b) replaced by particular values. We illustrate this by going through the proof above with particular values for a and b.

   Let us take $a = 30$, $b = 171$. In the statement of 1.1.4 we write b in the form $aq + r$, that is, we write 171 in the form $30q + r$. Let us take $q = 5$ so $r = 21$ and the equation in the statement of the lemma is $171 = 30 \cdot 5 + 21$ (but we do not have to take the form with smallest remainder r, we could have taken say $q = 3$ and $r = 81$, the conclusion of the lemma will still be true with those choices).

   The proof begins by assigning d to be $(30, 171)$. Then (says the proof) d divides both 30 and 171 so it divides the right-hand side of the rearranged equation $21 = 171 - 30 \cdot 5$ hence d divides the left-hand side, that is d divides 21. So d is a common divisor of 30 and 21. Therefore, by definition of the gcd $(30, 21)$ it must be that d divides $(30, 21)$.

Similarly, since $(30, 21)$ divides both 30 and 21 and since $171 = 30 \cdot 5 + 21$ it must be that $(30, 21)$ divides 171 and so is a common divisor of 30 and 171. Therefore, by definition of $d = (30, 171)$ we must have that $(30, 21)$ divides d.

Therefore d and $(30, 21)$ are positive integers which divide each other. The conclusion is that they must be equal: $(30, 171) = (30, 21)$. (Of course, you can compute the actual values of the gcd to check this but the point is that you do not need to do the computation to know that they are equal. In fact, the lemma that we have just proved is the basis of the practical method for computing greatest common divisors, so to say that we do not need this lemma because we can always compute the values completely misses the point!)

The next result appears in Euclid's *Elements* (Book VII Propositions 1 and 2) and so goes back as far as 300 BC. The proof here is essentially that given in Euclid (it also appears in the Chinese *Jiŭ zhāng sùan shù* (*Nine Chapters on the Mathematical Art*) which was written no later than the first century AD). Observe that the proof uses 1.1.1, and hence depends on the well-ordering principle (which was used in the proof of 1.1.1). Indeed it also uses the well-ordering principle directly. The (very useful) 1.1.3 is not explicit in Euclid.

Theorem 1.1.5 (Euclidean algorithm)   *Let a and b be positive integers. If a divides b then a is the greatest common divisor of a and b. Otherwise, applying 1.1.1 repeatedly, define a sequence of positive integers $r_1, r_2, \ldots, r_n$ by*

$$b = aq_1 + r_1 \qquad (0 < r_1 < a),$$
$$a = r_1q_2 + r_2 \qquad (0 < r_2 < r_1),$$
$$\vdots$$
$$r_{n-2} = r_{n-1}q_n + r_n \qquad (0 < r_n < r_{n-1}),$$
$$r_{n-1} = r_nq_{n+1}.$$

Then r_n is the greatest common divisor of a and b.

Proof   Apply Theorem 1.1.1, writing $r_1, r_2, \ldots, r_n$ for the successive *non-zero* remainders. Since $a, r_1, r_2, \ldots$ is a decreasing sequence of positive integers, it must eventually stop, terminating with an integer r_n which, because no non-zero remainder 'r_{n+1}' is produced must, therefore, divide r_{n-1}. Then, applying 1.1.4 to the second-to-last equation gives $(r_{n-2}, r_{n-1}) = (r_{n-1}, r_n)$ which, we have just observed, is r_n. Repeated application of Lemma 1.1.4, working back through the equations, shows that r_n is the greatest common divisor of a and b. $\square$

Example   Take $a = 30, b = 171$.

$$171 = 5 \cdot 30 + 21 \qquad \text{so } r_1 = 21 \qquad \text{and } (171, 30) = (30, 21);$$
$$30 = 21 + 9 \qquad \text{so } r_2 = 9 \qquad \text{and } (30, 21) = (21, 9);$$
$$21 = 2 \cdot 9 + 3 \qquad \text{so } r_3 = 3 \qquad \text{and } (21, 9) = (9, 3);$$
$$9 = 3 \cdot 3.$$

Hence

$$(171, 30) = (30, 21) = (21, 9) = (9, 3) = 3.$$

If we wish to write the gcd in the form $171s + 30t$, we can use the above equations to 'solve' for the remainders as follows.

$$3 = 21 - 2 \cdot 9$$
$$= 21 - 2(30 - 21)$$
$$= 3 \cdot 21 - 2 \cdot 30$$
$$= 3(171 - 5 \cdot 30) - 2 \cdot 30$$
$$= 3 \cdot 171 - 17 \cdot 30.$$

The calculation may be conveniently arranged in a matrix format.

To find (a, b) as a linear combination of a and b, set up the partitioned matrix

$$\begin{pmatrix} 1 & 0 & \Big| & b \\ 0 & 1 & \Big| & a \end{pmatrix}$$

(this may be thought of as representing the equations: '$x = b$' and '$y = a$'). Set $b = aq_1 + r_1$ with $0 \le r_1 < a$. If $r_1 = 0$ then we may stop since then $a = (a, b)$. If r_1 is non-zero, subtract q_1 times the bottom row from the top row to get (noting that $b - aq_1 = r_1$)

$$\begin{pmatrix} 1 & -q_1 & \Big| & r_1 \\ 0 & 1 & \Big| & a \end{pmatrix}.$$

Now write $a = r_1 q_2 + r_2$ with $0 \le r_2 < r_1$. We may stop if $r_2 = 0$ since r_1 is then the gcd of a and r_1, and hence by 1.1.4 is the gcd of a and b. Furthermore, the row of the matrix which contains r_1 allows us to read off r_1 as a combination of a and b: namely $1 \cdot b + (-q_1) \cdot a = r_1$.

If r_2 is non-zero then we continue. Thus, if at some stage one of the rows is

$$n_i \quad m_i \mid r_i \quad (*)$$

representing the equation

$$bn_i + am_i = r_i,$$

and if the other row reads

$$n_{i+1} \quad m_{i+1} \mid r_{i+1} \quad (**)$$

then we set

$$r_i = r_{i+1}q_{i+2} + r_{i+2} \text{ with } 0 \le r_{i+2} < r_{i+1}$$

and we subtract q_{i+2} times the second of these rows from the first and replace $(*)$ with the result.

Observe that these operations reduce the size of the (non-negative) numbers in the right-hand column, and so eventually the process will stop. When it stops we will have the gcd: moreover if the row containing the gcd reads

$$n \quad m \mid d$$

then we have the expression,

$$bn + am = d,$$

of d as an integral linear combination of a and b.

Example 1 We repeat the above example in matrix form: so $a = 30$ and $b = 171$.

$$\left(\begin{array}{cc|c} 1 & 0 & 171 \\ 0 & 1 & 30 \end{array} \right) \rightarrow \left(\begin{array}{cc|c} 1 & -5 & 21 \\ 0 & 1 & 30 \end{array} \right) \rightarrow \left(\begin{array}{cc|c} 1 & -5 & 21 \\ -1 & 6 & 9 \end{array} \right)$$

$$\rightarrow \left(\begin{array}{cc|c} 3 & -17 & 3 \\ -1 & 6 & 9 \end{array} \right) \rightarrow \left(\begin{array}{cc|c} 3 & -17 & 3 \\ -10 & 57 & 0 \end{array} \right).$$

So $(171, 30) = 3 = 3 \cdot 171 - 17 \cdot 30$.

Example 2 Take b to be 507 and a to be 391.

$$\left(\begin{array}{cc|c} 1 & 0 & 507 \\ 0 & 1 & 391 \end{array} \right) \rightarrow \left(\begin{array}{cc|c} 1 & -1 & 116 \\ 0 & 1 & 391 \end{array} \right) \rightarrow \left(\begin{array}{cc|c} 1 & -1 & 116 \\ -3 & 4 & 43 \end{array} \right)$$

$$\rightarrow \left(\begin{array}{cc|c} 7 & -9 & 30 \\ -3 & 4 & 43 \end{array} \right) \rightarrow \left(\begin{array}{cc|c} 7 & -9 & 30 \\ -10 & 13 & 13 \end{array} \right)$$

$$\rightarrow \left(\begin{array}{cc|c} 25 & -35 & 4 \\ -10 & 13 & 13 \end{array} \right) \rightarrow \left(\begin{array}{cc|c} 25 & -35 & 4 \\ -91 & 118 & 1 \end{array} \right)$$

$(507, 391) = 1 = -91 \cdot 507 + 118 \cdot 391$.

You may use whichever method you prefer for calculating gcds: the methods are essentially the same and it is only in the order of the calculations that they differ. The advantages of the matrix method are that there is less to write down and, at any stage, the calculation can be checked for correctness, since a row $u \quad v \mid w$ represents the equation $bu + av = w$. A disadvantage is that one has to put more reliance on mental arithmetic. Therefore it is especially important that, after finishing a calculation like those above, you should check the correctness of the final equation as a safeguard against errors in arithmetic.

A good exercise (if you have the necessary background) is to write a program (in pseudocode) which, given any two positive integers, finds their gcd as an integral linear combination. If you attempt this exercise you will find that any gaps in your understanding of the method will be highlighted.

The definition of greatest common divisor may be extended as follows.

Definition Let $a_1, \ldots, a_n$ be positive integers. Then their **greatest common divisor** $(a_1, \ldots, a_n)$, also written $\gcd(a_1, \ldots, a_n)$, is the positive integer m with the property that $m \mid a_i$ for each i and, whenever c is an integer with $c \mid a_i$ for each i, we have $c \mid m$.

This exists, and can be calculated, by using the case $n = 2$ 'and induction'. We discuss induction at length in the next section but here we will give a somewhat informal indication of how it is used.

We claim that $(a_1, \ldots, a_n) = ((a_1, \ldots, a_{n-1}), a_n)$. In other words, we can compute the gcd of n numbers, $a_1, \ldots, a_n$ by computing the gcd of the first $n - 1$ of them and then computing the gcd of *that* and the last number a_n. As for computing the gcd of $a_1, \ldots, a_{n-1}$ we compute *that* by computing the gcd of the first $n - 2$ numbers and then computing the gcd of that with a_{n-1}. Etc. So, in the end, all we need is to be able to compute the gcd of *two* numbers. Here is an example.

Suppose that we wish to compute $(24, 60, 30, 8)$. We claim that this is equal to $((24, 60, 30), 8)$ and that this is equal to $(((24, 60), 30), 8)$. Now we do some arithmetic and find that $(24, 60) = 12$, then $(12, 30) = 6$ and then $(6, 8) = 2$, so we conclude $(24, 60, 30, 8) = 2$.

After you have read the section on induction you can try, as an exercise, to give a formal proof that, for all n and integers $a_1, \ldots, a_n$, we have $(a_1, \ldots, a_n) = ((a_1, \ldots, a_{n-1}), a_n)$.

Definition Two positive integers a and b are said to be **relatively prime** (or **coprime**) if their greatest common divisor is 1: $(a, b) = 1$. Example 2 above shows that 507 and 391 are relatively prime.

We now give some properties of relatively prime integers. You are probably aware of these properties though you may not have seen them stated formally. A special case of (i) below is the deduction that since 15 and 8 are relatively prime and since 15 divides $8 \cdot 30 = 240$ it must be that 15 divides 30. A special case of (ii) is that since 15 and 8 are relatively prime and since 15 divides 360 and 8 divides 360 we must have that $15 \cdot 8 = 120$ divides 360. Perhaps by giving numerical values to a, b, and c in this way it all seems rather obvious but, beware: neither (i) nor (ii) is true without the assumption that a and b are relatively prime. If we were to replace 15 and 8 by, say 6 and 9, the statements (i) and (ii) would be false for some values of c. See Exercises 1.1.4 and 1.1.5 at the end of the section.

Theorem 1.1.6   *Let a, b, c be positive integers with a and b relatively prime. Then*

(i) *if a divides bc then a divides c,*
(ii) *if a divides c and b divides c then ab divides c.*

Proof   (i) Since a and b are relatively prime there are, by 1.1.3, integers r and s such that

$$1 = ar + bs.$$

Multiply both sides of this equation by c to get

$$c = car + cbs. \qquad (*)$$

Since a divides bc, it divides the right-hand side of the equation and hence divides c.

(ii) With the above notation, consider equation $(*)$. Since a divides c, ab divides cbs and, since b divides c, ab divides car. Thus ab divides c as required. $\square$

Comment   Note how using 1.1.3 gives a beautifully simple argument – surely not the argument one would first think of trying.

The results of this section may be extended in fairly obvious ways to include negative integers. For example, to apply Theorem 1.1.1 with b negative and a positive it makes best sense to demand that the remainder 'r' still satisfy the inequality $0 \leq r < a$. This means that in order to divide the negative number b by a we do *not* simply divide the positive number $-b$ by a and then put a minus sign in front of everything.

Example To divide -9 by 4: find the multiple of 4 which is just below -9 (that is $-12 = 4(-3)$) and then write:

$$-9 = 4(-3) + 3,$$

noting that the remainder 3 satisfies $0 \leq 3 < 4$. (If we wrote $-9 = 4(-2) + -1$, then the remainder -1 would not satisfy the inequality $0 \leq r < 4$.)

So remember that the remainder should always be positive or zero.

A similar remark applies to Theorem 1.1.5: we require that the greatest common divisor always be positive.

Example The greatest common divisor of -24 and -102 equals the greatest common divisor of 24 and 102. To express it as a linear combination of -24 and -102, either we use the matrix method or we proceed as follows, remembering that remainders must always be non-negative:

$$-102 = -24 \cdot 5 + 18$$
$$-24 = 18(-2) + 12$$
$$18 = 12 \cdot 1 + 6$$
$$12 = 6 \cdot 2.$$

Hence the gcd of -24 and -102 is 6 and 6 is $-1(-102) + 4(-24)$.

To conclude this section, we note that there is the notion of **least common multiple** or **lcm** of integers a and b. This is defined to be the positive integer m such that both a and b divide m (so m is a common multiple of a and b), and such that m divides every common multiple of a and b. It is denoted by $\mathrm{lcm}(a, b)$. The proof that such an integer m does exist, and is unique, is left as an exercise.

More generally, given non-zero integers $a_1, \ldots, a_n$, we define their **least common multiple**, $\mathrm{lcm}(a_1, \ldots, a_n)$, to be the (unique) positive integer m which satisfies $a_i | m$ for all i and, whenever an integer c satisfies $a_i \mid c$ for all i, we have $m \mid c$.

For instance $\mathrm{lcm}(6, 15, 4) = \mathrm{lcm}(\mathrm{lcm}(6, 15), 4) = \mathrm{lcm}(30, 4) = 60$.

We shall see in Section 1.3 how to interpret both the greatest common divisor and the least common multiple of integers a and b in terms of the decomposition of a and b as products of primes.

All of the concepts and most of the results of this section are to be found in the *Elements* of Euclid (who flourished around 300 BC). Euclid's origins are unknown but he was one of the scholars called to the Museum of Alexandria. The Museum was a centre of scholarship and research established by Ptolemy, a general of Alexander the Great, who, after the latter's death in 323 BC, gained control of the Egyptian part of the empire.

The *Elements* probably was a textbook covering all the elementary mathematics of the time. It was not the first such 'elements' but its success was such that it drove its predecessors into oblivion. It is not known how much of the mathematics of the *Elements* originated with Euclid: perhaps he added no new results; but the organisation, the attention to rigour and, no doubt, some of the proofs, were his. It is generally thought that the algebra in Euclid originated considerably earlier.

No original manuscript of the *Elements* survives, and modern editions have been reconstructed from various recensions (revised editions) and commentaries by other authors.

Exercises 1.1

1. For each of the following pairs a, b of integers, find the greatest common divisor d of a and b and express d in the form $ar + bs$:
   (i) $a = 7$ and $b = 11$;   (ii) $a = -28$ and $b = -63$;
   (iii) $a = 91$ and $b = 126$;   (iv) $a = 630$ and $b = 132$;
   (v) $a = 7245$ and $b = 4784$;   (vi) $a = 6499$ and $b = 4288$.
2. Find the gcd of 6, 14 and 21 and express it in the form $6r + 14s + 21t$ for some integers r, s and t.
   [Hint: compute the gcd of two numbers at a time.]
3. Let a and b be relatively prime integers and let k be any integer. Show that b and $a + bk$ are relatively prime.
4. Give an example of integers a, b and c such that a divides bc but a divides neither b nor c.
5. Give an example of integers a, b and c such that a divides c and b divides c but ab does not divide c.
6. Show that if $(a, c) = 1 = (b, c)$ then $(ab, c) = 1$ (this is Proposition 24 of Book VII in Euclid's *Elements*).
7. Explain how to measure 8 units of water using only two jugs, one of which holds precisely 12 units, the other holding precisely 17 units of water.

1.2   Mathematical induction

We can regard the positive integers as having been constructed in the following way. Start with the number 1. Then add 1 to get 2. Then add 1 to get 3. Add 1 again to get 4. And so on. That is, we start with a certain base, 1, and then, again and again without ending, we add 1. In this way we generate the positive integers. A construction of this sort (begin with a base case then apply a process again and again) is described as an **inductive construction**. Here is another example.

ex. counting
base 1 add 1 and soon

Define a sequence of integers as follows. Set $a_1 = 2$, define a_{n+1} inductively by the formula $a_{n+1} = 2a_n + 1$. So $a_2 = 2a_1 + 1 = 2 \cdot 2 + 1 = 5$, $a_3 = 2a_2 + 1 = 2 \cdot 5 + 1 = 11$, $a_4 = 2a_3 + 1 = 2 \cdot 11 + 1 = 23$, and so on. You might notice that if we add 1 to any of the numbers that we have generated so far we obtain a multiple of 3. You might check a few more values and see that this still seems to be a property of the numbers generated in this sequence. But how can we *prove* that this holds for every number in the sequence? Obviously we cannot check each one, because the sequence never ends. What we can do is use a proof by induction. Essentially this is a proof that uses the way that the sequence is generated by a base number together with a rule ($a_{n+1} = 2a_n + 1$) which is applied again and again. At the base case we can just compute: adding 1 to a_1 gives $1 + 2 = 3$ which is certainly divisible by 3. At the 'inductive step', where we go from a_n to a_{n+1}, we argue as follows. Suppose that we know that $a_n + 1$ is a multiple of 3, say $a_n + 1 = 3k$. Then $a_{n+1} + 1 = (2a_n + 1) + 1 = 2a_n + 2 = 2(a_n + 1) = 2(3k)$: which is certainly a multiple of 3. It follows that for every n it is true that $a_n + 1$ is a multiple of 3.

Use of the induction principle can take very complicated forms but, at base, is the fact that the positive integers are constructed by starting somewhere and then applying a 'rule' again and again.

Here is a very abstract statement of the induction principle. In the statement, '$P(n)$' is any mathematical assertion involving the positive integer n (think of 'n' as standing for an integer variable, as in the assertion '$\frac{1}{2} + \frac{1}{3} + \cdots + \frac{1}{n}$ is not an integer').

Induction principle   Let $P(n)$ be an assertion involving the positive integer variable n. If

(a) $P(1)$ holds and
(b) whenever $P(k)$ holds so also does $P(k + 1)$

then $P(n)$ holds for every positive integer n.

That is, if we can prove the 'base case' at $n = 1$ and then, if we have an argument that proves the $k + 1$ case from the k case, then we have the result for all positive integers.

The typical structure of a proof by induction is as follows.
Base case – show $P(1)$;
Induction step – *assume* that $P(k)$ holds (this assumption is the **induction hypothesis**) and *deduce* that $P(k + 1)$ follows.

Then the **conclusion** (by the induction principle) is that $P(n)$ holds for all positive integers n.

Example Show that the sum $1 + 2 + \cdots + n$ of the first n positive integers is $n(n + 1)/2$.

This may be proved using the induction principle: for the assertion $P(n)$ we take '$1 + 2 + \cdots + n = n(n + 1)/2$'.

First, the *base case* holds because when $n = 1$ the left-hand side and right-hand side of the formula are both equal to 1: so the formula is valid for $n = 1$.

For the *induction step*, the induction hypothesis, $P(k)$, is that

$$1 + 2 + \cdots + k = k(k + 1)/2$$

(so we *assume* this, and have to *prove* $P(k + 1)$).

The statement $P(k + 1)$ concerns the sum of the first $k + 1$ positive integers, so let us try writing down this sum and then using the above equation to replace the sum of the first k terms. We get

$$1 + 2 + \cdots + k + (k + 1) = k(k + 1)/2 + (k + 1).$$

Simplifying the right-hand side gives

$$k(k + 1)/2 + (k + 1) = (k + 1)(k/2 + 1) = (k + 1)(k + 2)/2.$$

Thus we have deduced

$$1 + 2 + \cdots + k + (k + 1) = (k + 1)(k + 2)/2 \qquad (= (k + 1)\{(k + 1) + 1\}/2)$$

which is the required assertion, $P(k + 1)$.

It follows by induction that the formula is valid for every $n \geq 1$.

Example For each positive integer n let

$$a_n = 4^{2n-1} + 3^{n+1}.$$

We will show that, for all positive integers n, a_n is divisible by 13.

For the proposition $P(n)$ we take: 'a_n is divisible by 13.'

The *base case* is $n = 1$. In that case a_n equals $4 + 9 = 13$, which certainly is divisible by 13.

For the *induction step* we assume the induction hypothesis, that $4^{2k-1} + 3^{k+1}$ is divisible by 13: so $4^{2k-1} + 3^{k+1} = 13r$ for some integer r. We must deduce that 13 divides

$$4^{2(k+1)-1} + 3^{(k+1)+1} = 4^{2k+1} + 3^{k+2}.$$

To see this we note that

$$4^{2k+1} + 3^{k+2} = 4^2(4^{2k-1}) + 3(3^{k+1})$$
$$= 16(4^{2k-1} + 3^{k+1}) - 16(3^{k+1}) + 3(3^{k+1})$$
$$= 16(13r) - 13(3^{k+1}).$$

It is clear that 13 divides the right-hand side of this expression and so 13 divides $4^{2k+1} + 3^{k+2}$, as required.

It follows by induction that $4^{2n-1} + 3^{n+1}$ is divisible by 13 for every positive integer n.

We should be clear about the following point. Induction is a *form* of argument: if we want to use it then we have to assume $P(k)$ and try to prove $P(k + 1)$, but induction does not tell us *how* to do that. In the examples we have given above, we just had to rearrange equations a bit: but it is not always so easy!

We say a bit more about **definition by induction** (sometimes termed **definition by recursion**). This is even used, for example, in defining the positive powers of an integer a. Informally one says: '$a^1 = a, a^2 = a \cdot a, a^3 = a \cdot a \cdot a$, and so on'. More formally, one proceeds by setting $a^1 = a$ (the 'base case') and then inductively defining $a^{k+1} = a^k \cdot a$ (think of a^k as being already defined). Another example of this occurs in defining the factorial symbol $n!$ Here $0!$ is defined to be 1 and, inductively, $(n + 1)!$ is defined to be $(n + 1) \times n!$ (Thus $4!$ is $4 \cdot 3! = 4 \cdot 3 \cdot 2! = 4 \cdot 3 \cdot 2 \cdot 1! = 4 \cdot 3 \cdot 2 \cdot 1 \cdot 0! = 4 \cdot 3 \cdot 2 \cdot 1 \cdot 1 = 24$.) An informally presented definition by induction is usually signalled by use of '...' or a phrase such as 'and so on'. For other examples of definition by induction see Exercises 1.2.3 and 1.2.9.

As another example of proof by induction, we establish the binomial theorem, 1.2.1 below. Supposing that you have not seen this before, stated in this generality, how can you make sense of the statement of the theorem which, at first sight, might look rather complicated?

Try substituting in values: in this case giving values to all of n, x and y would probably obscure what is being said (remember that one reason for using letters to stand for numbers is that it allows clearer statements!). We will leave x and y as variables but try out giving values to n.

For $n = 1$ you should check that the statement becomes $(x + y)^1 = 1 \cdot x^1 + 1 \cdot y^1$, not very exciting!

For $n = 2$ we obtain $(x + y)^2 = 1 \cdot x^2 + 2 \cdot x^1 y^1 + 1 \cdot y^2$ and for $n = 3$ the statement becomes $(x + y)^3 = 1 \cdot x^3 + 3 \cdot x^2 y^1 + 3 \cdot x^1 y^2 + 1 \cdot y^3$. You should write out the corresponding statements for $n = 4$ and $n = 5$. This exhibits the theorem as a very general statement covering some familiar special cases. You might also notice that the coefficients occurring are those seen in Pascal's Triangle, the first few rows of which are shown below.

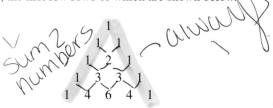

Pascal's Triangle is formed by adding pairs of adjacent numbers in one row to give the numbers in the next row: you should look out for where that rule occurs in the proof.

Theorem 1.2.1 *Let n be a positive integer and let x, y be any numbers. Then*

$$(x + y)^n = \binom{n}{0} x^n + \cdots + \binom{n}{i} x^{n-i} y^i + \cdots + \binom{n}{n} y^n$$

where for $0 \leq k \leq n$, $\binom{n}{k}$ is defined to be $\frac{n!}{k!(n-k)!}$ (and is known as a binomial coefficient).

Proof Observe that, for any $n \geq 1$, $\binom{n}{n} = \frac{n!}{n!0!} = 1$ and $\binom{n}{0} = \frac{n!}{0!n!} = 1$. For the base case, $n = 1$, the theorem asserts that $(x + y)^1 = \binom{1}{0}x^1 + \binom{1}{1}y^1$ which, by the observation just made, is true. Now suppose that the result holds for $n = k$ (induction hypothesis). Then, using the induction hypothesis, we have

$$(x + y)^{k+1} = (x + y)(x + y)^k$$
$$= (x + y)\left(\binom{k}{0}x^k + \binom{k}{1}x^{k-1}y^1 + \cdots + \binom{k}{k-1}x^1y^{k-1} + \binom{k}{k}y^k\right).$$

When we multiply this out, the term involving x^{k+1} is

$$\binom{k}{0}x^{k+1} = x^{k+1} = \binom{k+1}{0}x^{k+1},$$

and that involving y^{k+1} is

$$\binom{k}{k}y^{k+1} = y^{k+1} = \binom{k+1}{k+1}y^{k+1}.$$

The term involving $x^{k+1-i}y^i$ $(1 \leq i \leq k)$ is obtained as the sum of two terms, namely

$$x \binom{k}{i} x^{k-i} y^i + y \binom{k}{i-1} x^{k-(i-1)} y^{i-1}.$$

This simplifies to

$$\left(\binom{k}{i} + \binom{k}{i-1} \right) x^{k+1-1} y^i.$$

We must show that the coefficient, $\binom{k}{i} + \binom{k}{i-1}$, of $x^{k+1-i} y^i$ is $\binom{k+1}{i}$. We have

$$\binom{k}{i} + \binom{k}{i-1} = \frac{k!}{i!(k-1)!} + \frac{k!}{(i-1)!(k-(i-1))!}$$

$$= \frac{k!}{i!(k-i)!} + \frac{k!}{(i-1)!(k-i+1)!}$$

$$= \frac{k!}{i \cdot (i-1)!(k-i)!} + \frac{k!}{(i-1)!(k-i+1) \cdot (k-i)!}$$

$$= \frac{k!}{(i-1)!(k-i)!} \left(\frac{1}{i} + \frac{1}{k-i+1} \right)$$

$$= \frac{k!}{(i-1)!(k-i)!} \cdot \frac{k+1}{i \cdot (k-i+1)}$$

$$= \frac{(k+1) \cdot k!}{i \cdot (i-1)! \cdot (k-i+1) \cdot (k-i)!} = \frac{(k+1)!}{i!(k+1-i)!}$$

$$= \binom{k+1}{i}$$

as required. $\square$

(The rule for forming Pascal's Triangle was the last part of the proof, where we showed that $\binom{k}{i} + \binom{k}{i-1} = \binom{k+1}{i}$.)

Next we show that the principle of mathematical induction may be deduced from the well-ordering principle. There is no harm in skipping this proof in your first reading.

Theorem 1.2.2 *The well-ordering principle implies the principle of mathematical induction.*

Proof Suppose that the assertion $P(n)$ satisfies the conditions for the induction principle: so $P(1)$ holds and whenever $P(k)$ holds, so also does $P(k+1)$. Let S be the set of positive integers m for which $P(m)$ is false. There are two cases:

either S is the empty set (that is the set with no elements) or else S is non-empty. We will see that the second case leads to a contradiction.

If S is not empty then we can apply the well-ordering principle, to deduce that S has a least element, which we call t. Since $P(1)$ holds we know that 1 is not in S and so t must be greater than 1. Hence $t - 1$ is positive. The definition of t as the least element of S implies that $P(t - 1)$ does hold. Since $t = (t - 1) + 1$ it follows, by our assumption on P (take $k = t - 1$), that $P(t)$ holds. This is a contradiction to the fact that t is in S.

Thus the hypothesis that S is non-empty allows us to derive a contradiction, and so it must be the case that S is the empty set. In other words, $P(n)$ is true for every positive integer n. □

Comment Note that this is essentially a proof by contradiction: we set S to be the set of positive integers where $P(n)$ is false; we showed that, if S is non-empty, then one can derive a contradiction; so we concluded that S must be empty, in other words, we concluded that $P(n)$ is true for all $n > 0$.

In fact, the converse of the above result is also true: the well-ordering principle can be deduced from the principle of mathematical induction. So the two principles are logically equivalent. We do not need this fact but we indicate its proof in Exercise 1.2.10.

There are some useful variations of the induction principle: let $P(n)$ be an assertion as before.

(a) If P holds for an integer n_0 and if, for every integer $k \geq n_0$, $P(k)$ implies $P(k + 1)$, then P holds for all integers $k \geq n_0$.
(b) If $P(0)$ holds and if, for each $k \geq 0$, from the hypothesis that P holds for all non-negative $m \leq k$ one may deduce that $P(k + 1)$ holds, then $P(n)$ holds for every natural number n.

The first variation simply says that the induction need not start at $n = 1$: for example, it may be appropriate to start with the base case being at $n = 0$.

The second of these variations is known variously as **strong induction**, **complete induction** or **course of values induction** and is a very commonly used form of the induction principle (of course the '0' in its statement could as well be replaced by any integer 'n_0' as in (a) and the conclusion would be modified accordingly). Several examples of its use will occur later (in the proof of 1.3.3, for example). This variation takes note of the fact that, if in an induction we have reached the stage where we have that $P(k)$ is true then, in getting there, we also showed that $P(k - 1)$, $P(k - 2)$, ... (down to the base case) are true, so it is legitimate to use all this information (not just the fact that $P(k)$ is true) in trying to prove that $P(k + 1)$ holds.

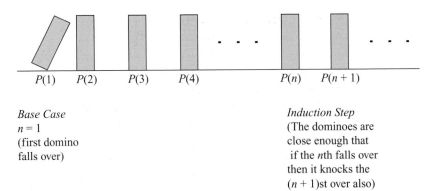

$P(1)$ $P(2)$ $P(3)$ $P(4)$ $P(n)$ $P(n+1)$

Base Case   *Induction Step*
$n = 1$ (The dominoes are
(first domino   close enough that
falls over)   if the nth falls over
                                               then it knocks the
                                               $(n+1)$st over also)

Fig. 1.1

Fig. 1.1 sometimes is used to illustrate the idea of proof by induction.

Imagine a straight line of dominoes all standing on end: these correspond to the integers. They are sufficiently close together that if any one of them falls, then it will knock over the domino next to it: that corresponds to the induction step (from $P(k)$ we get $P(k+1)$). One of the dominoes is pushed over: that corresponds to the base case. Now imagine what happens.

In these terms, the principle of strong induction says that to knock over the $(k+1)$st domino we are not restricted to using just the force of the kth domino: we can also, if we can, use the fact that *all* the previous dominoes have fallen over.

The well-ordering principle was explicitly recognised long before the principle of induction.

Since the well-ordering principle expresses an 'obvious' property of the positive integers, one would not expect to see it stated until there was some recognition of the need for mathematical assertions to be backed up by proofs from more or less clearly stated axioms. There is a perfectly explicit statement of it in Euclid's *Elements*. It is not, however, stated as one of his axioms but, rather, is presented as an obvious fact in the course of one of the proofs (Book VII, proof of Proposition 31, our Theorem 1.3.3), which is by no means the first proof in the *Elements* where it is used (e.g. it is implicit in the proofs of Propositions 1 and 2 in Book VII: our 1.1.4 and 1.1.5). In Euclid, the principle is of course applied to the set of positive numbers rather than to the set of natural numbers, for it was to be many centuries before zero would be recognised as a number (especially in Europe).

There are instances in Euclid's *Elements* of something approaching proof by induction, though not in a form that would be recognised today as correct.

It is not unusual for a student new to the idea of proof by induction to 'prove' that $P(n)$ is true, by just checking it for the first few values of n and then claiming that it necessarily holds also for all greater values of n. In fact, up into the seventeenth century this was not an uncommon method of 'proof'. For example, Wallis in his *Arithmetica Infinitorum* of 1655–6 made much use of such procedures, and he was heavily criticised (in 1657) by Fermat for doing so. By 1636 Fermat had used the principle of induction in a way we would now regard as valid, and Blaise Pascal in his *Triangle Arithmétique* of 1653 spells out the details of a proof by induction. As Fermat points out in his criticism of Wallis' methods, one may manufacture an assertion ($P(n)$) which is true for small values (of n) but which fails at some large value (also see Exercise 1.2.11 below). Actually, many (but not all!) of these early 'proofs' by induction are easily modified to give rigorous proofs because, although their authors used particular numbers, their arguments often apply equally well to an arbitrary positive integer.

Exercises 1.2

1. Define a sequence a_n ($n \geq 1$) of integers by $a_1 = 1$, $a_{n+1} = 2a_n + 1$ for $n \geq 2$. Compute the values of a_i for $i = 1, \ldots, 5$. Prove by induction that for all $n \geq 1$, $a_n + 1$ is a power of 2.

2. Prove that for all positive integers n,

$$1 + 4 + \cdots + n^2 = n(n+1)(2n+1)/6 = \frac{1}{3}n^3 + \frac{1}{2}n^2 + \frac{1}{6}n.$$

3. The **Fibonacci sequence** is the sequence 1, 1, 2, 3, 5, 8, 13,... where each term is the sum of the two preceding terms. Show that every two successive terms of the Fibonacci sequence are relatively prime.
   [Hint: write down an explicit definition (by induction) of this sequence.]

4. We saw in the text that the sum of the first n positive integers is given by a quadratic (degree 2) polynomial in n. From Exercise 1.2.2 above you see that the sum of the squares of the first n positive integers is given by a polynomial in n of degree 3. Given the information that the formula for the sum of the cubes of the first n positive integers is given by a polynomial in n of degree 4, find this polynomial. [Hint: suppose that the polynomial is of the form $an^4 + bn^3 + cn^2 + dn + e$ for certain constants $a, \ldots, e$, then express the sum of the first $n + 1$ cubes in two different ways.]

5. Prove that for all positive integers n,

$$\frac{1}{3} + \frac{1}{15} + \cdots + \frac{1}{(2n-1)(2n+1)} = \frac{n}{2n+1}.$$

6. Find a formula for the sum of the first n odd positive integers.

7. Prove that if x is not equal to 1 and n is any positive integer then

$$1 + x + x^2 + \cdots + x^n = \frac{1 - x^{n+1}}{1 - x}.$$

8. (i) Show that, for every positive integer n, $n^5 - n$ is divisible by 5.
   (ii) Show that, for every positive integer n, $3^{2n} - 1$ is divisible by 8.

9. Given that $x_0 = 2$, $x_1 = 5$ and

$$x_{n+2} = 5x_{n+1} - 3x_n$$

for n greater than or equal to 0, prove that

$$2^n x_n = (5 + \sqrt{13})^n + (5 - \sqrt{13})^n$$

for every natural number n.

10. Show that the principle of induction implies the well-ordering principle.
    [Hint: let X be a set of positive integers which contains no least element;
    we must show that X is empty. Define L to be the set of all positive
    integers, n, such that n is not greater than or equal to any element in X.
    Show by induction that L is the set of all positive integers, and hence that
    X is indeed empty.]

11. Consider the assertion: (∗) 'for every prime number n, $2^n - 1$ is a prime
    number' (a positive integer is prime if it cannot be written as a product of
    two strictly smaller positive integers). Taking n to be 2, 3, 5 in turn, the
    corresponding values of $2^n - 1$ are 3, 7, 31 and these certainly are prime.
    Is (∗) true? (Also see Exercise 1.3.6.)

12. The following arguments purport to be proofs by induction: are they valid?
    (a) This argument shows that all people have the same height. More
    formally, it is shown that if X is any set of people then each person in X
    has the same height as every person in X. The proof is by induction on n
    the number of people in X.
    Base case $n = 1$. This is clear, since if the set X contains just one person
    then that person certainly has the same height as him / herself.
    Induction step. We assume that the result is true for every set of $k - 1$
    people (the induction hypothesis), and deduce that it is true for any set X
    containing exactly k people.
    Choose any person a in X, and let Y be the set X with a removed. Then Y
    contains exactly $k - 1$ people who, by the induction hypothesis, must all
    be of the same height: h metres, say.
    Choose any other person b (say) in X, and let Z be the original set X with b
    removed. Since Z has just $k - 1$ people, the induction hypothesis applies,
    to give that the people in Z all have the same height, let us say k metres.
    Now let c be any person in X other than a or b. Since b and c both are in Y,

each is h metres tall. Since a and c both are in Z, each is k metres tall. So (consider the height of c) $h = k$. But that means that a and b are of the same height.

Therefore, since a and b were arbitrary members of X, it follows that the people in X all have the same height. Thus the induction step is complete, and so the initial assertion follows by induction.

(b) To establish the formula $1 + 2 + \cdots + n = \frac{n^2}{2} + \frac{n}{2} + 1$.

Assume inductively that the formula holds for $n = k$; thus

$$1 + 2 + \cdots + k = \frac{k^2}{2} + \frac{k}{2} + 1.$$

Add $k + 1$ to each side to obtain

$$1 + 2 + \cdots + k + (k + 1) = \frac{k^2}{2} + \frac{k}{2} + 1 + (k + 1).$$

The term on the left-hand side is $1 + 2 + \cdots + (k + 1)$, and the term on the right-hand side is easily seen to be equal to

$$\frac{(k + 1)^2}{2} + \frac{k + 1}{2} + 1.$$

Thus the induction step has been established and so the formula is correct for all values of n.

1.3 Primes and the Unique Factorisation Theorem

Definition A positive integer p is **prime** if p has exactly two positive divisors, namely 1 and p.

Thus, for example, 5 is prime since its only positive divisors are itself and 1, whereas 4 is not prime since it is divisible by 1, 4 and 2.

Notes (i) The definition implies that 1 is not prime since it does not have two distinct positive divisors.

(ii) The smallest prime number is therefore 2 and this is the only even prime number, since any other even positive integer n has at least three distinct divisors (namely 1, 2 and n).

(iii) We may begin listing the primes in ascending order:

2, 3, 5, 7, 11, 13, 17, 19, 23, 29, ...

If one wishes to continue this list beyond the first few primes, then it is not very efficient to check each number in turn for **primality** (the property of being prime). A fairly efficient, and very old, method for generating the list of primes is the Sieve of Eratosthenes, described below.

Eratosthenes of Cyrene (c.280–c.194 BC) is probably more widely known for his estimate of the size of the earth: he obtained a circumference of 250 000 stades (believed to be about 25 000 miles); the actual value varies between 24 860 and 24 902 miles.

The Sieve of Eratosthenes To find the primes less than some number n, prepare an array of the integers from 2 to n. Save 2 and then delete all multiples of 2. Now look for the next undeleted integer (which will be 3), save it and delete all its multiples. The smallest undeleted number will be the next prime, 5. Continue in this way to find all the primes up to n. In fact, it will turn out that you can stop this process once you have reached the greatest integer which is less than or equal to the square root of n, in the sense that any integers left undeleted at this stage will be prime. (You will be asked to think about this in Exercise 1.3.2 at the end of the section.)

As an exercise, you might like to write a computer program which, given a positive integer n, will use the sieve of Eratosthenes to find all the prime numbers up to n. In fact, such a program is one of the standard benchmark tests which is used to evaluate the speed of a computer. Also you should use the sieve to find all prime numbers less than or equal to $n = 50$ (you will be asked to do this for a larger value of n in the exercises).

The first result of this section describes a very useful property of primes: the property is characteristic of these numbers and is sometimes used as the definition of prime. The theorem occurs as Proposition 30 in Book VII of Euclid's *Elements*.

Theorem 1.3.1 *Let p be a prime integer and suppose that a and b are integers such that p divides ab. Then p divides either a or b.*

Proof Since the only positive divisors of p are 1 and p, it follows that the greatest common divisor of p and a is either p, in which case p divides a, or 1. So, if p does not divide a, the greatest common divisor of a and p must be 1. The result now follows by applying Theorem 1.1.6(i) (with p, a, b in place of a, b, c). □

Comment This is short but quite subtle: there is not an immediately obvious connection between the property of being prime and having the property expressed in 1.3.1, but probably you were already aware that primes have that property (just through experience with numbers). But how to prove it? The concept of greatest common divisor is the key to the short and simple proof above.

It is not difficult to see that any integer p which has the property expressed in Theorem 1.3.1 must be prime.

To see how the statement of Theorem 1.3.1 can fail if p is not prime consider: 4 divides $6 \cdot 2 = 12$ yet 4 divides neither 6 nor 2.

Notice that, as is usual in mathematics, the term 'or' is used in the inclusive sense: so the conclusion of Theorem 1.3.1 is more fully expressed as 'p divides a or b or both'.

The next result is an extension of Theorem 1.3.1. It provides an illustration of one kind of use of the principle of mathematical induction.

Lemma 1.3.2   *Let p be a prime and suppose that p divides the product*

$$a_1 a_2 \ldots a_r.$$

Then p divides at least one of $a_1, a_2, \ldots, a_r$.

Proof   The proof is by induction on the number, r, of factors a_i. The base case is trivial since the 'product' is then just a_1. We therefore suppose inductively that if p divides a product of the form

$$b_1 b_2 \ldots b_{r-1}$$

then p divides at least one of $b_1, b_2, \ldots, b_{r-1}$.

Suppose then that p divides the product $a_1 \ldots a_r$. We want to write this as a product, $b_1 \ldots b_{r-1}$, of $r - 1$ integers. All we have to do is multiply the last two together. So define b_i to be equal to a_i for $i \leq r - 2$ and let b_{r-1} be the product $a_{r-1}a_r$: thus we think of bracketing the product of the a_i in the following way:

$$a_1 a_2 \ldots a_{r-2}(a_{r-1}a_r)$$

as a product of $r - 1$ integers. It follows by induction that either p divides one of $a_1, a_2, \ldots a_{r-2}$ or p divides $a_{r-1}a_r$ and, in the latter case, Theorem 1.3.1 implies that p divides a_{r-1} or a_r. So, either way, we conclude that p divides one of the original integers $a_1, \ldots, a_r$. $\square$

Discussion   The key to this, which is the 'obvious' (but nevertheless has to be proved) extension of 1.3.1 is temporarily to bracket together and multiply two of the numbers so that the product of r integers is 'reduced' to a product of $r - 1$ integers. That allowed us to apply the induction hypothesis. Then, we were able to get back to the original list of r integers because p was prime so, by 1.3.1, if it divided the product of the last two numbers, it must have been a divisor of at least one of those numbers.

The following result is sometimes referred to as the Fundamental Theorem of Arithmetic. It says that, in some sense, the primes are the multiplicative building blocks from which every (positive) integer may be produced in a unique way.

Therefore positive integers, other than 1, which are not prime are referred to as **composite**. The distinction between prime and composite numbers, and the importance of this distinction, was recognised at least as early as the time of Philolaus (who died around 390 BC).

Theorem 1.3.3 (The Unique Factorisation Theorem for Integers)   *Every positive integer n greater than or equal to 2 may be written in the form*

$$n = p_1 p_2 \ldots p_r$$

where the integers $p_1, p_2, \ldots, p_r$ are prime numbers (which need not be distinct) and $r \geq 1$. This factorisation is unique in the sense that if also

$$n = q_1 q_2 \ldots q_s$$

where $q_1, q_2, \ldots, q_s$ are primes, then $r = s$ and we can renumber the q_i so that $q_i = p_i$ for $i = 1, 2, \ldots, r$. In other words, up to rearrangement, there is just one way of writing a positive integer as a product of primes.

Proof   The proof is in two parts. We show in this first part, using strong induction, that every positive integer greater than or equal to 2 has a factorisation as a product of primes.

The base case holds because 2 is prime. If n is greater than 2, then either n is prime, in which case n has a factorisation (with just one factor) of the required form, or n can be written as a product ab where $1 < a < n$ and $1 < b < n$. In this latter case, apply the inductive hypothesis to deduce that both a and b have factorisations into primes: so juxtaposing the factorisations of a and b (i.e. putting them next to each other), we obtain a factorisation of n as a product of primes.

For the second part of the proof, we use the standard form of mathematical induction, this time on r, the number of prime factors, to show that any positive integer which has a factorisation into a product of r primes has a *unique* factorisation.

To establish the base case ($r = 1$ so n is prime) let us suppose that n is a prime which also may be expressed as:

$$n = q_1 q_2 \ldots q_s.$$

If we had $s \geq 2$ then n would have distinct divisors $1, q_1, q_1 q_2$, contradicting that it is prime. So $s = 1$, and the base case is proved.

Now take as induction hypothesis the statement 'any positive integer greater than 2 which has a factorisation into $r - 1$ primes has a unique factorisation (in the above sense)'. Suppose that

$$n = p_1 p_2 \cdots p_r = q_1 q_2 \cdots q_s$$

are two prime factorisations of n. We show that, up to rearrangement, they are the same.

Since p_1 divides n it divides one of $q_1, q_2, \ldots, q_s$ by Lemma 1.3.2. It is harmless to renumber the q_i so that it is q_1 which p_1 divides. Since q_1 is prime it must be that p_1 and q_1 are equal. We may therefore cancel $p_1 = q_1$ from each side to get

$$p_2 p_3 \cdots p_r = q_2 q_3 \cdots q_s.$$

Since the integer on the left-hand side is a product of $r - 1$ primes, the induction hypothesis allows us to conclude that $r - 1$ is equal to $s - 1$, and hence that r is equal to s, and also that, after renumbering, $p_i = q_i$ for $i = 2, \ldots, r$ and hence, since we already have $p_1 = q_1$, for $i = 1, \ldots, r$. $\square$

The first part of Theorem 1.3.3 (existence of the decomposition) occurs, stated in a some what weaker form, as Proposition 31 in Book VII of the *Elements*. It is in the proof of this that Euclid clearly asserts the well-ordering principle. Euclid's argument is in essence the same as that given above.

Comment   The first part of the proof was done efficiently but that, to some extent, obscures the simplicity of the idea, which is as follows.

If n is not prime, then factor it, as ab. If a is not prime then factor it, similarly for b. Continue: that is, keep splitting any factors which are not prime. This cannot go on forever – the integers we produce are decreasing and each factorisation gives strictly smaller numbers. So eventually the process stops, with a product of primes.

For instance $588 = 4 \times 147 = (2 \times 2) \times (7 \times 21) = (2 \times 2) \times (7 \times (3 \times 7))$ (see Fig. 1.2).

The second part of the proof is based on the observation that if a prime divides one side of the equation then it divides the other, so we can cancel it from each side, then we continue this process, which can only halt with the equation $1 = 1$. Therefore there must have been the same number of primes, and indeed the same primes, on each side of the original equation. In giving a more formal proof, we rearranged this idea and used induction.

The following result, that there are infinitely many primes, and its elegantly simple proof appear in Euclid's *Elements* (Book IX, Proposition 20).

Corollary 1.3.4   *There are infinitely many prime integers.*

Proof   Choose any positive integer n and suppose that $p_1, p_2, \ldots, p_n$ are the first n prime numbers. We will show that there is a prime number different from

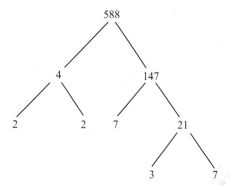

Fig. 1.2 Factorisation of 588 by splitting.

each of $p_1, p_2, \ldots, p_n$. Since n may be chosen as large as we like, this will show that there are not just finitely many primes.

Define the number N as follows:

$$N = (p_1 p_2 \ldots p_n) + 1.$$

Note that N has remainder 1 when divided by each of $p_1, p_2, \ldots, p_n$: in particular, none of $p_1, p_2, \ldots, p_n$ divides N exactly. By Theorem 1.3.3, N has a prime divisor p. Since p divides N, p cannot be equal to any of $p_1, p_2, \ldots, p_n$: thus we have shown that there exists a prime which is not on our original list, as required. □

Comment This is a beautiful and clever proof. The scheme of the proof is to show how, given any finite set of primes, we can construct a number (≥ 2) which is not divisible by any of them (and which, therefore, must have a prime factor not in our original set).

The integer N defined in the proof need not itself be prime: we simply showed that it has a prime divisor not equal to any of $p_1, p_2, \ldots, p_n$. In principle, one may find a 'p' as in the proof by factorising N. So the proof is, in principle, a recipe which, given any finite list of primes, will produce a 'new' prime (i.e. one not on the list).

When we write an integer as a product of prime numbers it is often convenient to group together the occurrences of the same prime: for example, rather than writing $72 = 2 \times 2 \times 2 \times 3 \times 3$ one writes $72 = 2^3 \times 3^2$.

The following characterisation of greatest common divisor and least common multiple is easily obtained.

Corollary 1.3.5 *Let a and b be positive integers. Let*

$$a = (p_1)^{n_1}(p_2)^{n_2} \ldots (p_r)^{n_r}$$
$$b = (p_1)^{m_1}(p_2)^{m_2} \ldots (p_r)^{m_r}$$

be the prime factorisations of a and b, where $p_1, p_2 \ldots, p_r$ are distinct primes and $n_1, n_2, \ldots, n_r, m_1, m_2, \ldots, m_r$ are non-negative integers (some perhaps zero, in order to allow a common list of primes to be used, see the example which follows the proof).

Then the greatest common divisor, d, of a and b is given by

$$d = (p_1)^{k_1}(p_2)^{k_2} \ldots (p_r)^{k_r},$$

where, for each i, k_i is the smaller of n_i and m_i, and the least common multiple, f, of a and b is given by

$$f = (p_1)^{t_1}(p_2)^{t_2} \ldots (p_r)^{t_r},$$

where, for each i, t_i is the larger of n_i and m_i.

Proof   The characterisation of the greatest common divisor follows using Theorem 1.1.6(ii) since any number of the form p^n, with p prime, divides the greatest common divisor (a, b) exactly if it divides both a and b.

Similarly the characterisation of the lowest common multiple follows since a prime power, p^n, is a factor of a or of b exactly if it is a factor of their lowest common multiple.   □

Comment   We have been a bit brief here, leaving out the detailed argument but pointing out what you need to use. This is the first place where we have written 'Proof' yet have not really given a proof, only an indication of how the proof goes.

If you want to see a detailed argument then, in this case, probably it is better to write it down yourself rather than read it. Once you see what the result is saying (if you do not, try it with some numbers in place of the letters) and have understood the relevance of the statements we made in the 'proof', you will probably be able to see how a proof could go, though writing it down would take some organisation and, depending on how you do it, could be a little tedious.

The above result is often of practical use in calculating the greatest common divisor d of two integers, provided we do not need to express d as a linear combination of the numbers, and provided we can find the prime factorisations of the numbers quickly.

Example   To find the greatest common divisor of 135 and 639 we factorise these to obtain that 135 is 5 times $27 = 3^3$ and that 639 is $9 = 3^2$ times 71. So $135 = 3^3 \cdot 5^1 \cdot 71^0$ and $639 = 3^2 \cdot 5^0 \cdot 71^1$. It follows that the greatest common divisor is $3^2 \cdot 5^0 \cdot 71^0 = 3^2 = 9$.

Example To find the lowest common multiple of 84 and 56 observe that $84 = 2^2 \cdot 3 \cdot 7$ and that $56 = 2^3 \cdot 7$. Therefore lcm $(84, 56) = 2^3 \cdot 3 \cdot 7 = 168$.

Along with geometry, the study of the arithmetic properties of integers (which, until rather late on, meant the positive integers) forms the most ancient part of mathematics. Significant discoveries were made in many of the early civilisations around the Mediterranean, in the Near East, in Asia and in South America but undoubtedly the greatest discoveries in ancient times were made by the Greeks. Probably the main factor in accounting for this is that their interest in numbers was motivated less by practical motives (such as commerce and astrological calculations) than by philosophical considerations. This relative freedom from particular applications gave them a rather abstract viewpoint from which, perhaps, they were more likely to discover general properties.

Almost all of what we have covered in Sections 1.1 and 1.3 may be found in Euclid's *Elements* and probably was of earlier origin. It should be remarked, however, that the presentation of these results in Euclid is very different from their presentation above, There are two main differences.

The first difference is peculiar to the Greek mathematicians, and it is that numbers were treated by them as lengths of line segments. Thus, for example, they would often represent the product of two numbers a and b as the area of a rectangle with sides of length a and b respectively.

Euclid describes the process of finding the greatest common divisor of a and b in terms of starting with two line segments, one of length a and the other of length b ($\neq a$); from the longer line one removes a segment of length equal to the length of the shorter line segment; one continues this process, always subtracting the current shorter length from the current longer one. Provided that the starting lengths, a and b, are integers this process will terminate, in the sense that at some stage one reaches two lines of equal length: this length is the 'common measure' (greatest common divisor) of a and b. The process described in 1.1.5 is just a somewhat telescoped version of this.

Actually, for a and b to have a common measure in the above sense it is not necessary that they be integers: it is enough that they be rational numbers (fractions). The earlier Greek mathematicians believed that *any* two line segments have a common measure in this sense, and an intellectual crisis arose when it was discovered that, on the contrary, the side of a square does not have a common measure with the diagonal of the square (or, as we would put it, the square root of 2 is irrational).

The second difference was the lack of a good algebraic notation. This was a weakness to a greater or lesser degree of all early mathematics, although the Indian mathematicians adopted a relatively symbolic notation quite early on.

In Europe Viète (1540–1603) was largely responsible for the beginnings of a reasonable symbolic notation.

In this connection, it is worthwhile pointing out that Euclid's proof of the infinity of primes (Corollary 1.3.4) goes (in modern terminology) more or less as follows.

Suppose that there are only finitely many primes, say a, b and c. Consider the product $abc + 1$. This number has a prime divisor d. Since none of a, b, c divides $abc + 1$, d is a prime different from each of a, b, c. This is a contradiction. Hence the number of primes is not finite.

Nowadays, this proof would be criticised since it derives a contradiction only in the special case that there are just three primes (a, b, c): we would say 'suppose there were only finitely many primes $p_1, \ldots, p_n$'. But how could Euclid even say that in the absence of a notation for indices or subscripts? Since he did not even have the notation with which to express the general case, Euclid had to resort to a particular instance, but his readers would have understood that the argument itself was perfectly general. (Note, by the way, that Euclid's argument is presented as a proof by contradiction.)

Perhaps the high point of Greek work in number theory was the *Arithmetica* of Diophantus (who flourished around AD 250). Originally there were thirteen books comprising this work but only six have survived: it is not even known what kinds of problems were treated in the seven missing books. A major concern of the *Arithmetica* was the finding of integer or rational solutions to equations of various sorts. The methods were presented in the form of solutions to problems and, because of the inadequacy of the notation, in any given problem every unknown but one would be replaced by a particular numerical value (and the generality of the method would then have to be inferred). Many problems raised in that work are still unsolved today, despite the attention of some of the greatest mathematicians. On the other hand, work on these problems has given rise to extremely deep mathematics, and this has led to many successes, such as results of Gerd Faltings in 1983 answering old questions on integer solutions to equations and, in particular, Andrew Wiles' proof of 'Fermat's Last Theorem', which we discuss at the end of Section 1.6.

After the work of the Greeks, very little advance in number theory was made in Europe until interest was rekindled by Fermat and later by Euler. This was in contrast to the continuing advances made by Arab, Chinese and Indian mathematicians.

One problem which Fermat (1601–65) considered was that of finding various methods to generate sequences of prime numbers. He considered numbers of the form $2^n + 1$: such a number cannot be prime unless n is a power of 2 (see Exercise 1.3.7). Setting $F(k) = 2^{2^k} + 1$ one has that $F(0), F(1), \ldots, F(6)$ are

$$3, 5, 17, 257, 65\,537, 4\,294\,967\,297, 18\,446\,744\,073\,709\,551\,617.$$

In a letter of 1640 to Frénicle, Fermat lists the above numbers and expresses his belief that all are prime, and he conjectures that the sequence of integers $F(k)$ might be a sequence of primes. In fact, although $F(0)$, $F(1)$, $F(2)$, $F(3)$, $F(4)$ all are primes, $F(5)$ and $F(6)$ are not. It is rather surprising that Fermat and Frénicle failed to discover that $F(5)$ is not prime since, although this number is rather large, it is possible to find a factor by using an argument similar to that used by Fermat when he showed that $2^{37} - 1$ is not prime (see Exercise 1.6.10 below). In fact, such an argument was used by Euler almost a century later to show that $F(5)$ is not prime (in the process Euler rediscovered Fermat's Theorem, 1.6.3 below). Fermat persisted in his belief that $F(5)$ was prime, though he later added that he did not have a full proof. Actually no new **Fermat primes** (that is, numbers of the form $F(k)$ which are prime) have subsequently been discovered.

A better source of primes is provided by the Mersenne sequence $M(n)$, of numbers of the form $2^n - 1$. It may be shown that $M(n)$ can only be prime if n itself is prime (Exercise 1.3.6). The converse is false, that is, there are prime values of n for which $M(n)$ is not prime. One such value is $n = 37$ (another, as you may check is $n = 11$). Fermat showed that $M(37)$, which equals $137\,438\,953\,471$, is not prime, by an argument using the theorem which bears his name (Theorem 1.6.3 below). Exercise 1.6.10 asks you to do the same. There are currently 39 **Mersenne primes** known, the last 27 having been discovered (i.e. shown to be prime) by computer (now most commonly by networks of computers linked over the internet). The largest to date is $M(13\,466\,917)$, an integer whose decimal expansion has over four million digits! Discovered in 2001, it is currently (2003) the largest known prime.

Perhaps the most famous unsolved questions concerning prime numbers are the following.

(i) Are there an infinite number of prime pairs, that is, numbers of the form p, $p + 2$ with both numbers prime?

(ii) (Goldbach's conjecture) Can every integer greater than 2 be written as a sum of two primes?

The answers to these simply stated problems are unknown.

The second is stated as a question but the **conjecture** (what Goldbach expected to be true) is that every integer greater than 2 *can* be written as a sum of two primes. How can such a conjecture be verified (or shown to be wrong)? Of course we can check for 'small' values: it is easy enough to check that (say) each of the first hundred even integers greater than 2 may be written as a sum of two primes. With the aid of a computer one may extend one's search for counterexamples to much larger numbers. A **counterexample** to Goldbach's conjecture would be a number greater than 2 which cannot be written as a sum of two primes. So far, no counterexample to Goldbach's conjecture has been found. On the other hand still there is no general proof of its validity. So it could be that tomorrow some computer search will turn up a counterexample (or someone will find a proof that it is correct).

One of the attractions of number theory lies in the fact that such simply stated questions are still unanswered.

Exercises 1.3

1. Use the Sieve of Eratosthenes to find all prime numbers less than 250.
2. Show why, when using the sieve method to find all primes less than n, you need only strike out multiples of the primes whose square is less than or equal to n.
3. (a) Find the prime factorisations for the following integers (a calculator will be useful for the larger values): 136, 150, 255, 713, 3549, 4591.
   (b) Use your answers to find the greatest common divisor and least common multiple of each of the pairs: 136 and 150; 255 and 3549.
4. Let $p_1 = 2, p_2 = 3, \ldots$ be the list of primes, in increasing order. Consider products of the form

$$(p_1 \times p_2 \times \cdots \times p_n) + 1$$

(compare with the proof of Corollary 1.3.4).
Show that this number is prime for $n = 1, \ldots, 5$.
Show that when $n = 6$ this number is not prime. [Use your answer to Exercise 1.3.1. A calculator will speed the work of checking divisibility.]
5. By considering the prime decomposition of (ab, n), show that if a, b and n are integers with n relatively prime to each of a and b, then n is relatively prime to ab.
6. Show that if $2^n - 1$ is prime, then n must be prime.
7. Show that if $2^n + 1$ is prime, where $n \geq 1$, then n must be of the form 2^k for some positive integer k.
8. Prove that there are infinitely many primes of the form $4k + 3$. Argue by contradiction: supposing that there are only finitely many primes

$p_1 = 3, p_2 = 7, \ldots, p_n$ of that form, consider

$$N = 4(p_2 \times \cdots \times p_n) + 3.$$

9. Show that for any non-zero integers a and b

$$ab = \gcd(a, b)\mathrm{lcm}(a, b).$$

1.4 Congruence classes

Some of the problems in Diophantus' *Arithmetica* (see above, end of Section 1.3) concerned questions such as 'When may an integer be expressed as a sum of two squares'? (This is a natural question in view of the Greeks' geometric treatment of algebra, and Pythagoras' Theorem.) One of the first results of Fermat's reading of Diophantus was his proof that no number of the form $4k + 3$ can be a sum of two squares (although he was by no means the first to discover this, for example Bachet and Descartes already knew it).

The result is not difficult for us to prove: we could give the proof now, but it will be much easier to describe after the following definition and observations.

The main concepts in this section are the idea of integers being congruent modulo some fixed integer and the notion of congruence class. The first concept was fairly explicit in the work of Fermat and his contemporaries, and both concepts occur in Euler's later works in the mid-eighteenth century, but the notation which we use now was introduced by Carl Friedrich Gauss (1777–1855) in his *Disquisitiones Arithmeticae* published in 1801, which begins with a thorough treatment of these ideas.

Definition Suppose that n is an integer greater than 1, and let a, b be integers. We say that a is **congruent** to b **mod(ulo)** n if a and b have the same remainder when divided (according to 1.1.1) by n. We write

$$a \equiv b \bmod n$$

if this is so.

The definition may be more usefully formulated as follows:

$$a \equiv b \bmod n \text{ if and only if } n \text{ divides } a - b.$$

Examples $-1 \equiv 4 \bmod 5$

$$6 \equiv 18 \bmod 12$$

$$19 \equiv -5 \bmod 12.$$

The notion of two integers being congruent modulo some fixed integer is actually one with which we are familiar from special cases in everyday life. For example, if we count days from now, then day k and day m will be the same day of the week if, when divided by 7, k and m have the same remainder, that is, if $k \equiv m \bmod 7$. Similarly a clock works, in hours, modulo 12 (or 24). Christmas in 1988 fell on a Sunday: therefore Christmas 1989 fell on a Monday since there were 365 days in 1989 (not a leap year), and 365 is congruent to 1 modulo 7: therefore the day of the week on which Christmas fell moved one day forward. For another example, take measurements of angles, where it is often appropriate to work modulo 360 degrees.

Notes (i) The condition 'n divides a' can be written as '$a \equiv 0 \bmod n$.'

(ii) The properties of congruence '$\equiv$' are very similar to those of the usual equality sign '$=$'. For example it is permissible to add to, or subtract from, both sides of any congruence the same quantity, or to multiply both sides by a constant. Thus if a, b and c are integers and if

$$a \equiv b \bmod n$$

so, by definition, a and b have the same remainder when divided by n, then

$$a + c \equiv b + c \bmod n$$
$$a - c \equiv b - c \bmod n, \quad \text{and}$$
$$ca \equiv cb \bmod n.$$

However, the situation for division is more complicated, as we shall see.

Now we may return to the problem at the beginning of this section. Let us take any integer m and square it: what are the possibilities for m^2 modulo 4? It is an easy consequence of the rules above that the value of m^2 modulo 4 depends only on the value of m modulo 4. For example, if $m \equiv 3 \bmod 4$ so $m = 4k + 3$ for some k in $\mathbb{Z}$, then

$$m^2 = (4k + 3)^2 = 16k^2 + 24k + 9 = 4(4k^2 + 6k) + 9 \equiv 9 \equiv 1 \bmod 4.$$

If m is respectively congruent to 0, 1, 2, 3 modulo 4 then m^2 is respectively congruent to 0, 1, 4, 9 modulo 4, and these are in turn congruent to 0, 1, 0, 1 modulo 4. Therefore if an integer k is the sum of two squares, say $k = n^2 + m^2$, then, modulo 4, the possibilities for k are

$$(0 \text{ or } 1) + (0 \text{ or } 1).$$

In particular, it is impossible for k to be congruent to 3 modulo 4. In other words, a sum of two squares cannot be of the form $4k + 3$.

Consider 'equations' involving this notion: such equations are called **congruences**. For a specific example take

$$2x \equiv 0 \bmod 4.$$

What should be meant by a *solution* to this congruence?

Notice that there are infinitely many integer values for 'x' which will solve it:

$$\ldots, -4, -2, 0, 2, 4, 6, \ldots.$$

These 'solutions' may, however, be divided into two classes, namely:

$$\ldots, -8, -4, 0, 4, 8, \ldots \text{ and} \ldots, -6, -2, 2, 6, 10, \ldots$$

where, within each class, all the integers are congruent to each other modulo 4, but no integer in the one class is congruent modulo 4 to any integer in the other class. So in some sense the congruence

$$2x \equiv 0 \bmod 4$$

may be thought of as having essentially two solutions, where each solution is a 'congruence class' of integers. We make the following definition.

Definition   Fix an integer n greater than 1 and let a be any integer. The **congruence class** of a **modulo** n is the set of all integers which are congruent to a modulo n:

$$[a]_n = \{b : b \equiv a \bmod n\}.$$

The set of all congruence classes modulo n is referred to as the **set of integers modulo** n and is denoted $\mathbb{Z}_n$. Observe that this is a set with n elements, for there are exactly n possibilities for the remainder when an integer is divided by n. By the **zero congruence class** we mean the congruence class of 0 (that is, the congruence class consisting of all multiples of n).

Note that

$$[a]_n = [b]_n \text{ if and only if } a \equiv b \bmod n.$$

Example 1   When n is 2, there are two congruence classes namely $[0]_2$, which is the set of even integers and $[1]_2$ (the set of odd integers).

Example 2   When n is 10, the positive integers in a given congruence class are those which have the same last digit when written (as usual) in base 10.

The solutions to

$$2x \equiv 0 \bmod 4$$

are, therefore, the congruence classes $[0]_4$ and $[2]_4$.

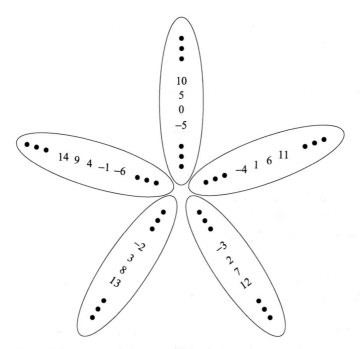

Fig. 1.3 Congruence classes in $\mathbb{Z}_5$.

There are many ways of representing a given congruence class: for example we could equally well have written any of $\ldots, [-4]_4, [4]_4, [8]_4, \ldots$ in place of $[0]_4$; similarly $\ldots, [-2]_4, [2]_4, [6]_4, \ldots$ all equal $[2]_4$.

Since every element of $\mathbb{Z}_n$ may be represented in infinitely many ways it is useful to fix a set of standard representatives (by a **representative** of a congruence class we mean any integer in that class): these are usually taken to be the integers from 0 to $n - 1$. Thus

$$\mathbb{Z}_n = \{[0]_n, [1]_n, [2]_n, \ldots, [n-2]_n, [n-1]_n\}.$$

For example

$$\mathbb{Z}_5 = \{[0]_5, [1]_5, [2]_5, [3]_5, [4]_5\} \qquad \text{(see Fig. 1.3)},$$
$$\mathbb{Z}_2 = \{[0]_2, [1]_2\}.$$

We may drop the subscript 'n' when doing so leads to no ambiguity. Also for convenience sometimes we denote the congruence class $[a]_n$ simply by a, provided it is clear from the context that $[a]_n$ is meant.

One may say that the notions of congruence modulo n and congruence classes were implicit in early work in number theory in the sense that if one refers to, say, 'integers of the form $4n + 3$' then one is implicitly referring to the congruence class $[3]_4$. These notions only became explicit with Euler: as his work in number theory developed, he became increasingly aware that he was working not with numbers, but with certain sets of numbers. However, the *Tractatus de Numerorum Doctrina*, in which he systematically developed these notions (around 1750), was not published by him, although he did incorporate many of the results in various of his papers. The *Tractatus* was printed posthumously in 1830 but by then it had been superseded by Gauss' *Disquisitiones Arithmeticae* (1801). In that work Gauss went considerably further than Euler had. The notations that we use here for congruence and for congruence classes were introduced by Gauss.

Consideration of $\mathbb{Z}_n$ would be rather pointless if we could not do arithmetic modulo n: in fact $\mathbb{Z}_n$ inherits the arithmetic operations of $\mathbb{Z}$, as follows.

Definition   Fix an integer n greater than 1 and let a, b be any integers. Define the **sum** and **product** of the congruence classes of a and b by:

$$[a]_n + [b]_n = [a + b]_n,$$
$$[a]_n \times [b]_n = [a \times b]_n.$$

(As usual $[a]_n \cdot [b]_n$ or just $[a]_n[b]_n$ may also be used to denote the product.)

There is a potential problem with this definition. We have defined the sum (and product) of two congruence classes by reference to particular representatives of the classes. How can we be sure that if we chose to represent $[a]_n$ in some other form (say as $[a + 99n]_n$) then we would get the same congruence class for the sum? Well, we can check.

Before giving the general proof, let us illustrate this point with an example. Suppose that we take $n = 6$ and we wish to compute $[3]_6 + [5]_6$. By the definition above, this is $[3 + 5]_6 = [8]_6 = [2]_6$. But $[3]_6 = [21]_6$ so we certainly want to have $[3]_6 + [5]_6 = [21]_6 + [5]_6$. We have just seen that the term on the left is equal to $[2]_6$, so if our definition of addition of congruence classes is a good one then the term on the right-hand side should turn out to be the same. We check: $[21]_6 + [5]_6 = [26]_6 = [2]_6$, so no problem has appeared. Of course we also have $[3]_6 = [-9]_6$, so it should also be that $[-9]_6 + [5]_6 = [2]_6$, and you may check that this is so. These are just two cases checked: but there are infinitely many representatives for $[3]_6$ (and for $[5]_6$). So we need a general proof that the definitions are good: such a proof is given next.

Theorem 1.4.1   *Let n be an integer greater than 1 and let a, b and c be any integers. Suppose that*

$$[a]_n = [c]_n.$$

Then:

(i) $[a + b]_n = [c + b]_n$, *and*
(ii) $[ab]_n = [cb]_n$.

Proof (i) Since $[a]_n = [c]_n$, n divides $c - a$. So we can write

$$c = a + kn$$

for some integer k. Therefore

$$[c + b]_n = [a + kn + b]_n$$
$$= [a + b + kn]_n$$
$$= [a + b]_n \text{ (by definition of congruence class)}$$

as required.

(ii) With the above notation, we have that

$$[cb]_n = [(a + kn)b]_n$$
$$= [ab + nkb]_n$$
$$= [ab]_n. \quad \square$$

Comment The proof itself is, we hope, easy to follow line by line. In the discussion before the result we tried to explain the purpose of the theorem and proof. Experience suggests, however, that students often find this rather baffling so we say just a little more.

We are going to make $\mathbb{Z}_n$ into an algebraic structure: in particular we want to add and multiply congruence *classes*. In the statement of Theorem 1.4.1 we started by saying 'Suppose that $[a]_n = [c]_n$', in other words, suppose that a and c belong to the same congruence class. Then in (i) we take an element b in a possibly different class and add it to each of a and c. The assertion we prove is that the two resulting integers, $a + b$ and $c + b$, belong to the same class. Part (ii) says the corresponding thing for multiplication.

By symmetry we can also replace b by an element d in the same class as b, so the next result is an immediate corollary.

Corollary 1.4.2 *If*

$$[a]_n = [c]_n \text{ and } [b]_n = [d]_n$$

then

(i) $[a + b]_n = [c + d]_n$, *and*
(ii) $[ab]_n = [cd]_n$.

Therefore we may write

$$[a]_n + [b]_n = [a+b]_n, \text{ and}$$
$$[a]_n[b]_n = [ab]_n$$

without ambiguity.

Example Show that 11 divides $10! + 1$ (recall from Section 1.2 that $10! = 10 \times 9 \times 8 \times 7 \times 6 \times 5 \times 4 \times 3 \times 2 \times 1$). It is not necessary to compute $10! + 1$ and then find the remainder modulo 11, rather we reduce modulo 11 as we go along:

$$2 \times 3 \times 4 = 24 \equiv 2 \bmod 11,$$

$$6! = (2 \times 3 \times 4) \times 5 \times 6 \equiv 2 \times 5 \times 6 \bmod 11$$
$$\equiv 60 \bmod 11$$
$$\equiv 5 \bmod 11,$$
$$6! \times 7 \equiv 5 \times 7 \bmod 11$$
$$\equiv 35 \bmod 11$$
$$\equiv 2 \bmod 11,$$
$$7! \times 8 \equiv 2 \times 8 \bmod 11$$
$$\equiv 16 \bmod 11$$
$$\equiv 5 \bmod 11,$$
$$8! \times 9 \times 10 \equiv 5 \times 9 \times 10 \bmod 11$$
$$\equiv 5 \times (-2) \times (-1) \bmod 11$$
$$\equiv 10 \bmod 11.$$

Therefore $10! + 1 \equiv 10 + 1 \equiv 0 \bmod 11$, as required.

Example In the last stage of the computation above, we simplified by replacing 9 and 10 mod 11 by -2 and -1 respectively. Similarly, if we wish to compute the standard representative of, say $([13]_{18})^3$ then we can make use of the fact that $13 \equiv -5 \bmod 18$:

$$13^3 \equiv (-5)^3 \equiv 25 \times (-5) \equiv 7 \times (-5) \equiv -35 \equiv 1 \bmod 18$$

(for the last step we added a suitable multiple of 18, in this case 36).

We can make addition and multiplication tables for $\mathbb{Z}_n$ as given below when n is 8, where the entry in the intersection of the a-row and b-column is $[a]_n + [b]_n$ (or $[a]_n \times [b]_n$, as appropriate). Note that we abbreviate $[a]_8$ to a in these tables.

Addition and multiplication tables for $\mathbb{Z}_8$

+	0	1	2	3	4	5	6	7
0	0	1	2	3	4	5	6	7
1	1	2	3	4	5	6	7	0
2	2	3	4	5	6	7	0	1
3	3	4	5	6	7	0	1	2
4	4	5	6	7	0	1	2	3
5	5	6	7	0	1	2	3	4
6	6	7	0	1	2	3	4	5
7	7	0	1	2	3	4	5	6

×	0	1	2	3	4	5	6	7
0	0	0	0	0	0	0	0	0
1	0	1	2	3	4	5	6	7
2	0	2	4	6	0	2	4	6
3	0	3	6	1	4	7	2	5
4	0	4	0	4	0	4	0	4
5	0	5	2	7	4	1	6	3
6	0	6	4	2	0	6	4	2
7	0	7	6	5	4	3	2	1

Definition   Fix an integer n greater than 1, and let a be any integer. We say that $[a]_n$ is **invertible** (or a is **invertible modulo** n) if there is an integer b such that

$$[a]_n[b]_n = [1]_n$$

(that is, such that $ab \equiv 1 \bmod n$), in which case $[b]_n$ is the **inverse** of $[a]_n$, and we write $[b]_n = [a]_n^{-1}$. We say that a non-zero congruence class $[a]_n$ is a **zero-divisor** if there exists an integer b with

$$[b]_n \neq [0]_n \text{ and } [a]_n[b]_n = [0]_n$$

(in which case, note, $[b]_n$ also is a zero-divisor).

Example   In $\mathbb{Z}_8$ there are elements, such as $[5]_8$, other than $\pm[1]_8$ with multiplicative inverses. Also there are elements other than $[0]_8$ which are zero-divisors, for instance $[2]_8$ (because $[2]_8[4]_8 = [0]_8$).

How do we tell if a given congruence class has an inverse? and if the class is invertible how may we set about finding its inverse?

Theorem 1.4.3 *Let n be an integer greater than or equal to 2, and let a be any integer. Then $[a]_n$ has an inverse if and only if the greatest common divisor of a and n is 1. In fact, if r and s are integers such that*

$$ar + ns = 1$$

then the inverse of $[a]_n$ is $[r]_n$.

Proof   Since n is fixed, we will leave off the subscripts from congruence classes. Suppose first that $[a]$ has an inverse, $[k]$ say. So $[ak]$ is equal to $[1]$. Hence

$$ak \equiv 1 \bmod n,$$

that is, n divides $ak - 1$. Therefore, for some integer t,

$$ak - 1 = nt.$$

Hence

$$ak - nt = 1,$$

which, by Corollary 1.1.3, means that the greatest common divisor, (a, n), of a and n is 1.

Suppose, conversely, that (a, n) is 1 and that r and s are integers such that

$$ar + ns = 1.$$

It follows that $ar - 1$ is divisible by n, and so

$$ar \equiv 1 \bmod n,$$

that is,

$$[a][r] = [1],$$

as required. □

Comment   The important thing here is to look at the equation $ar + ns = 1$ and to realise that if it is 'reduced modulo n' then, because n becomes 0, the term ns disappears and we are left with the equation $[ar]_n = [1]_n$, that is $[a]_n[r]_n = [1]_n$ so there, plainly before us, is an inverse for $[a]_n$.

It is 'if and only if' because the argument just outlined (that if the gcd is 1 then a has an inverse modulo n) reverses: from the conclusion, that a has an inverse mod n, we can work back to the assumption that $(a, n) = 1$.

Since we already have a method for expressing the greatest common divisor of two integers as an integral linear combination of them, the above theorem provides us with a practical method for finding out if a congruence class is invertible and, at the same time, calculating its inverse.

Example 1   We saw in Section 1.1 that

$$1 = -91 \cdot 507 + 118 \cdot 391$$

and so the inverse of 391 modulo 507 is 118.

Example 2   If a is 215 and n is 795, since 5 divides both these integers, their greatest common divisor is not 1 and so 215 has no inverse modulo 795.

Example 3 Let a be 23 and let n be 73. The matrix method for finding the gcd of a and n gives

$$\begin{pmatrix} 1 & 0 & \vline & 73 \\ 0 & 1 & \vline & 23 \end{pmatrix} \rightarrow \begin{pmatrix} 1 & -3 & \vline & 4 \\ 0 & 1 & \vline & 23 \end{pmatrix} \rightarrow \begin{pmatrix} 1 & -3 & \vline & 4 \\ -5 & 16 & \vline & 3 \end{pmatrix} \rightarrow \begin{pmatrix} 6 & -19 & \vline & 1 \\ -5 & 16 & \vline & 3 \end{pmatrix}.$$

From the top row we have

$$6 \cdot 73 - 19 \cdot 23 = 1.$$

'Reduce this equation modulo 73' to obtain

$$[-19]_{73}[23]_{73} = [1]_{73}$$

and so the inverse of 23 modulo 73 is -19.

It is usual to express the answer using the standard representative, and so normally we would say that the inverse of 23 modulo 73 is 54 $(= -19 + 73)$ and write $[23]_{73}^{-1} = [54]_{73}$.

Example 4 When the numbers involved are small it can be cumbersome to use the matrix method, and inverses can often be found quite easily by inspection. For example, if we wish to find the inverse of 8 modulo 11, then we are looking for an integer multiple of 8 which has remainder 1 when divided by 11, so we can inspect multiples of 11, plus 1, for divisibility by 8: one observes that

$$55 + 1 = 56 = 7 \times 8,$$

and so it follows that the inverse of 8 modulo 11 is 7. Similarly, observing that

$$11^2 = 121 \equiv 1 \bmod 20$$

one sees that $[11]_{20}$ is its own inverse (is 'self-inverse'):

$$[11]_{20}^{-1} = [11]_{20}.$$

A method for finding inverses modulo n (when they exist) is found in Bachet's *Problèmes plaisants et délectables* (1612), but Brahmagupta, who flourished about AD 628, had already given the general solution.

We now give three results which may be regarded as consequences of Theorem 1.4.3. The first of these considers the problem of cancelling in congruences.

Corollary 1.4.4 *Let n be an integer greater than or equal to 2, and let a, b, c be any integers. If n and c are relatively prime and if*

$$ac \equiv bc \bmod n,$$

then

$$a \equiv b \bmod n.$$

Proof   The congruence may be written as the equation

$$[a]_n[c]_n = [b]_n[c]_n.$$

Since n and c are relatively prime, it follows by Theorem 1.4.3 that $[c]_n^{-1}$ exists.
So we multiply each side of the equation on the right by $[c]_n^{-1}$ to obtain

$$[a]_n[c]_n[c]_n^{-1} = [b]_n[c]_n[c]_n^{-1}.$$

Hence $[a]_n[1]_n = [b]_n[1]_n$
and so $[a]_n = [b]_n.$
Therefore $a \equiv b \bmod n$ as required.   □

Comment   The idea of dividing each side of an equation by the same thing
is surely familiar and is used elsewhere in this book (e.g. in the last part of
the proof of the next result). Care must be taken, however, because dividing
by something is really multiplying by the inverse of that thing and not every
congruence class has an inverse. Dividing can be hazardous – it is easy, if you
are not experienced, to 'divide by 0': better to multiply by the inverse since
doing that explicitly points up the issue of whether the inverse exists.

Note   The assumption in 1.4.4 that $(c, n) = 1$ is needed. For example $30 \equiv$
$6 \bmod 8$, but if we try to divide both sides by 2 (which is *not* relatively prime
to 8) then we get '$15 \equiv 3 \bmod 8$', which is false. On the other hand since $(3,8)$
$= 1$ we *can* divide both sides by 3 to obtain the congruence $10 \equiv 2 \bmod 8$.

Corollary 1.4.5   *Let n be an integer greater than 1. Then each non-zero element
of $\mathbb{Z}_n$ is either invertible or a zero-divisor, but not both.*

Proof   Suppose that $[a]_n$ is not invertible. So, by Theorem 1.4.3, the greatest
common divisor, d, of n and a is greater than 1. Since d divides a and n we have
that $a = kd$ for some k and also $n = td$ where t is a positive integer necessarily
less than n. It follows that $at = ktd$ is divisible by n. Hence

$$[a]_n[t]_n = [0]_n$$

and so, since $[t]_n \neq [0]_n$, $[a]_n$ is indeed a zero-divisor.
    To see that an element cannot be both invertible and a zero-divisor, suppose
that $[a]_n$ is invertible. Then, given any equation $[a]_n[b]_n = [0]_n$, we can multiply
both sides by $[a]_n^{-1}$ and simplify to obtain $[b]_n = [0]_n$, so, from the definition,
$[a]_n$ is not a zero-divisor.   □

Comment   If the first part of the argument is not clear to you then run through
it with some numbers in place of n and a (and hence d).

Example   This corollary implies that every non-zero congruence class in, for example, $\mathbb{Z}_{14}$ is invertible or a zero-divisor, but not both. By Theorem 1.4.3 the invertible congruence classes are $[1]_{14}, [3]_{14}, [5]_{14}, [9]_{14}, [11]_{14}$ and $[13]_{14}$. By Corollary 1.4.5 all the rest are zero-divisors. We can see all this explicitly: for invertibility we have $[1]_{14}^2 = [1]_{14}, [3]_{14} \cdot [5]_{14} = [1]_{14}, [9]_{14} \cdot [11]_{14} = [1]_{14}, [13]_{14}^2 = [-1]_{14}^2 = [1]_{14}$; also $[2]_{14} \cdot [7]_{14} = [0]_{14}$ and so each of $[4]_{14}, [6]_{14}, [8]_{14}, [10]_{14}, [12]_{14}$, multiplied by $[7]_{14}$ gives $[0]_{14}$ and hence is a zero-divisor.

The result shows that these new arithmetic structures $\mathbb{Z}_n$ can be rather strange: they can have elements which are not zero but which multiply together to give zero, so working in them requires some care.

The next result says that if we are working modulo a prime then things are better (but we still have to remember that non-zero elements can *add* together to give zero: there is no concept in $\mathbb{Z}_n$ of an element being 'greater than zero').

This next result, in essentially this form, was given by Euler.

Corollary 1.4.6   *Let p be a prime. Then every non-zero element of $\mathbb{Z}_p$ is invertible.*

Proof   If $[a]_p$ is non-zero then p does not divide a and so a and p are relatively prime. Then the result follows by Theorem 1.4.3.   □

To conclude this section, we consider a special subset of $\mathbb{Z}_n$.

Definition   Let n be an integer greater than 1. We denote by G_n (some authors write $\mathbb{Z}_n^*$) the set of invertible congruence classes of $\mathbb{Z}_n$. By 1.4.3 $[a]_n$ is in G_n if and only if a is relatively prime to n.

Theorem 1.4.7   *Let n be an integer greater than or equal to 2. The product of any two elements of G_n is in G_n.*

Proof   Suppose that $[a]$ and $[b]$ are in G_n. So each of a and b is relatively prime to n. Since any prime divisor, p, of ab must also divide one of a or b (by 1.3.1) it follows that ab and n have no prime common factor and hence no common factor greater than 1. Therefore, by 1.4.3, ab is invertible modulo n.   □

Example   When n is 20, G_n consists of the classes

$$[1], [3], [7], [9], [11], [13], [17], [19].$$

We can form the multiplication table for G_{20} as follows, where we write $[a]_{20}$ more simply as a.

	1	3	7	9	11	13	17	19
1	1	3	7	9	11	13	17	19
3	3	9	1	7	13	19	11	17
7	7	1	9	3	17	11	19	13
9	9	7	3	1	19	17	13	11
11	11	13	17	19	1	3	7	9
13	13	19	11	17	3	9	1	7
17	17	11	19	13	7	1	9	3
19	19	17	13	11	9	7	3	1

Observe the way in which the above 8 by 8 table breaks into four 4 by 4 blocks. We shall see later (in Section 5.3) why this happens.

Of course, 1.4.7 may be extended (by induction) to the statement that the product of any finite number of, possibly repeated, elements of G_n lies in G_n. A particular case of this is obtained when all the elements are equal: that is, if a is any member of G_n then every positive power a^k is in G_n. It is easy to show (again, by induction) that the inverse of a^k is $(a^{-1})^k$: for this, the notation a^{-k} is employed.

Exercises 1.4

1. Determine which of the following are true (a calculator will be useful for the larger numbers):
   (i) $8 \equiv 48 \bmod 14$,   (ii) $-8 \equiv 48 \bmod 14$,
   (iii) $10 \equiv 0 \bmod 100$,   (iv) $7754 \equiv 357482 \bmod 3643$,
   (v) $16023 \equiv 1325227 \bmod 25177$,   (vi) $4015 \equiv 33303 \bmod 1295$.
2. Construct the addition and multiplication tables for $\mathbb{Z}_n$ when n is 6 and when n is 7.
3. Find the following inverses, if they exist:
   (i)   the inverse of 7 modulo 11;
   (ii) the inverse of 10 modulo 26;
   (iii) the inverse of 11 modulo 31;
   (iv) the inverse of 23 modulo 31;
   (v)   the inverse of 91 modulo 237.
4. Write down the multiplication table for G_n when n is 16 and when n is 15.
5. Show that no integer of the form $8n + 7$ can be written as a sum of three squares.
6. Let p be a prime number. Show that the equation $x^2 = [1]_p$ has just two solutions in $\mathbb{Z}_p$.
7. Let p be a prime number. Show that

$$(p - 1)! \equiv -1 \bmod p.$$

8. Choose a value of n and count the number of elements in G_n. Try this with various values of n. Can you discover any rules governing the relation between n and the number of elements in G_n? [In Section 1.6 below we give rules for computing the number of elements in G_n directly from n.]

9. The observation that $10 \equiv 1 \bmod 9$ is the basis for the procedure of 'casting out nines'. The method is as follows.

Given an integer X written in base 10 (as is usual), compute the sum of the digits of X: call the result the **digit sum** of X. If the digit sum is greater than 9, we form the digit sum again. Continue in this way to obtain the **iterated digit sum** which is at most 9. (Thus 5734 has digit sum 19 which has digit sum 10 which has digit sum 1, so the iterated digit sum of 5734 is 1.)

Now suppose that we have a calculation which we want to check by hand: say, for example, someone claims that

$$873\,985 \times 79\,041 = 69\,069\,967\,565.$$

Compute the iterated digit sums of $873\,985$ and $79\,041$ (these are 4 and 3 respectively), multiply these together (to get 12), and form the iterated digit sum of the product (which is 3). Then the result should equal the iterated digit sum of $69\,069\,967\,565$ (which is 5). Since it does not, the 'equality' is incorrect. If the results had been equal then all we could say would be that no error was detected.

(i)   Using the method of casting out nines what can you say about the following computations?

$$56\,563 \times 9961 = 563\,454\,043;$$
$$1234 \times 5678 \times 901 = 6\,213\,993\,452;$$
$$333 \times 666 \times 999 = 221\,556\,222.$$

(ii)   The following equation is false but you are told that only the underlined digit is in error. What is the correct value for that digit?

$$674\,532 \times 9764 = 6\,586\,1\underline{4}0\,448.$$

(iii)   Justify the method of casting out nines.

1.5   Solving linear congruences

A **linear congruence** is an 'equation' of the form

$$ax \equiv b \bmod n$$

where x is an integer variable. Written in terms of congruence classes this becomes the equation

$$[a]_n X = [b]_n$$

where a solution X is now to be a congruence class.

Such an equation may have

(i) no solution (as, for example, $2x \equiv 1 \bmod 4$),
(ii) exactly one solution (for example $2x \equiv 1 \bmod 5$), or
(iii) more than one solution (for example the congruence $2x \equiv 0 \bmod 4$ discussed at the beginning of Section 1.4).

The first result shows how to distinguish between these cases and how to find all solutions for such a congruence (if there are any). This result was first given by Brahmagupta (c. 628). Of course he did not express it as we have done: rather he gave the criterion for solvability of, and the general solution of, $ak + nt = b$, where a, n, b are fixed integers and k and t are integer unknowns. (Note that if we have solved $ax \equiv b \bmod n$ then if k is a solution for x we have that n divides $ak - b$, that is, $ak - b = ns$ for some integer s so, writing t for $-s$, we have $ak + nt = b$. Since a, k, n and b are known we compute t from this equation. Therefore solving $ak + nt = b$ for k and t is equivalent to solving the congruence $ax \equiv b \bmod n$ for x.) An equation of the form $ak + nt = b$ is 'indeterminate' in the sense that, since it is just one equation with two unknowns, it has infinitely many solutions if it has any at all. One sees, however, that the solutions form themselves into complete congruence classes.

Theorem 1.5.1   *The linear congruence*

$$ax \equiv b \bmod n$$

has solutions if and only if the greatest common divisor, d, of a and n divides b. If d does divide b there are d solutions up to congruence modulo n, and these solutions are all congruent modulo n/d.

Proof   Suppose that there is a solution, c say, to

$$ax \equiv b \bmod n.$$

Then, since

$$ac \equiv b \bmod n,$$

we have that n divides $ac - b$; say

$$ac - b = nk.$$

Rearrange this to obtain

$$b = ac - nk.$$

The greatest common divisor d of a and n divides both terms on the right-hand side of this equation, and hence we deduce that d divides b, as claimed.

Conversely, suppose d divides b, say $b = de$. Write d as a linear combination of a and n; say

$$d = ak + nt.$$

Multiply this by e to obtain

$$b = ake + nte.$$

This gives

$$a(ke) \equiv b \bmod n,$$

and so the congruence has a solution, ke, as required. Therefore the first assertion of the theorem has been proved.

Suppose now that c is a solution of

$$ax \equiv b \bmod n.$$

So as before we have

$$ac = b + nk$$

for some integer k. By the above, d divides b and hence we may divide this equation by d to get the equation in integers

$$(a/d)c = b/d + (n/d)k.$$

Thus

$$(a/d)c \equiv b/d \bmod (n/d).$$

That is, every solution of the original congruence is also a solution of the congruence

$$(a/d)x \equiv b/d \bmod (n/d).$$

Conversely it is easy to see (by reversing the steps) that every solution to this second congruence is also a solution to the original one. So the solution is really a congruence class modulo n/d. Such a congruence class splits into d distinct congruence classes modulo n. Namely if c is a solution then the congruence classes of

$$c, c + (n/d), c + 2(n/d), c + 3(n/d), \ldots, c + (d-1)(n/d)$$

are distinct solutions modulo n, and are all the solutions modulo n. □

Comment We strongly suggest working through the above proof with particular values for a, b and n (say, the values from Example 3 (or 4) below). Try running the proof with particular numbers parallel to the proof with letters to see how the general and special cases relate to each other.

This yields the following method for solving a linear congruence.

To find all solutions of the linear congruence $ax \equiv b \bmod n$.

1. Calculate $d = (a, n)$.
2. Test whether d divides b.
    (a) If d does not divide b then there is no solution.
    (b) If d divides b then there are d solutions mod n.
3. To find the solutions in case (b), 'divide the congruence throughout by d' to get

$$(a/d)\, x \equiv (b/d) \bmod (n/d).$$

Notice that since a/d and n/d have greatest common divisor 1, this congruence will have a unique solution.
4. Calculate the inverse $[e]_{n/d}$ of $[a/d]_{n/d}$ (by inspection or by the matrix method).
5. Multiply to get

$$[x]_{n/d} = [e]_{n/d}[b/d]_{n/d}$$

and calculate a solution, c, for x.
6. The solutions to the original congruence will be the classes modulo n of

$$c, c + (n/d), \ldots, c + (d - 1)(n/d).$$

Example 1 Solve the congruence

$$6x \equiv 5 \bmod 17.$$

Since $(6, 17) = 1$ and 1 divides 5 there is, by Theorem 1.5.1, a unique solution modulo 17. It is found by calculating $[6]_{17}^{-1}$ (found by inspection to be $[3]_{17}$) and multiplying both sides by this inverse. We obtain

$$x \equiv 3 \times 5 \equiv 15 \bmod 17$$

as the solution (unique up to congruence mod 17). (Therefore the values of x which are solutions are $\ldots, -19, -2, 15, 32, \ldots$.)

Example 2 To solve

$$6x \equiv 5 \bmod 15$$

note that $(6, 15) = 3$ and 3 does not divide 5, so by 1.5.1 there is no solution.

Example 3 In the congruence

$$6x \equiv 9 \bmod 15,$$

$(6, 15) = 3$ and 3 divides 9, so by 1.5.1 there are three solutions up to congruence modulo 15.

  To find these we find the solutions up to congruence modulo 5 ($5 = 15/3$), and we do this by dividing the whole congruence by the greatest common divisor of 6 and 15. This gives

$$2x \equiv 3 \bmod 5.$$

Now, $(2, 5) = 1$ so there is a unique solution (this is the point of dividing through by the gcd). One quickly sees that

$$x \equiv 4 \bmod 5$$

is the unique solution mod 5. The proof of 1.5.1 shows that the solutions of the original congruence are therefore the members of the congruence class $[4]_5$. In order to describe the solutions in terms of congruence classes modulo 15, we note that $[4]_5$ splits up as

$$[4]_{15}, [4 + 5]_{15}, [4 + 10]_{15}$$

that is, as

$$[4]_{15}, [9]_{15}, [14]_{15}.$$

Example 4 Solve the congruence

$$432x \equiv 12 \bmod 546.$$

The first task is to calculate the greatest common divisor of 432 and 546. Since we do not need to express this as a linear combination of 432 and 546 it is not necessary to use the matrix method: it is enough to factorise these numbers. We have that 432 is 6 times 72 while 546 is 6 times 91: since 91 is 7×13 and 72 is 8×9 one sees that 432 and 546 have no common factor greater than 6. Dividing the congruence by 6 gives

$$72x \equiv 2 \bmod 91.$$

  The next task is to find the inverse of 72 modulo 91, and unless the reader is unusually gifted at arithmetic calculations, this is best done using the matrix method:

$$\left(\begin{array}{cc|c} 1 & 0 & 91 \\ 0 & 1 & 72 \end{array} \right) \rightarrow \left(\begin{array}{cc|c} 1 & -1 & 19 \\ 0 & 1 & 72 \end{array} \right) \rightarrow \left(\begin{array}{cc|c} 1 & -1 & 19 \\ -3 & 4 & 15 \end{array} \right) \rightarrow \left(\begin{array}{cc|c} 4 & -5 & 4 \\ -3 & 4 & 15 \end{array} \right)$$

$$\rightarrow \left(\begin{array}{cc|c} 4 & -5 & 4 \\ -15 & 19 & 3 \end{array} \right) \rightarrow \left(\begin{array}{cc|c} 19 & -24 & 1 \\ -15 & 19 & 3 \end{array} \right).$$

The top line of this matrix corresponds to the equation $19 \cdot 91 - 24 \cdot 72 = 1$ so it follows that the inverse of 72 modulo 91 is -24, or 67. So multiply both sides of the congruence by 67 to obtain

$$x \equiv 2 \times 67 \bmod 91$$
$$\equiv 134 \bmod 91$$
$$\equiv 43 \bmod 91.$$

Finally, to describe the solutions in terms of congruence classes modulo 546, we have that $[43]_{91}$ splits into six congruence classes modulo 546, namely

$$[43]_{546}, [134]_{546}, [225]_{546}, [316]_{546}, [407]_{546}, [498]_{546}.$$

Next we consider how to solve systems of linear congruences.

Suppose that we wish to find an integer which, when divided by 7 has a remainder of 3, and when divided by 25 has a remainder of 6. Is there such an integer? and if so how does one find it?

This question may be formulated in terms of congruences as: find an integer x that satisfies

$$x \equiv 3 \bmod 7 \text{ and } x \equiv 6 \bmod 25.$$

The next theorem implies that there is a simultaneous solution to these congruences, and its proof tells us how to find a solution.

The theorem may have been known to the eighth century Buddhist monk Yi Xing. Certainly it appears in Qín Jiǔsháo's *Shù shū jiǔ zhāng* (*Mathematical Treatise in Nine Sections*) of 1247.

Theorem 1.5.2 (Chinese Remainder Theorem) *Suppose that $m \geq 2$ and $n \geq 2$ are relatively prime integers and that a and b are any integers. Then there is a simultaneous solution to the congruences*

$$x \equiv a \bmod m,$$
$$x \equiv b \bmod n.$$

The solution is unique up to congruence mod mn.

Proof Since m and n are relatively prime, there exist integers k and t such that

$$mk + nt = 1. \qquad (*)$$

Then it is easily checked that $c = bmk + ant$ is a simultaneous solution for the congruences. For,

$$c \equiv ant \bmod m$$

and, from equation ($*$)

$$nt \equiv 1 \bmod m.$$

Hence

$$c \equiv a \times 1 = a \bmod m.$$

The proof that c is congruent to b modulo n is similar.

To show that the solution is unique up to congruence modulo mn, suppose that each of c, d is a solution to both congruences. Then

$$c \equiv a \bmod m \text{ and } d \equiv a \bmod m.$$

Hence

$$c - d \equiv 0 \bmod m.$$

Similarly

$$c - d \equiv 0 \bmod n.$$

That is, $c - d$ is divisible by both m and n. Since m and n are relatively prime it follows by Theorem 1.1.6(ii) that $c - d$ is divisible by mn, and hence c and d lie in the same congruence class mod mn.

Conversely, if c is a solution to both congruences and if

$$d \equiv c \bmod mn$$

then d is of the form $c + kmn$, and so the remainder when d is divided by m or n is the same as the remainder when c is divided by m or n. So d solves both congruences, as required. □

Comment   For the first part of the proof (existence of a solution) notice that the equation $mk + nt = 1$, when reduced modulo n, becomes $[mk]_n = [1]_n$ so, if we multiply both sides by $[b]_n$ we obtain $[bmk]_n = [b]_n$. That is where the term bmk in $c = bmk + ant$ comes from, similarly (reducing mod m) for the other term.

Example   Consider the problem, posed before the statement of Theorem 1.5.2, of finding a solution to the congruences

$$x \equiv 3 \bmod 7 \text{ and } x \equiv 6 \bmod 25.$$

First, find a combination of 7 and 25 which is 1: one such combination is

$$7(-7) + 25 \times 2 = 1.$$

Then we multiply these two terms by 6 and 3 respectively. (Note the 'swop over'!) This gives us

$$6 \cdot 7 \cdot (-7) + 3 \cdot 25 \cdot 2 = -144.$$

So the solution is $[-144]_{175}$ $(175 = 7 \cdot 25)$. We should put this in standard form by adding a suitable multiple of 175: we obtain that the solution is $[31]_{175}$.

Alternatively, there is a method for solving this type of problem which does not involve having to remember how to construct the solution. We repeat the above example to illustrate this method.

A solution of the first congruence is of the form

$$x = 3 + 7k,$$

so if x satisfies the second congruence, we have

$$3 + 7k \equiv 6 \bmod 25.$$

Now solve this congruence for k: we have

$$7k \equiv 3 \bmod 25.$$

The inverse of 7 modulo 25 is 18 (by inspection), so

$$k \equiv 3 \times 18 \bmod 25$$
$$\equiv 4 \bmod 25.$$

Thus, for some integer r,

$$x = 3 + 7(4 + 25r)$$
$$= 3 + 28 + 175r$$
$$= 31 + 175r$$

as before.

Each of these methods allows us to solve systems of more than two congruences, so long as the 'moduli' are pairwise relatively prime, by solving two congruences at a time. Actually in the *Mathematical Treatise in Nine Sections* there are examples to show that the idea behind the method may sometimes be applied even if the moduli are not all pairwise relatively prime (see [Needham, Section 19 (i) (4)] or [Li Yan and Du Shiran, p. 165]).

Example   Solve the simultaneous congruences

$$x \equiv 2 \bmod 7$$
$$x \equiv 0 \bmod 9$$
$$2x \equiv 6 \bmod 8.$$

Observe that the third congruence is not in an immediately usable form, so we first solve it to obtain the two (since $(2, 8) = 2$) solutions:

$$x \equiv 3 \bmod 8 \text{ and } x \equiv 7 \bmod 8.$$

So now we have two sets of three congruences to solve, and we could treat these as entirely separate problems, only combining the solutions at the end. We may note, however, that there is no need to separate the solution for the third congruence into two solutions modulo 8, since the solution is really just the congruence class $[3]_4$. Thus we reduce to solving the simultaneous congruences

$$x \equiv 2 \bmod 7$$
$$x \equiv 0 \bmod 9$$
$$x \equiv 3 \bmod 4.$$

Since $(7, 9) = 1 = (7, 4) = (9, 4)$ we will be able to apply 1.5.2. Take (say) the first two congruences to solve together. We have

$$7 \cdot (-5) + 9 \cdot 4 = 1,$$

so a solution to the first two is:

$$0 \cdot 7(-5) + 2 \cdot 9 \cdot 4 = 72 \bmod 7 \cdot 9.$$

This simplifies to 9 mod 63. So now the problem has been reduced to solving

$$x \equiv 9 \bmod 63$$
$$x \equiv 3 \bmod 4.$$

We have

$$16 \cdot 4 - 1 \cdot 63 = 1.$$

This gives

$$9 \cdot 16 \cdot 4 - 3 \cdot 1 \cdot 63 \bmod 63 \cdot 4$$

as the solution. This simplifies to 135 mod 252.

Finally, in this section, we briefly consider solving non-linear congruences. There are many deep and difficult problems here and to give a reasonable account would take us very far afield. So we content ourselves with merely indicating a few points (below, and in the exercises).

Example Consider the quadratic equation

$$x^2 + 1 \equiv 0 \bmod n.$$

The existence of solutions, as well as the number of solutions, depends on n. For example, when n is 3, we can substitute the three congruence classes $[0]_3$, $[1]_3$ and $[2]_3$ into the equation to see that $x^2 + 1$ is never $[0]_3$. When n is 5, it can be seen that $[2]_5$ and $[3]_5$ are solutions. If n is 65, it can be checked that $[8]_{65}$, $[-8]_{65}$, $[18]_{65}$ and $[-18]_{65}$ are all solutions, and this leads to the (different) factorisations

$$x^2 + 1 \equiv (x + 8)(x - 8) \bmod 65$$
$$\equiv (x + 18)(x - 18) \bmod 65.$$

When n is a prime, however, to the extent that a polynomial can be factorised, the factorisation is unique.

Example Consider the polynomial $x^3 - x^2 + x + 1$: does it have any integer roots? Suppose that it had an integer root k: then we would have $k^3 - k^2 + k + 1 = 0$. Let n be any integer greater than 1, and reduce this equation modulo n to obtain

$$[k]_n^3 - [k]_n^2 + [k]_n + [1]_n = [0]_n.$$

So we would have that the polynomial $X^3 - X^2 + X + [1]_n$ with coefficients from $\mathbb{Z}_n$ has a root in $\mathbb{Z}_n$. This would be true for *every* n.

Let us take $n = 2$: so reducing $x^3 - x^2 + x + 1 = 0$ modulo 2 gives $X^3 - X^2 + X + [1]_2$. It is straightforward to check whether or not this equation has a solution in $\mathbb{Z}_2$: all we have to do is to substitute $[0]_2$ and $[1]_2$ in turn. Doing this, we find that $[1]_2$ is a root. This tells us nothing about whether or not the original polynomial has a root.

So we try taking $n = 3$: reduced modulo 3, the polynomial becomes $X^3 - X^2 + X + [1]_3$. Let us see whether this has a root in $\mathbb{Z}_3$. Substituting in turn $[0]_3$, $[1]_3$ and $[2]_3$ for X we get the values $[1]_3$, $[2]_3$ and $[1]_3$ for the polynomial. In particular none of these is zero, so the polynomial has no root modulo 3. Therefore the original polynomial has no integer root (for by the argument above, if it did, then it would also have to have a root modulo 3). In Chapter 6 we look again at polynomials with coefficients which are congruence classes.

Exercises 1.5

1. Find all the solutions (when there are any) of the following linear congruences:
   (i) $3x \equiv 1 \bmod 12$;
   (ii) $3x \equiv 1 \bmod 11$;
   (iii) $64x \equiv 32 \bmod 84$;

(iv) $15x \equiv 5 \bmod 17$;

(v) $15x \equiv 5 \bmod 18$;

(vi) $15x \equiv 5 \bmod 100$;

(vii) $23x \equiv 16 \bmod 107$.

2. Solve the following sets of simultaneous linear congruences:

   (i) $x \equiv 4 \bmod 24$ and $x \equiv 7 \bmod 11$;

   (ii) $3x \equiv 1 \bmod 5$ and $2x \equiv 6 \bmod 8$;

   (iii) $x \equiv 3 \bmod 5$, $2x \equiv 1 \bmod 7$ and $x \equiv 3 \bmod 8$.

3. Find the smallest positive integer whose remainder when divided by 11 is 8, which has last digit 4 and is divisible by 27.

4. (i) Show that the polynomial $x^4 + x^2 + 1$ has no integer roots, but that it has a root modulo 3, and factorise it over $\mathbb{Z}_3$.

   (ii) Show that the equation $7x^3 - 6x^2 + 2x - 1 = 0$ has no integer solutions.

5. A hoard of gold pieces 'comes into the possession of' a band of 15 pirates. When they come to divide up the coins, they find that three are left over. Their discussion of what to do with these extra coins becomes animated, and by the time some semblance of order returns there remain only 7 pirates capable of making an effective claim on the hoard. When, however, the hoard is divided between these seven it is found that two pieces are left over. There ensues an unfortunate repetition of the earlier disagreement, but this does at least have the consequence that the four pirates who remain are able to divide up the hoard evenly between them. What is the minimum number of gold pieces which could have been in the hoard?

1.6 Euler's Theorem and public key codes

Suppose that we are interested in the behaviour of integers modulo 20. Fix an integer a and then form the successive powers of its congruence class:

$$[a]_{20}, [a]_{20}^2, [a]_{20}^3, \ldots, [a]_{20}^n, \ldots$$

What can happen? Let us try some examples (write '[3]' for '$[3]_{20}$' etc.). Taking $a = 3$ we obtain

$$[3]^1 = [3], \ [3]^2 = [9], \ [3]^3 = [27] = [7],$$
$$[3]^4 = [3]^3[3] = [7][3] = [21] = [1], \ [3]^5 = [3]^4[3] = [1][3] = [3],$$
$$[3]^6 = [3]^2 = [9], \ [3]^7 = [3]^3 = [7], \ [3]^8 = [1], \ [3]^9 = [3], \ldots$$

Observe that the successive powers are different until we reach [1] and then the pattern starts to repeat.

If we take $a = 4$ then the pattern of powers is somewhat different, in that [1] is never reached:

$$[4]^1 = [4], \quad [4]^2 = [16], \quad [4]^3 = [64] = [4], \quad [4]^4 = [16], \ldots$$

Taking $a = 10$ the sequence of powers of $[10]_{20}$ is:

$$[10], \quad [100] = [0], \quad [0], \quad [0], \ldots$$

If we take $a = 11$ then the behaviour is similar to that when $a = 3$; we reach [1] and then the pattern repeats from the beginning:

$$[11], \quad [1], \quad [11], \quad [1], \quad [11], \ldots$$

Those congruence classes, like $[3]_{20}$ and $[11]_{20}$, which have some power equal to the class of 1 are of particular significance. In this section we will give a criterion for a congruence class to be of this form and we examine the behaviour of such classes.

Definition   Let n be a positive integer greater than 1. The integer a is said to have **finite multiplicative order modulo** n if there is a positive integer k such that

$$[a]_n^k \; (= [a^k]_n) = [1]_n.$$

Thus $[3]_{20}$ and $[11]_{20}$ have finite multiplicative order, but $[4]_{20}$ and $[10]_{20}$ do not. Similarly if n is 6, then for all k

$$[3]_6^k = [3]_6$$

and so 3 does not have finite multiplicative order modulo 6.

Going beyond examples, we now give a general result which explains what can happen with finite multiplicative order.

Theorem 1.6.1   *The integer a has finite multiplicative order modulo n if, and only if, a is relatively prime to n.*

Proof   Fix a and n and suppose that there is a positive integer k such that

$$[1]_n = [a^k]_n = [a^{k-1}]_n \, [a]_n.$$

It follows that $[a]_n$ has an inverse, namely $[a^{k-1}]_n$, and so, by Theorem 1.4.3, a is relatively prime to n.

Conversely, suppose that a and n are relatively prime so, by 1.4.3, $[a]_n$ has an inverse and hence, by 1.4.7 (and the comment after that), all its powers $[a]_n^k$

have inverses. Now consider the $n + 1$ terms

$$[a], [a]^2, \ldots, [a]^{n+1}.$$

Since $\mathbb{Z}_n$ has only n distinct elements, at least two of these powers are equal as elements of $\mathbb{Z}_n$: say

$$[a]^k = [a]^t \text{ where } 1 \le k < t \le n + 1$$

(so note that $1 \le t - k$).

Take both terms to the same side and factorise to get

$$[a]^k([1] - [a]^{t-k}) = [0].$$

Multiplying both sides of the equation by $[a]^{-k}$ and simplifying, we obtain

$$[1] - [a]^{t-k} = [0].$$

This may be rewritten as

$$[a]^{t-k} = [1]$$

and so a does have finite multiplicative order modulo n. $\square$

Definition   If a has finite multiplicative order modulo n, then the **order** of a modulo n is the smallest positive integer k such that

$$[a]_n^k = [1]_n,$$

(or in terms of congruences $a^k \equiv 1 \bmod n$.)

We also say, in this case, that the **order** of the congruence class $[a]_n$ is k.

Example 1   The discussion at the beginning of the section shows that the order of 3 modulo 20 is 4 and the order of 11 modulo 20 is 2.

Example 2   Since the first three powers of 2 are 2, 4 and 8, it follows that 2 has order 3 modulo 7. Similarly it can be seen that 3 has order 6 modulo 7.

Example 3   When n is 17, we see that

$$2^4 \equiv -1 \bmod 17$$

and so it follows that $[2]_{17}$ has order 8. (To see this, square each side to obtain $2^8 \equiv 1 \equiv 2^0 \bmod 17$. It then follows, by 1.6.2 below, that the order of $[2]_4$ is a divisor of 8. It cannot be 1, 2 or 4 because $2^4 \not\equiv 1 \bmod 17$, so must be 8.) Also, since

$$13^2 = 169 \equiv -1 \bmod 17,$$

so

$$13^4 \equiv (-1)^2 \equiv 1 \bmod 17,$$

we deduce that $[13]_{17}$ has order 4. Notice that this also implies that the inverse of 13 modulo 17 is 4 since

$$13^3 \equiv -1 \cdot 13 = -13 \equiv 4 \bmod 17,$$

and so

$$13 \cdot 4 \equiv 13 \cdot 13^3 \equiv 13^4 \equiv 1 \bmod 17.$$

The next theorem explains the periodic behaviour of the powers of 3 and 11 mod 20, seen at the beginning of this section.

Theorem 1.6.2   *Suppose that a has order k modulo n. Then*

$$a^r \equiv a^s \bmod n$$

if, and only if,

$$r \equiv s \bmod k.$$

Proof   If

$$r \equiv s \bmod k$$

then r has the form $s + kt$ for some integer t, and so

$$\begin{aligned}
a^r &= a^{s+kt} \\
&= a^s(a^{kt}) \\
&= a^s(a^k)^t \\
&\equiv a^s(1)^t \bmod n \\
&\equiv a^s \bmod n.
\end{aligned}$$

Conversely, if

$$a^r \equiv a^s \bmod n,$$

then suppose, without loss of generality, that r is less than or equal to s. Since, by Theorem 1.6.1, a is relatively prime to n, it follows, by Theorems 1.4.3 and 1.4.7, that a^r has an inverse modulo n. Multiplying both sides of the above congruence by this inverse gives

$$1 \equiv a^{s-r} \bmod n.$$

Now write $s - r$ in the form

$$s - r = qk + u$$

where u is a natural number less than k. It then follows, as in the proof of the first part, that

$$a^{s-r} \equiv a^u \bmod n,$$

and so

$$a^u \equiv 1 \bmod n.$$

The minimality of k forces u to be 0.
Hence

$$r \equiv s \bmod k. \qquad \square$$

We now turn our attention to the possible orders of elements in $\mathbb{Z}_n$, considering first the case when n is prime. This result was announced by Pierre de Fermat in a letter of 1640 to Frénicle de Bessy, in which Fermat writes that he has a proof. Fermat states his result in the following words: 'Given any prime p, and any geometric progression $1, a, a^2$, etc., p must divide some number $a^n - 1$ for which n divides $p - 1$: if then N is any multiple of the smallest n for which this is so, p divides also $a^N - 1$'. We use the language of congruence to restate this as follows.

Theorem 1.6.3 (Fermat's Theorem) *Let p be a prime and suppose that a is an integer not divisible by p. Then*

$$[a]_p^{p-1} = [1]_p.$$

That is,

$$a^{p-1} \equiv 1 \bmod p.$$

Therefore, for any integer a

$$a^p \equiv a \bmod p.$$

Proof Let G_p be the set of invertible elements of $\mathbb{Z}_p$, so by Corollary 1.4.6, G_p consists of the $p - 1$ elements

$$[1]_p, [2]_p, \ldots, [p - 1]_p.$$

Denote by $[a]G_p$ the set of all multiples of elements of G_p by $[a]$:

$$[a]G_p = \{[a][b] : [b] \text{ is in } G_p\}$$
$$= \{[a][1], [a][2], \ldots, [a][p-1]\}.$$

Since $[a]$ is in G_p it follows by Theorem 1.4.7 that every element in $[a]G_p$ is in G_p. No two elements $[a][b]$ and $[a][c]$ of $[a]G_p$ with $[b] \neq [c]$ are equal since, if

$$[a][b] = [a][c]$$

then, by Corollary 1.4.4,

$$[b] = [c].$$

It follows, since the sets $[a]G_p$ and G_p have the same finite number of elements, that the sets $[a]G_p$ and G_p are equal. Now, multiply all the elements of G_p together to obtain the element

$$[N] = [1][2] \cdots [p-1].$$

By Theorem 1.4.7, $[N]$ is in G_p. Since the set G_p is equal to the set $[a]G_p$ (though the elements might be written in a different order), multiplying all the elements of $[a]G_p$ together must give us the same result:

$$[1][2] \cdots [p-1] = [a][1] \times [a][2] \times \cdots [a][p-1].$$

Collecting together all the '$[a]$' terms shown on the right-hand side we deduce that

$$[N] = [a]^{p-1}[N].$$

Since $[N]$ is in G_p, it is invertible: so we may cancel, by Corollary 1.4.4, to obtain

$$[1] = [a]^{p-1},$$

as required.

Finally, notice that for any integer a, either a is divisible by p, in which case a^p is also divisible by p, or a is not divisible by p, in which case, as we have just shown,

$$a^{p-1} \equiv 1 \bmod p.$$

Thus, in either case,

$$a^p \equiv a \bmod p. \qquad \square$$

Comment Run through the proof with particular (small) values for a and p to see, first, how multiplication by $[a]$ just rearranges the non-zero congruence classes and, second, how the cancellation argument involving $[N]$ works. For the purpose of understanding the proof it is better to leave $[N]$ as a product of terms rather than calculate its value. (It is, however, an interesting exercise to calculate the value of $[N]$: to explain what you find see Exercise 1.4.7.)

Corollary 1.6.4 *Let p be a prime number and let a be any integer not divisible by p. Then the order of a mod p divides $p - 1$.*

Proof This follows directly from the above theorem and 1.6.2. □

Warning The corollary above does *not* say that the order of a *equals* $p - 1$: certainly one has

$$a^{p-1} = 1 \bmod p,$$

but $p - 1$ need not be the lowest positive power of a which is congruent to 1 modulo p.

For example, consider the elements of G_7. The orders of its elements, $[1]_7, [2]_7, [3]_7, [4]_7, [5]_7, [6]_7$, are, respectively, 1, 3, 6, 3, 6, 2 (all, in accordance with Corollary 1.6.4, divisors of $6 = 7 - 1$).

Example 1 Let p be 17: so $p - 1$ is 16. It follows by Theorem 1.6.2 that 2^{100} is congruent to 2^4 modulo 17 since $100 (= 6 \times 16 + 4)$ is congruent to 4 modulo 16. That is,

$$2^{100} \equiv 2^4 \bmod 17$$
$$\equiv 16 \bmod 17.$$

Example 2 When p is 101 we have, by the same sort of reasoning, that $15^{601} \equiv (15^{100})^6 \cdot 15 \equiv 1^6 \cdot 15 \equiv 15 \bmod 101$.

It is not known what Fermat's original proof of Theorem 1.6.3 was (it seems reasonable to suppose that he did in fact have a proof). The first published proof was due to Leibniz (1646–1716): it is very different from the proof we gave above, being based on the Binomial Theorem (see Exercise 1.6.4 for this alternative proof). In 1742 Euler found the same proof but, his interest in number theory having been aroused, he went on to discover (before 1750) a 'multiplicative proof' like that we gave above. In a sense, that proof is better since it deals only with the essential aspects of the situation and it generalises to give a proof of Euler's Theorem (below). Actually Euler's proof was closer to that we give for Lagrange's Theorem (Theorem 5.2.3).

By 1750 Euler had managed to generalise Fermat's Theorem to cover the case of any integer $n \geq 2$ in place of the prime p. The power $p - 1$ of Fermat's Theorem had to be interpreted correctly, since if n is an arbitrary integer then the order of an invertible element modulo n certainly need not divide $n - 1$. The point is that if p is prime then $p - 1$ is the number of invertible congruence classes modulo p: that is, the number of elements in G_p. The function which assigns to n the number of elements in G_n is referred to as **Euler's phi-function**. Euler introduced this function and described its elementary properties in his *Tractatus*.

Definition   The number of elements in G_n is denoted by $\phi(n)$. Thus, by Theorem 1.4.3, $\phi(n)$ equals the number of integers between 1 and n inclusive which are relatively prime to n. The symbol ϕ used here is the Greek letter corresponding to the letter f in the Roman alphabet: in the Roman alphabet it is written 'phi' and pronounced accordingly. We will occasionally use other Greek letters in this book.

Theorem 1.6.5   *Suppose that p is a prime and let n be any positive integer. Then*

$$\phi(p^n) = p^n - p^{n-1}.$$

Proof   The only integers beween 1 and p^n which have a factor in common with p^n are the integers which are divisible by p, namely

$$p, 2p, \ldots, p^2, \ldots, p^n = p^{n-1}p.$$

Thus there are p^{n-1} numbers in this range which are divisible by p and so there are $p^n - p^{n-1}$ numbers between 1 and p^n which are *not* divisible by p, i.e. which are relatively prime to p^n.   □

Examples $\phi(5) = 4;$
                        $\phi(25) = \phi(5^2) = 5^2 - 5^1 = 20;$
                        $\phi(4) = \phi(2^2) = 2^2 - 2^1 = 2;$
                        $\phi(81) = \phi(3^4) = 3^4 - 3^3 = 54.$

Theorem 1.6.6 *Let a and b be relatively prime integers. Then*

$$\phi(ab) = \phi(a)\,\phi(b).$$

Proof Let $[r]_a$ and $[s]_b$ be elements of G_a and G_b respectively. From $[r]_a$ and $[s]_b$ we will produce an element $[t]_{ab}$ which we will show lies in G_{ab}. By the Chinese Remainder Theorem (1.5.2) there is an integer t satisfying

$$t \equiv r \bmod a \text{ and}$$

$$t \equiv s \bmod b,$$

and t is uniquely determined up to congruence modulo ab. Now we show that the class $[t]_{ab}$ is invertible. Since we have $r = t + ka$ for some integer k, and since the gcd of r and a is 1, it follows by 1.1.4 that the gcd of t and a is 1. Similarly $(t, b) = 1$. Therefore, by Exercise 1.1.6 we may deduce that $(t, ab) = 1$. Hence $[t]_{ab}$ is in G_{ab}.

Next we show that every element $[t]_{ab}$ in G_{ab} comes from a pair consisting of an element $[r]_a$ in G_a and an element $[s]_b$ in G_b. So, given $[t]_{ab}$ in G_{ab}, let r be the standard representative for $[t]_a$. Since $(t, ab) = 1$, certainly $(t, a) = 1$. So, since t is of the form $r + ka$, we have (by 1.1.4) that $(r, a) = 1$, and hence $[r]_a$ is in G_a. Similarly if s is the standard representative for $[t]_b$, then $[s]_b$ is in G_b. It follows that each element $[t]_{ab}$ in G_{ab} determines (uniquely) a pair $([r]_a, [s]_b)$ where

$$t \equiv r \bmod a \text{ and}$$

$$t \equiv s \bmod b.$$

By the first paragraph (uniqueness of t up to congruence modulo ab), different elements of G_{ab} determine different pairs.

Now imagine writing down all the elements of G_{ab} in some order. Underneath each element $[t]_{ab}$ write the pair $([t]_a, [t]_b)$ $(([r]_a, [s]_b)$ in the notation used above). We have shown that the second row contains no repetitions, and also that it contains every possible pair of the form $([r]_a, [s]_a)$ with $[r]_a$ in G_a and $[s]_b$ in G_b (since every element of G_a can be paired with every element of G_b). Thus the numbers of elements in the two rows must be equal. The first row contains $\phi(ab)$ elements, and the second row contains $\phi(a)\,\phi(b)$ elements: thus $\phi(ab) = \phi(a)\,\phi(b)$, as required. $\square$

Comment The reader may feel a little unsure about some points in the above proof. Within that proof we implicitly introduced two ideas which will be discussed at greater length in Chapter 2. The first of these is the idea of the Cartesian product, $X \times Y$, of sets X and Y. This is the set of all pairs of the form (x, y) with x in X and y in Y (and should not be confused with product in the arithmetic sense). The number of elements in $X \times Y$ is the product of the number of elements in X and the number of elements in Y. The second idea arises in the way in which we showed that the number of elements in the two sets G_{ab} and $G_a \times G_b$ are equal. The 'matching' obtained by writing the elements of the sets in two rows, one above the other, is an illustration of a bijective function, as we shall see in Section 2.3. This, rather than making a count of the elements, is the most common way in which pure mathematicians show that two sets have the same number of elements!

Examples
$$\phi(100) = \phi(25)\phi(4) = 20 \cdot 2 = 40;$$
$$\phi(14) = \phi(2)\phi(7) = 6;$$
$$\phi(41) = 40.$$

Now we come to Euler's generalisation of Fermat's Theorem.

Theorem 1.6.7 (Euler's Theorem)   *Let n be greater than or equal to 2 and let a be relatively prime to n. Then*

$$[a]_n^{\phi(n)} = [1]_n$$

that is

$$a^{\phi(n)} \equiv 1 \bmod n.$$

Proof   The proof is a natural generalisation of that given for Fermat's Theorem. We have arranged matters so that we can repeat that proof almost unchanged.

Let G_n be the set of invertible elements of $\mathbb{Z}_n$. Denote by $[a]G_n$ the set of all multiples of elements of G_n by $[a]$:

$$[a]G_n = \{[a][b] : [b] \text{ is in } G_n\}.$$

Since $[a]$ is in G_n it follows by Theorem 1.4.7 that every element in $[a]G_n$ is in G_n. No two elements $[a][b]$ and $[a][c]$ of $[a]G_n$ with $[b] \neq [c]$ are equal, since if

$$[a][b] = [a][c]$$

then, by Corollary 1.4.4,

$$[b] = [c].$$

It follows that the sets $[a]G_n$ and G_n are equal.

Now multiply together all the elements of G_n to obtain an element $[N]$, say, and note that $[N]$ is in G_n by Theorem 1.4.7. Multiplying the elements of $[a]G_n$ together gives $[a]^{\phi(n)}[N]$ so, since the set G_n is equal to the set $[a]G_n$, we deduce that

$$[a]^{\phi(n)}[N] = [N].$$

Since $[N]$ is in G_n it is invertible and so, by Corollary 1.4.4, we may cancel to obtain

$$[a]^{\phi(n)} = [1],$$

as required. □

Corollary 1.6.8 *Suppose that n is a positive integer and let a be an integer relatively prime to n. Then the order of a mod n divides $\phi(n)$.*

Proof This follows directly from the theorem and 1.6.2. □

Example 1 Since $14 = 2 \cdot 7$, so $\phi(14) = \phi(2)\phi(7) = 6$, the value of 3^{19} modulo 14 is determined by the congruence class of 19 modulo 6 and so 3^{19} is congruent to 3^1, that is, to 3, modulo 14. More explicitly, $3^{19} \equiv (3^6)^3 \cdot 3^1 \equiv (1^3) \cdot 3^1 = 3 \bmod 14$ since $3^6 \equiv 1 \bmod 14$.

Example 2 Since the last two digits of a positive integer are determined by its congruence class modulo 100 and since $\phi(100) = \phi(2^2)\phi(5^2)$ is 40, we have that the last two digits of 3^{125} are 43, since

$$3^{125} \equiv (3^{40})^3 \times 3^5 \equiv (1)^3 \times 243 \equiv 43 \bmod 100.$$

Warning If the integers a and n are not relatively prime then, by 1.6.1 no power of a can be congruent to $1 \bmod n$, although it might happen that $a^{\phi(n)+1} \equiv a \bmod n$.

For instance, take $n = 100$ (so $\phi(n) = 40$) and $a = 5$: then it is easily seen (and proved by induction) that every power of a beyond the first is congruent to $25 \bmod 100$ and so, in particular, $5^{\phi(n)+1}$ is not congruent to 5.

On the other hand, if one takes n to be 50 (so $\phi(n) = 20$) and a to be 2 then it does turn out that $2^{\phi(n)+1} \equiv 2 \bmod 50$ (in Exercise 1.6.8 you are asked to explain this).

To conclude this chapter, we discuss the idea of public key codes and how such codes may be constructed.

The traditional way to transmit and receive sensitive information is to have both sender and receiver equipped with a 'code book' which enables the sender to encode information and the receiver to decode the resulting message. It is a general feature of such codes that if one knows how to *en*code a message then one can in practice *de*code an intercepted message. Thus, if one wishes to receive sensitive information from a number of different sources, one is confronted with obvious problems of security.

The idea of a public key code is somewhat different. Suppose that the 'receiver' R wishes to receive information from a number of different sources $S_1, S_2, \ldots$ (R could be a company or bank headquarters, a computer database containing medical records, an espionage headquarters,... with the S_i being correspondingly branches, hospitals, field operatives,...) Rather than equipping each S_i with a 'code book', R provides, in a fairly public way, certain information which allows the S_i to encode messages. These messages may then be sent over public channels. The code is designed so that if some third party T intercepts a message then T will find it impossible in practice to decode the message *even if T has access to the information that tells the S_i how to encode messages.*

That is, decoding a message is somehow inherently more difficult than encoding a message, even if one has access to the 'code book'.

Various ways of realising such a code in practice have been suggested (the idea of public key codes was put forward by Diffie and Hellman in 1976). One method is the 'knapsack method' (see [Salomaa, Section 7.3] for a description of this method). It seemed for a while that this provided a method for producing public key codes: it was however discovered by Shamir in 1982 that the method did not give an inherently 'safe' code, although it has since been modified to give what appears to be a safe code.

The mathematics of the method which we describe here is based on Euler's Theorem. It is generally believed to be 'inherently safe', but there is no proof of that, and so it is not impossible that it will have to be fundamentally modified or replaced. The method is referred to as the RSA system, after its inventors: Rivest, Shamir and Adleman (1978). It also transpired that a group at GCHQ in the UK had come up with the same idea somewhat earlier but, for security reasons, it was kept secret (see www.cesg.gov.uk/publications/media/nsecret/ellis.pdf for details). The (assumed) efficacy of this type of code depends on the inherent difficulty of factorising a (very large!) integer into a product of primes. It has been shown by Rabin that deciphering (a variant of) this system is as difficult as factorising integers.

Construction of the code   First one finds two very large primes (say about 100 decimal digits each). With the aid of a reasonably powerful computer it is very quickly checked whether a given number is prime or not, so what one could do in practice is to generate randomly a sequence of 100-digit numbers, check each in turn for primality, and stop when two primes have been found. Let us denote the chosen primes by p and q.

Set n equal to the product pq: this is one of the numbers, the **base**, which will be made public.

By Theorem 1.6.5 and Theorem 1.6.6, $\phi(n)$ is equal to $(p-1)(q-1)$. Now choose a number a, the **exponent**, which is relatively prime to $\phi(n)$. To do this, simply generate a large number randomly and test whether this number is relatively prime to $\phi(n)$ (using 1.1.5): if it is, then take it for a; if it is not, then try another number...(the chance of having to try out many numbers is very small). Using the methods of Section 1.1, find a linear combination of $\phi(n)$ and a which is 1:

$$ax + \phi(n)y = 1. \quad (*)$$

Note that x is, in particular, the inverse of a modulo $\phi(n)$.

Now one may publish the pair of numbers (n, a).

To encode a message   If the required message is not already in digital form then assign an integer to each letter of the alphabet and to each punctation mark according to some standard agreement, with all such letter–number equivalents having the same length (perhaps a $= 01$, b $= 02$ and so on). Break the digitised message into blocks of length less than the number of digits in either p or q (so if p and q are of 100 digits each then break the message into blocks each with length less than 100 digits).

Now encode each block β by calculating the standard representative m for β^a modulo n. Now send the sequence of encoded blocks with the beginning of each block clearly defined or marked in some way.

To decode   The constructor of the code now receives the message and breaks it up into its blocks. To decode a block m, simply calculate the standard representative of

$$m^x \bmod n,$$

where x is as in $(*)$. The result is the original block β of the message. To see that this is so, we recall that m was equal to $\beta^a \bmod n$. Therefore

$$m^x \equiv (\beta^a)^x = \beta^{ax} = \beta^{1-\phi(n)y} = \beta \cdot \left(\beta^{\phi(n)}\right)^{-y} \equiv \beta \cdot 1^{-y} = \beta \bmod n.$$

Here we are using Euler's Theorem (1.6.7) to give us that

$$\beta^{\phi(n)} \equiv 1 \bmod n.$$

This, of course, is only justified if β is relatively prime to n: but that is ensured by our choosing β to have fewer digits than either of the prime factors of n (actually the chance of an arbitrary integer β not being relatively prime to n is extremely small).

At first sight it might seem that this is not an effective code, for surely anyone who intercepts the message may perform the calculation above, and so decode the message. But notice that the number x is not made public. Very well you may say: an interceptor may simply calculate x. But how does one calculate x? One computes 1 as a linear combination of a and $\phi(n)$. And here is the point: although a and n are made public, $\phi(n)$ is not and, so far as is known, there is no way in which one may easily calculate $\phi(n)$ for such a large number n. Of course, one way to calculate $\phi(n)$ from n is to factorise n as the product of the two primes p and q, but factorisation of such large numbers seems to be an inherently difficult task. Certainly, at the moment, factorisation of such a number (of about 200 decimal digits) seems to be well beyond the range of any existing computer (unless one is prepared to hang around, waiting for an answer, for a few million years). See www.rsasecurity.com/rsalabs/challenges/factoring/index.html for up-to-date information.

It should be said that, in order to obtain a code which cannot easily be broken, there are a few more (easily met) conditions to impose on p, q and a: see [Salomaa] or the RSA Labs website www.rsasecurity.com/rsalabs/ for this, as well as for a more detailed discussion of these codes. For up-to-date information about this code and its uses see the RSA Labs website.

We give an example: we will of course choose small numbers for the purpose of illustrating the method, so our code would be very easy to break.

Example Take 3 and 41 to be our two primes p, q. So $n = 123$ and $\phi(n) = (3 - 1)(41 - 1) = 80$.

We choose an integer a relatively prime to 80: say $a = 27$. Express 1 as a linear combination of 80 and 27:

$$3 \cdot 27 - 1 \cdot 80 = 1$$

so 'x' is 3. We publish $(n, a) = (123, 27)$.

To encode a block β, the sender calculates $\beta^{27} \bmod 123$, and to decode a received block m, we calculate $m^3 \bmod 123$.

Thus, for example, to encode the message $\beta = 05$, the sender computes

$$5^{27} \bmod 123 \,(= (125)^9 \equiv 2^9 \equiv 4 \cdot 128 \equiv 4 \cdot 5 \equiv 20 \bmod 123)$$

and so sends $m = 20$. On receipt of this message, anyone who knows 'x' (the inverse of 27 mod 80) computes $20^3 \bmod 123$ which, you should check, is equal to the original message 05.

If now we use the number-to-letter equivalents:

$$G = 1, R = 2, A = 3, D = 4, U = 5, O = 6, S = 7, I = 8, T = 9, Y = 0,$$

and the received message is 10/04, the original message is decoded by calculating

$$10^3 = 1000$$
$$= 8 \cdot 123 + 16$$
$$\equiv 16 \bmod 123$$

and

$$4^3 = 64 \bmod 123.$$

Juxtaposing these blocks gives 1664, and so the message was the word G O O D.

(In this example we used small primes for purposes of illustration but, in doing so, violated the requirement that the number of digits in any block should be less than the number of digits in either of the primes chosen. Exercise 1.6.9 below asks you to discover what effect this has.)

Pierre Fermat was born in 1601 near Toulouse. In 1631 he became a magistrate in the 'Parlement' of Toulouse, and so became 'Pierre de Fermat'. He held this office until his death in 1665. Fermat's professional life was divided between Toulouse, where he had his main residence, and Castres, which was the seat of the 'Chambre' of the Parlement which dealt with relations between the Catholic and Protestant communities within the province.

Fermat's contact with other mathematicians was almost entirely by letter: his correspondence with Mersenne and others in Paris starts in 1636. In 1640 he was put in contact with one of his main correspondents, Frénicle de Bessy, by Mersenne. In fact, Fermat seems never to have ventured far from home, in contrast to most of his scientific contemporaries.

Fermat's name is perhaps best known in connection with what was, for many centuries, one of the most celebrated unsolved problems in mathematics. The equation

$$x^n + y^n = z^n$$

can be seen to have integer solutions when n is 1 or 2 (for example when n is 2, $x = 3$, $y = 4$ and $z = 5$ is an integer solution). Fermat claimed, in the margin of his copy of Diophantus' *Arithmetica*, that he could show that this equation never has a solution in positive integers when n is greater than 2. Fermat appended his note to Proposition 8 of Book II of the *Arithmetica*: 'To divide a given square number into two squares'. Fermat's note translates as

> On the other hand it is impossible to separate a cube into two cubes, or a biquadrate into two biquadrates, or generally any power except a square into two powers with the same exponent. I have discovered a truly marvellous proof of this, which however the margin is not large enough to contain.

However, until very recently, no-one had been able to supply a proof of 'Fermat's Last Theorem', and one may reasonably doubt whether Fermat did in fact have a correct proof.

Many attempts were made over the centuries to prove this result and various special cases were dealt with. In 1983 Faltings proved a general result which put strong limits on the number of solutions but the conjecture, that there are no solutions for $n \geq 3$, remained open. Then, in 1993, Andrew Wiles announced, in a lecture at the Isaac Newton Institute in Cambridge, that he had proved 'Fermat's Last Theorem'. As it turned out, however, there was a gap in the proof. It took over a year for Wiles, and a collaborator, Richard Taylor, to correct the proof. But, it was corrected and so, finally, after more than 400 years, Fermat's assertion has been proved to be correct.

Number theory, as opposed to many other parts of mathematics, had not enjoyed a renaissance before Fermat's time. For instance, the first Latin translation, by 'Xylander', of Diophantus' *Arithmetica* had only appeared in 1575, and the first edition to contain the full Greek text, with many of the corrupt passages corrected, was published by Bachet in 1621. It was in a copy of this edition that Fermat made marginal notes, including the (in?)famous one above.

Fermat had hoped to see a revival of interest in number theory, but towards the end of his life he despaired of the area being treated with the seriousness he felt it deserved. In fact, Fermat's work in number theory remained relatively unappreciated for almost a century, until Euler, having been referred to Fermat's works by Goldbach, found his interest aroused.

Exercises 1.6

1. Find the orders of
   (i)   2 modulo 31,
   (ii) 10 modulo 91,

     (iii) 7 modulo 51, and

     (iv) 2 modulo 41.

2. Find

     (i) 5^{20} mod 7,

     (ii) 2^{16} mod 8,

     (iii) 7^{1001} mod 11, and

     (iv) 6^{76} mod 13.

3. Prove that for every positive integer a, written in the base 10, a^5 and a have the same last digit.

4. This exercise indicates the 'additive proof' (see above) of Fermat's Theorem. Let p be a prime. Consider the expansion of $(x + y)^p$ using the Binomial Theorem. Replace each of x and y in this expansion by 1, and reduce modulo p to deduce that $2^p \equiv 2$ mod p, that is, Fermat's Theorem for the case $a = 2$.

    (This proof may be generalised to cover the case of an arbitrary a by writing a as a sum of a '1s', using the Multinomial Theorem expansion of $(x_1 + x_2 + \ldots + x_a)^p$ and deducing that p divides all the coefficients in the expanded expression except the first and the last. For the Multinomial Theorem, see [Biggs, p. 99] for example.)

5. Calculate $\phi(32)$, $\phi(21)$, $\phi(120)$ and $\phi(384)$.

6. Find

     (i) 2^{25} mod 21,

     (ii) 7^{66} mod 120 and

     (iii) the last two digits of $1 + 7^{162} + 5^{121} \cdot 3^{312}$.

7. Show that, for every integer n, $n^{13} - n$ is divisible by 2, 3, 5, 7 and 13.

8. Show that, if $n \geq 2$ and if p is a prime which divides n but is such that p^2 is not a factor of n, then $p^{\phi(n)+1} \equiv p$ mod n. Can you find and prove a generalisation of this?

    [Hint for first part: n may be written as pm, and $(p, m) = 1$; consider powers of p modulo m.

    Hint for second part: for example, you may check that, although $2^{\phi(100)+1}$ is not congruent to 2^1 mod 100, one does have $2^{\phi(100)+2}$ congruent to 2^2 mod 100; also $6^{20+1} \equiv 6$ mod 66.]

9. In the example at the end of this section we used small primes for purposes of illustration, and in doing so violated the requirement that the number of digits in any block should be less than the number of digits in either of the primes chosen. This means that certain blocks, such as 18, 39,... which we might wish to send, will not be relatively prime to 123. What happens if we attempt to encode and then decode such blocks?

[Hint: the previous exercise is relevant. You should assume that x in (∗) on p. 71 is positive. The argument is quite subtle.]

10. Recall that the Mersenne primes are those numbers of the form $M(n) = 2^n - 1$ that are prime. In Exercise 1.3.6 you were asked to show that if $M(n)$ is prime then n itself must be prime. The converse is false: there are primes p such that $M(p)$ is not prime. One such value is $p = 37$. A factorisation for $M(37) = 2^{37} - 1$ was found by Fermat: he used what is a special case of Fermat's Theorem, indeed it seems that this is what led him to discover the general case of 1.6.3. In this exercise we follow Fermat in finding a non-trivial proper factor of $2^{37} - 1$, which equals 137 438 953 471.

   (i) Show that if p is a prime and if $q(\neq 2)$ is a prime divisor of $2^p - 1$ then q is congruent to 1 mod p.
   [Hint: since q divides $2^p - 1$ we have $2^p \equiv 1 \bmod q$; apply Fermat's Theorem to deduce that p divides $q - 1$.]

   (ii) Apply part (i) with $p = 37$ to deduce that any prime divisor of $2^{37} - 1$ must have the form $37k + 1$ for some k. Indeed, since clearly 2 does not divide $2^{37} - 1$, any such prime divisor must have the form $74k + 1$ (why?). Hence find a proper factorisation of $2^{37} - 1$ and so deduce that $2^{37} - 1$ is not prime.
   [We have cut down the possibilities for a prime divisor to: 75 (which may be excluded since it is not prime), 149, 213, ... The arithmetic in this part may be a little daunting, but you will not have to search too far for a divisor (there is a factor below 500), provided your arithmetic is accurate! It would be a good idea to use 'casting out nines' (Exercise 1.4.9) to check your divisions.]

11. Use a method similar to that in the exercise just above to find a prime factor of $F(5) = 2^{32} + 1$ (see notes to Section 1.3).
   [Hint: as before, start with a prime divisor $q(\neq 2)$ of $2^{32} + 1$ and work modulo 32. You should be able to deduce that q has the form $64k + 1$. Eliminate non-primes such as 65 and 129. As before, be very careful in your arithmetic. There is a factor below 1000.]

12. A word has been broken into blocks of two letters and converted to two-digit numbers using the correspondence

   $a = 0, b = 1, c = 2, d = 3, o = 4, k = 5, f = 6, h = 7, l = 8, j = 9.$

   The blocks are then encoded using the public key code with base 87 and exponent 19. The coded message is 04/10. Find the word which was coded.

13. A public key code has base 143 and exponent 103. It uses the following letter-to-number equivalents:

$$J = 1, N = 2, R = 3, H = 4, D = 5, A = 6, S = 7, Y = 8,$$
$$T = 9, O = 0.$$

A message has been converted to numbers and broken into blocks. When coded using the above base and exponent the message sent is 10/03. Decode the message.

Summary of Chapter 1

We have investigated the divisibility relation on the set of integers. We defined the greatest common divisor of any two non-zero integers and showed that this is an integral linear combination of them. The notion of integers being relatively prime was introduced and was seen to play an important role in the investigation of congruence classes. We saw the prime numbers as being the 'building blocks' of integers under divisibility, in particular, we proved that every positive integer is, in an essentially unique way, a product of primes.

The additive structure on integers was used to define the set of congruence classes modulo n. It was shown that the set of congruence classes modulo n carries a natural arithmetic structure and a criterion for existence of inverses was established. We learned how to determine whether (sets of) congruences are solvable and, when they are, how to find the solutions.

Investigating the multiplicative structure of invertible congruence classes, we proved Fermat's Theorem and its generalisation, Euler's Theorem. The Euler phi-function was defined and we learned how to compute it. This was used to design public key codes.

We have also seen a variety of techniques of proof used. In particular, definition and proof by induction were introduced, as well as proof by contradiction.

2 Sets, functions and relations

In this chapter we set out some of the foundations of the mathematics described in the rest of the book. We begin by examining sets and the basic operations on them. This material will, at least in part, be familiar to many readers but if you do not feel entirely comfortable with set-theoretic notation and terminology you should work through the first section carefully. The second section discusses functions: a rigorous definition of 'function' is included and we present various elementary properties of functions that we will need. Relations are the topic of the third section. These include functions, but also encompass the important notions of partial order and equivalence relation.

The fourth section is a brief introduction to finite state machines.

2.1 Elementary set theory

The aim of this section is to familiarise readers with set-theoretic notation and terminology and also to point out that the set of all subsets of any given set forms a kind of algebraic structure under the usual set-theoretic operations.

A **set** is a collection of objects, known as its **members** or **elements**. The notation $x \in X$ will be used to mean that x is an element of the set X, and $x \notin X$ means that x is not an element of X. We will tend to use upper case letters as names for sets and lower case letters for their elements.

A set may be defined either by listing its elements or by giving some 'membership criterion' for an element to belong to the set. In listing the elements of a set, each element is listed only once and the order in which the elements are listed is unimportant. For example, the set

$$X = \{2, 3, 5, 7, 11, 13\} = \{3, 5, 7, 11, 13, 2\}$$

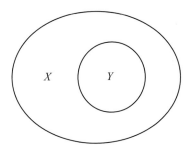

Fig. 2.1 $Y \subseteq X$

has just been defined by listing its elements, but it could also be specified as the set of positive integers that are prime and less than 15:

$$X = \{p \in \mathbb{P}: p \text{ is prime and } p < 15\}.$$

The colon in this formula is read as 'such that' and so the symbols are read as 'X is the set of positive integers p such that p is prime and is less than 15'. If the context makes our intended meaning clear, then we can define an infinite set by indicating the list of its members: for example

$$\mathbb{Z} = \{0, \pm 1, \pm 2, \pm 3, \ldots\},$$

where the sequence of three dots means 'and so on, in the same way'. The notation $\varnothing$ is used for the **empty set**: the set with no elements.

Two sets X, Y are said to be **equal** if they contain precisely the same elements. If every member of Y is also a member of X, then we say that Y is a **subset** of X and write $Y \subseteq X$ (or $X \supseteq Y$). Note that if $X = Y$ then $Y \subseteq X$. If we wish to emphasise that Y is a subset of X but not equal to X then we write $Y \subset X$ and say that Y is a **proper subset** of X. Observe that $X = Y$ if and only if $X \subseteq Y$ and $Y \subseteq X$.

Every set X has at least the subsets X and $\varnothing$; and these will be distinct unless X is itself the empty set.

We may illustrate relationships between sets by use of **Venn diagrams** (certain pictorial representations of such relationships). For instance, the Venn diagram in Fig. 2.1 illustrates the relationship '$Y \subseteq X$': all members of Y are to be thought of as inside the boundary shown for Y so, in the diagram, all members of Y are inside (the boundary corresponding to) X. The diagram is intended to leave open the possibility that there is nothing between X and Y (a region need not contain any elements), so it represents $Y \subseteq X$ rather than $Y \subset X$.

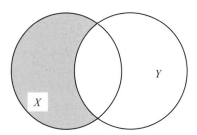

Fig. 2.2 $X \setminus Y$

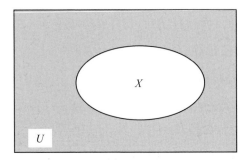

Fig. 2.3 X^c is the shaded area.

Venn, extending earlier systems of Euler and Leibniz, introduced these diagrams to represent logical relationships between defined sets in 1880. Dodgson, better known as Lewis Carroll, described a rather different system in 1896.

Given sets X and Y, we define the **relative complement** of Y in X to be the set of elements of X which do not lie in Y: we write

$$X \setminus Y = \{z: z \in X \text{ and } z \notin Y\}.$$

This set is represented by the shaded area in Fig. 2.2.

It is often the case that all sets that we are considering are subsets of some fixed set, which may be termed the **universal** set and is commonly denoted by U (note that the interpretation of U depends on the context). In this case the **complement** X^c of a set X is defined to be the set of all elements of U which are not in X: that is, $X^c = U \setminus X$ (see Fig. 2.3).

If X, Y are sets then the **intersection** of X and Y is defined to be the set of elements which lie in both X and Y:

$$X \cap Y = \{z: z \in X \text{ and } z \in Y\}$$

(see Fig. 2.4).

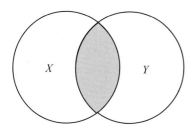

Fig. 2.4 $X \cap Y$

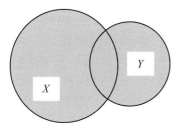

Fig. 2.5 $X \cup Y$

The sets X and Y are said to be **disjoint** if $X \cap Y = \emptyset$, that is, if no element lies in both X and Y.

Also we define the **union** of the sets X and Y to be the set of elements which lie in at least one of X and Y:

$$X \cup Y = \{z : z \in X \text{ or } z \in Y\}$$

(see Fig. 2.5).

There are various relationships between these operations which hold whatever the sets involved may be. For example, for any sets X, Y one has

$$(X \cup Y)^c = X^c \cap Y^c.$$

How does one establish such a general relationship? We noted above that two sets are equal if each is contained in the other. So to show that $(X \cup Y)^c = X^c \cap Y^c$ it will be enough to show that every element of $(X \cup Y)^c$ is in $X^c \cap Y^c$ and, conversely, that every element of $X^c \cap Y^c$ is in $(X \cup Y)^c$.

Suppose then that x is an element of $(X \cup Y)^c$: so x is not in $X \cup Y$. That is, x is not in X nor is it in Y. Said otherwise: x is in X^c and also in Y^c. Thus x is in $X^c \cap Y^c$. So we have established $(X \cup Y)^c \subseteq X^c \cap Y^c$.

Suppose, conversely, that x is in $X^c \cap Y^c$. Thus x is in X^c and x is in Y^c. That is, x is not in X and also not in Y: in other words, x is not in $X \cup Y$, so x is in $(X \cup Y)^c$. Hence $X^c \cap Y^c \subseteq (X \cup Y)^c$.

Thus we have shown that $(X \cup Y)^c = X^c \cap Y^c$.

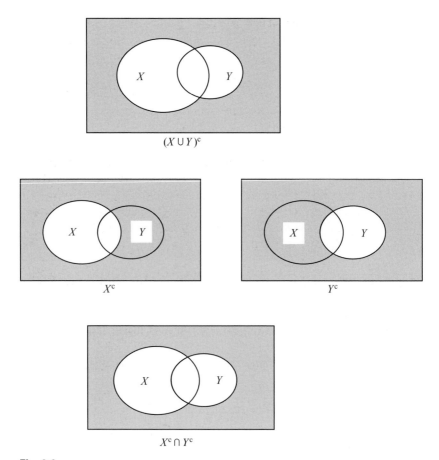

$(X \cup Y)^c$

X^c

Y^c

$X^c \cap Y^c$

Fig. 2.6

You may have observed how, in this proof, we used basic properties of the words 'or', 'and' and 'not'. Indeed, we replaced the set-theoretic operations union, intersection and complementation by use of these words and then applied elementary logic. For an explanation of this (general) feature, see Section 3.1 below.

One may picture the relationship expressed by the equation $(X \cup Y)^c = X^c \cap Y^c$ by using Venn diagrams (Fig. 2.6).

This sequence of pictures probably makes it more obvious why the equation $(X \cup Y)^c = X^c \cap Y^c$ is true. But do not mistake the sequence of pictures for a rigorous proof. For there may be hidden assumptions introduced by the way in which the pictures have been drawn. For example, does the sequence of pictures deal with the possibility that X is a subset of Y? (Pictures may be helpful in finding relationships in the first place or in understanding why they are true.)

The algebra of sets Let X be a set: we denote by $P(X)$ the set of all subsets of X. Thus, if X is the set with two elements x and y, $P(X)$ consists of the empty set, $\emptyset$, together with the sets $\{x\}$, $\{y\}$ and $X = \{x, y\}$ itself.

We will think of $P(X)$ as being equipped with the operations of intersection, union and complementation. Just as the integers with addition and multiplication obey certain laws (such as $x + y = y + x$) from which the other algebraic laws may be deduced, so $P(X)$ with these operations obeys certain laws (or 'axioms'). Some of these are listed in the next result. They are all easily established by the method that was used above to show $(X \cup Y)^c = X^c \cap Y^c$.

Theorem 2.1.1 *For any sets, X, Y and Z (contained in some 'universal set' U) we have*

$X \cap X = X$ *and*
$X \cup X = X$ *idempotence;*
$X \cap X^c = \emptyset$ *and*
$X \cup X^c = U$ *complementation;*
$X \cap Y = Y \cap X$ *and*
$X \cup Y = Y \cup X$ *commutativity;*
$X \cap (Y \cap Z) = (X \cap Y) \cap Z$ *and*
$X \cup (Y \cup Z) = (X \cup Y) \cup Z$ *associativity;*
$(X \cap Y)^c = X^c \cup Y^c$ *and*
$(X \cup Y)^c = X^c \cap Y^c$ *De Morgan laws;*
$X \cap (Y \cup Z) = (X \cap Y) \cup (X \cap Z)$ *and*
$X \cup (Y \cap Z) = (X \cup Y) \cap (X \cup Z)$ *distributivity;*
$(X^c)^c = X$ *double complement;*
$X \cap \emptyset = \emptyset$ *and*
$X \cup \emptyset = X$ *properties of empty set;*
$X \cap U = X$ *and*
$X \cup U = U$ *properties of universal set;*
$X \cap (X \cup Y) = X$ *and*
$X \cup (X \cap Y) = X$ *absorption laws.*

One may list a similar set of basic properties of the integers. In that case one would include rules such as the distributive law $a \times (b + c) = a \times b + a \times c$

and the law for identity $a \times 1 = a$. One could also include the law $a \times (b + (a + 1)) = a \times b + (a \times a + a)$. However, it is not necessary to do so because it already follows from two applications of the distributive law and one application of the law for identity:

$$a \times (b + (a + 1)) = a \times b + a \times (a + 1) \ (by \ distributivity)$$
$$= a \times b + (a \times a + a \times 1) \ (by \ distributivity)$$
$$= a \times b + (a \times a + a) \ (by \ identity).$$

Thus the inclusion of the above law would be redundant. Similarly, the list of laws in Theorem 2.1.1 has some redundancy.

For example, by properties of complement, $X \cup U$ is equal to $X \cup (X \cup X^c)$ which, by associativity, is equal to $(X \cup X) \cup X^c$ which, by idempotence, is equal to $X \cup X^c$; then, by another appeal to the properties of complement, this is equal to U. Thus the equality $X \cup U = U$ follows from some of the others.

You should work out proofs for the laws above: either verifications as with $(X \cup Y)^c = X^c \cap Y^c$ or, in appropriate cases, derivations from laws which you have already established (but avoid circular argument in such derivations).

So we are thinking of the set $P(X)$ of all subsets of X, equipped with the operations of '$\cap$', '$\cup$' and '^{c}', as being some kind of 'algebraic structure'. In fact it is an example of what is termed a 'Boolean algebra'. We say more about these in Section 4.4. Let us say that a **Boolean algebra of sets** is a subset B of the set $P(X)$ of all subsets of a set X, which contains at least the empty set $\emptyset$ and X, and also B must be closed under the 'Boolean' operations, $\cap$, $\cup$ and ^{c}, in the sense that if Y and Z are in B then so are $Y \cap Z$, $Y \cup Z$ and Y^c.

Example Let X be the set $\{0, 1, 2, 3\}$. Then $P(X)$ has $2^4 = 16$ elements. Take B to be the set $\{\emptyset, \{0, 2\}, \{1, 3\}, X\}$. You should check that B is closed under the operations and hence forms a Boolean algebra of sets.

The operations '$\cap$', '$\cup$' and '^{c}' produce new sets from existing ones. Here is a rather different way of producing new sets from old.

Definition The (**Cartesian**, named after Descartes) **product** of two sets X, Y is defined to be the set of all ordered pairs whose first entry comes from X and whose second entry comes from Y:

$$X \times Y = \{(x, y): x \in X \ and \ y \in Y\}.$$

Recall that ordered pairs have the property that $(x, y) = (x', y')$ exactly if $x = x'$ and $y = y'$. The product of a set X with itself is often denoted X^2.

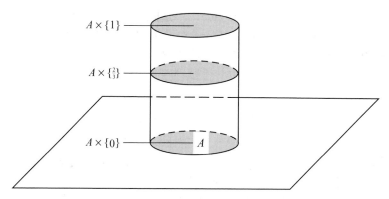

Fig. 2.7

Example 1 Let X be the set $\{0, 1, 2\}$ and let Y be the set $\{5, 6\}$. Then $X \times Y$ is a set with six elements: $X \times Y = \{(0, 5), (0, 6), (1, 5), (1, 6), (2, 5), (2, 6)\}$.

Example 2 Let $\mathbb{R}$ be the set of real numbers (conceived of as an infinite line) then $\mathbb{R} \times \mathbb{R}$ (also written $\mathbb{R}^2$) may be thought of as the real plane, where a point of this plane is identified with its ordered pair of coordinates (x, y) (with respect to the first and second coordinate axes $\mathbb{R} \times \{0\} = \{(a, 0): a \in \mathbb{R}\}$ and $\{0\} \times \mathbb{R} = \{(0, b): b \in \mathbb{R}\}$).

Example 3 Let X be any set. Then $X \times \emptyset$ is the empty set $\emptyset$ (since $\emptyset$ has no members).

Example 4 One may think of Euclidean 3-space $\mathbb{R}^3$ as being $\mathbb{R}^2 \times \mathbb{R}$ (that is, as the product of a plane with a line). Let A be a disc in the plane $\mathbb{R}^2$ and let $[0, 1]$ be the interval $\{x \in \mathbb{R} : 0 \leq x \leq 1\}$. Then $A \times [0, 1]$, regarded as a geometric object, is a vertical solid cylinder of height 1 and with base lying on the plane $\mathbb{R}^2 \times \{0\}$ (Fig. 2.7).

The ideas in this section go back mainly to Boole and Cantor. Cantor introduced the abstract notion of a set (in the context of infinite sets of real numbers). The 'algebra of sets' is due mainly to Boole – at least, in the equivalent form of the 'algebra of propositions' (for which, see Section 3.1).

In this section, we have introduced set theory simply to provide a convenient language in which to couch mathematical assertions. But there is much more to it than this: it can also be used as a foundation for mathematics. For this aspect, see discussion of the work of Cantor and Zermelo in the historical references.

Exercises 2.1

1. Which among the following sets are equal to one another?
   $X = \{x \in \mathbb{Z}: x^3 = x\}$;
   $Y = \{x \in \mathbb{Z}: x^2 = x\}$;
   $Z = \{x \in \mathbb{Z}: x^2 \leq 2\}$;
   $W = \{0, 1, -1\}$;
   $V = \{1, 0\}$.

2. List all the subsets of the set $X = \{a, b, c\}$. How many are there? Next, try with $X = \{a, b, c, d\}$. Now suppose that the set X has n elements: how many subsets does X have? Try to justify your answer.

3. Show that $X \backslash Y = X \cap Y^c$ (where the complement may be taken with respect to $X \cup Y$: that is, you may take the universal set U to be $X \cup Y$).

4. Define the **symmetric difference**, $A \triangle B$, of two sets A and B to be $A \triangle B = (A \backslash B) \cup (B \backslash A)$. Draw a Venn diagram showing the relation of this set to A and B. Show that this operation on sets is associative: for all sets A, B, C one has $(A \triangle B) \triangle C = A \triangle (B \triangle C)$.

5. Prove the parts of 2.1.1 that you have not yet checked.

6. List all the elements in the set $X \times Y$, where $X = \{0, 1\}$ and $Y = \{2, 3\}$. List all the subsets of $X \times Y$.

7. Let A, B, C, D be any sets. Are the following true? (In each case give a proof that the equality is true or a counterexample which shows that it is false.)
   (i) $(A \times C) \cap (B \times D) = (A \cap B) \times (C \cap D)$;
   (ii) $(A \times C) \cup (B \times D) = (A \cup B) \times (C \cup D)$.

8. Suppose that the set X has m members and the set Y has n members. How many members does the product set $X \times Y$ have?
   [Hint: try Exercise 2.1.6 first, then try with X having, say, three members and Y having four, ..., and so on – until you see the pattern. Then justify your answer.]

9. Give an example to show that if X is a subset of $A \times B$ then X does not need to be of the form $C \times D$ where C is a subset of A and D is a subset of B.

2.2 Functions

In this section we discuss functions and how they may be combined. The notion of a function is one of the most basic in mathematics, yet the way in which mathematicians have understood the term has changed considerably over the ages. In particular, history has shown that it is unwise to restrict the methods by which functions may be specified. Therefore the definition of function which we give may seem rather abstract since it concentrates on the end result – the

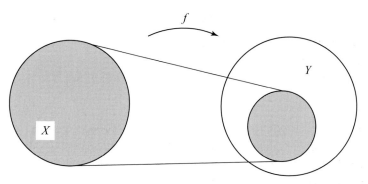

Fig. 2.8

function – rather than any way in which the function may be defined. For more on the development of the notion of function, see the notes at the end of the section.

As a first approximation, one may say that a function from the set X to the set Y is a rule which assigns to each element x of X an element of Y. Certainly something of the sort $f(x) = x^2 + 1$ serves to define a value $f(x)$ whenever the real number x is given and so this 'rule' defines a function from $\mathbb{R}$ to $\mathbb{R}$. But the term 'rule' is problematic since, in trying to specify what one means by a 'rule', one may exclude quite reasonable 'functions'.

Therefore, in order to bypass this difficulty, we will be rather less specific in our terminology and simply define a **function** from the set X to the set Y to be an assigment: to each element x of X is assigned an element of Y which is denoted by $f(x)$ and called the **image** of x. We refer to X as the **domain** of the function and Y is called the **codomain** of the function. The **image** of the function f is $\{f(x) : x \in X\}$, a subset of Y. The words **map** and **mapping** are also used instead of 'function'. The notation '$f : X \to Y$' indicates that f is a function from X to Y. A way of picturing this situation is shown in Fig. 2.8.

In the definition above, we replaced the word 'rule' by the rather more vague word 'assignment', in order to emphasise that a function need not be given by an explicit (or implicit) rule. In order to free our definition from the subtleties of the English language, we give a rigorous definition of function below.

You may find it helpful to think of a function $f : X \to Y$ as a 'black box' which takes inputs from X and yields outputs in Y and which, when fed with a particular value $x \in X$ outputs the value $f(x) \in Y$. Our rather free definition of 'function' means that we are saying nothing about how the 'black box' 'operates' (see Fig. 2.9).

Let us consider the following example.

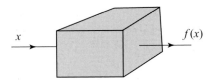

Fig. 2.9

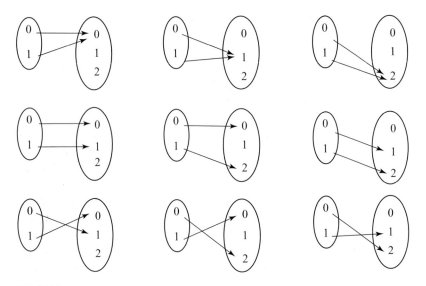

Fig. 2.10

Example We find all the functions from the set $\{0, 1\}$ to the set $\{0, 1, 2\}$. If one is used to functions given by rules such as $f(x) = x^2$, then one is tempted to spend rather a lot of time and energy in trying to describe the functions from X to Y by rules of that sort (e.g. $f(x) = x + 1$, $g(x) = 2 - x$, ...); but that is not what is asked for. All one needs to do to describe a particular function from X to Y is to specify, in an unambiguous way, for each element of X, an element of Y. So we can give an example of a function simply by saying that to the element 0 of X is assigned the element 2 of Y and to the element 1 of X is assigned the element 0 of Y. There is no need to explain this assignment in any other way.

Describing such functions in words is rather tedious: there are better ways. Figure 2.10 shows the nine ($=3^2$) possible functions from X to Y. (There are three choices for where to send $0 \in X$ and, for each of these three choices, three choices for the image of $1 \in X$: hence $3 \times 3 = 9$ in all.)

Another way of describing a function is simply to write down all pairs of the form $(x, f(x))$ with $x \in X$. So, the first function in the figure is completely described under this convention by the set $\{(0, 0), (1, 0)\}$ and the second function is described by $\{(0, 0), (1, 1)\}$. The set of all such ordered pairs is called the graph of the function. We define this formally.

Definition The **graph** of a function $f: X \to Y$ is defined to be the following subset of $X \times Y$:

$$\text{Gr}(f) = \{(x, y): x \in X \text{ and } y = f(x)\}; \text{ that is}$$
$$\text{Gr}(f) = \{(x, f(x)): x \in X\}.$$

Notice that, since a function takes only one value at each point of its domain, the graph of a function f has the property that for each x in X there is precisely one y in Y such that (x, y) is in $\text{Gr}(f)$.

Since a function and its graph each determines the other, we may now give an entirely rigorous definition of 'function' by saying that a function is a subset, G, of $X \times Y$ which satisfies the condition that for each element x of X there exists exactly one y in Y such that (x, y) is in G. (That is, we identify a function with its graph.)

It follows that two functions f, g are equal if they have the same domain and codomain and if, for every x in the common domain, $f(x) = g(x)$.

Example Suppose $X = Y = \mathbb{R}$ and let $f: \mathbb{R} \to \mathbb{R}$ be the function which takes any real number x to x^2. Then $\text{Gr}(f)$ is the set of those points in the real plane $\mathbb{R} \times \mathbb{R}$ which have the form (x, x^2) for some real number x. Think of this set geometrically to see why the term 'graph of a function' is appropriate for the notion defined above.

We may think of (the graph of) a function f as inducing a correspondence or relation from its domain X to its codomain Y. Since f is defined at each point of its domain, each element of X is related to at least one element of Y. Since the value of f at an element of X is uniquely defined, no element of X may be related to more than one element of Y. Therefore Fig. 2.11 *cannot* correspond to a function.

It is however possible that (a) there is some element of Y that is *not the image of any element* of X, (b) some element of Y is *the image of more than one element* of X. So Fig. 2.12 *may* arise from a function.

For instance, take $X = \mathbb{R} = Y$ and let $f(x) = x^2$. For an example of (a), consider -4 in Y (there is no $x \in \mathbb{R}$ with $x^2 = -4$); for an example of (b), consider 4 in Y (there are two values, $x = 2$, $x = -2$, from $\mathbb{R}$ with $x^2 = 4$). A function which avoids '(a)' is said to be surjective; one which avoids '(b)' is said to be injective; a function which avoids both '(a)' and '(b)' is said to be bijective. More formally, we have the following.

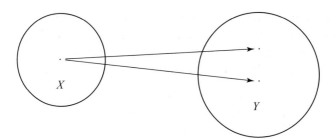

Fig. 2.11

(a)

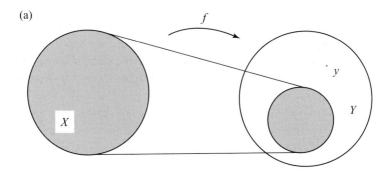

(b)

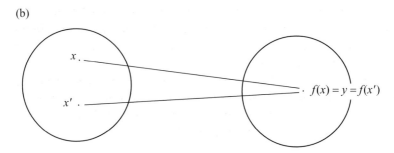

Fig. 2.12

Definition   Let $f\colon X \to Y$ be a function. We say that f is **surjective** (or **onto**) if for each y in Y there exists (at least one) x in X such that $f(x) = y$ (that is, every element of Y is the image of element of X). The function f is **injective** (or **one-to-one**, also written '1-1') if for x, x' in X the equality $f(x) \equiv f(x')$ implies $x = x'$ (that is, distinct elements of X cannot have the same image in Y).   Finally, f is **bijective** if f is both injective and surjective. A **surjection** is a function which is surjective; similarly with **injection** and **bijection**. A **permutation** of a set is a bijection from that set to itself. We will study the structure of permutations of finite sets in Chapter 4.

Example 1 The function $f: \mathbb{R} \to \mathbb{R}$ given by $f(x) = x^4$ is neither injective nor surjective. It is not injective since, for example, $f(2) = 16 = f(-2)$ but $2 \neq -2$. The fact that it is not surjective is shown by the fact that -1 (for example) is not in the image of f: it is not the fourth power of any real number.

Example 2 The function $s: \mathbb{P} \to \mathbb{P}$ defined by $s(n) = n + 1$ is injective but not surjective. It is not surjective because the equation $s(n) = 1$ has no solution in $\mathbb{P}$, that is, there is no $n \in \mathbb{P}$ with $n + 1 = 1$.

To show that s is injective, suppose that $s(n) = s(m)$, then $n + 1 = m + 1$. Then $n = m$. Turning this round (i.e. the 'contrapositive' statement – see p. 132) we have shown that if $n \neq m$ then $s(n) = s(m)$.

Example 3 The function $g: \mathbb{R} \to \mathbb{R}$ defined by $f(x) = x^5$ is bijective. To prove this, one may proceed as follows.

Surjective. To say that this function g is surjective is precisely to say that every real number has a real fifth root – an assertion which is true and, we assume, known to you.

Injective. Suppose that $f(x) = f(y)$: that is, $x^5 = y^5$. Thus $x^5 - y^5 = 0$. Factorising this gives

$$(x - y) \cdot (x^4 + x^3 y + x^2 y^2 + xy^3 + y^4) = 0.$$

If we can show that the second factor $t = x^4 + x^3 y + x^2 y^2 + xy^3 + y^4$ is never zero except when $x = y = 0$ then it will follow that $x^5 - y^5$ equals 0 only in the case that $x = y$: in other words, it will follow that the function f is injective. Now, there are various ways of showing that the factor t is zero only if $x = y = 0$: perhaps the most elementary is the following.

We intend to use the fact that a sum of squares of real numbers is zero only if the terms in the sum are individually zero. Notice that the term $x^3 y + xy^3$ equals $xy(x^2 + y^2)$. This suggests considering the term $(x + y)^2(x^2 + y^2)$ or at least, half of it. Following this up, we obtain:

$$t = \frac{1}{2}(x + y)^2(x^2 + y^2) + \frac{1}{2}x^4 + \frac{1}{2}y^4,$$

and this can be written as

$$\left(\left(\frac{1}{\sqrt{2}} \right) \cdot (x + y) \cdot x \right)^2 + \left(\left(\frac{1}{\sqrt{2}} \right) \cdot (x + y) \cdot y \right)^2$$
$$+ \left(\left(\frac{1}{\sqrt{2}} \right) x^2 \right)^2 + \left(\left(\frac{1}{\sqrt{2}} \right) y^2 \right)^2.$$

Thus t is indeed a sum of squares and we can see that this sum is zero only if $x = y = 0$, as required.

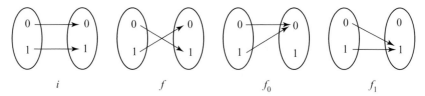

Fig. 2.13

In fact, for any function $f: \mathbb{R} \to \mathbb{R}$, $y = f(x)$, we can interpret the ideas of injective and surjective in terms of the graph of f ('graph' in the pictorial sense, drawn with the x-axis horizontal). Thus f is injective if and only if every horizontal line meets the graph in *at most one* point. Similarly, f is surjective if and only if every horizontal line meets the graph of f in *at least one* point. Using these ideas, it is easy to see that the function $f(x) = x^3$ is injective, that the function $h: \mathbb{R} \to \mathbb{R}$ given by $h(x) = x^3 - x$ is surjective but not injective, that the function $k: \mathbb{R} \to \mathbb{R}$ given by $k(x) = e^x$ is injective but not surjective.

We can also express these ideas in terms of solvability of equations. To say that $f: X \to Y$ is surjective is to say that for every $b \in Y$ the equation $f(x) = b$ *has a solution*. To say that f is injective is to say that for every $b \in Y$ the equation $f(x) = b$ has at most one solution. To say that f is bijective is to say that for every $b \in Y$ the equation $f(x) = b$ has exactly one solution.

Example 1   Consider the possible functions from $\{0, 1\}$ to itself. There are four possible functions (two choices for the value of f at 0; then, for each of these, two choices for $f(1)$). Their actions can be shown as in Fig. 2.13. The functions i and f are bijections and f_0 and f_1 are neither injections nor surjections.

Example 2   Refer back to the first example of this section. There are no surjections from $\{0, 1\}$ to $\{0, 1, 2\}$ and hence there are no bijections. But the function f defined by $f(0) = 2$ and $f(1) = 0$ is an example of an injection (you may check that six of the nine functions are injective).

Definitions   If X is a set then the function $\mathrm{id}_x: X \to X$ which takes every element to itself ($\mathrm{id}_x(x) = x$ for all x in X) is the **identity function** on X. If X and Y are sets and c is in Y then we may define the **constant function from X to Y with value** c by setting $f(x) = c$ for every x in X. (Do not confuse the identity function on a set with, when it makes sense, a function with constant value '1'.)

Definition   Suppose that $f: X \to Y$ and $g: Y \to Z$ are functions. We can take any element x of X, apply f to it and then apply g to the result (since the result is in Y). Thus we end up with an element $g(f(x))$ of Z. What we have just done is

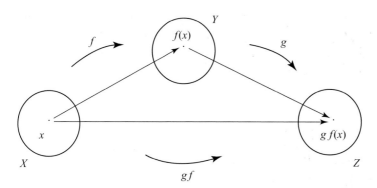

Fig. 2.14

to define a new function from X to Z: it is denoted by $gf: X \to Z$, is defined by $gf(x) = g(f(x))$, and is called the **composition** of f and g (note the reversal of order: gf means 'do f first and then apply g to the result'). See Fig. 2.14.

In the case that $X = Y = Z$ and $f = g$, the composition of f with itself is often denoted f^2 rather than ff (similarly for $f^3, \ldots$).

Example 1 Let f and g be the functions from $\mathbb{R}$ to $\mathbb{R}$ defined by $f(x) = x + 1$ and $g(x) = x^2$. The composite function fg is given by

$$fg(x) = f(g(x)) = f(x^2) = x^2 + 1.$$

Note that the composition gf is given by

$$gf(x) = g(x + 1) = (x + 1)^2 = x^2 + 2x + 1.$$

Thus, even if both functions gf and fg are defined, they need not be equal.

Example 2 Let $f, g: \mathbb{R} \to \mathbb{R}$ be defined by $f(x) = 4x - 3$ and $g(x) = (x + 3)/4$.
Then

$$fg(x) = f((x + 3)/4) = 4((x + 3)/4) - 3 = x + 3 - 3 = x$$

and

$$gf(x) = g(4x - 3) = ((4x - 3) + 3)/4 = 4x/4 = x.$$

In this case, it does turn out that fg and gf are the same function, namely the identity function on $\mathbb{R}$.

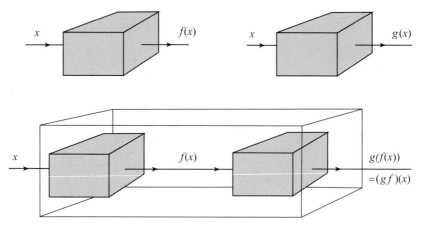

Fig. 2.15

Example 3 Suppose that F and G are computer programs, each of which takes integer inputs and produces integer outputs. To F we may associate the function f which is defined by $f(n) = $ that integer which is output by F if it is given input n. Similarly, let g be the function which associates to any integer n the output of G if G is given input n. We may connect these programs in series as shown in Fig. 2.15: thus the output of F becomes the input of G.

Regard this combination as a single program: the function which is associated to it is precisely the composition gf. If one thinks of a function as a 'black box', as indicated after the definition of function, then the picture above suggests a way of thinking about the composition of two functions.

Example 4 Let $f\colon X \to Y$ be any function. Then $f\,\mathrm{id}_x(x) = f(\mathrm{id}_x(x)) = f(x)$, so $f\,\mathrm{id}_x = f$. Similarly, $\mathrm{id}_Y\,f = f$.

Suppose now that we have functions $f\colon X \to Y$, $g\colon Y \to Z$ and $h\colon Z \to W$. Then we may form the composition gf and then compose this with h to get $h(gf)$. Alternatively we may form hg first and then apply this, having already applied f, to obtain $(hg)f$. The first result of this section says that the result is the same: $h(gf) = (hg)f$. This is the associative law for composition of functions.

Theorem 2.2.1 *If $f\colon X \to Y$, $g\colon Y \to Z$, $h\colon Z \to W$ are functions then $h(gf) = (hg)f$ and so this function from X to W may be denoted unambiguously by $hgf\colon X \to W$.*

Proof Consider Fig. 2.16. We see that the element x of X is sent to the same element w of W by the two routes. The first applies f and then the composite $hg\colon (hg)f$. The second applies the composite of gf and then $h\colon h(gf)$. □

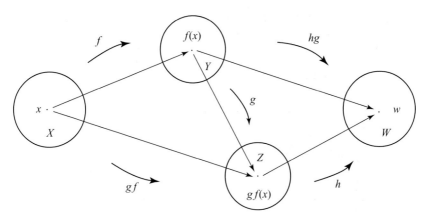

Fig. 2.16

Consider the function $f: \mathbb{R} \to \mathbb{R}$ defined by $f(x) = x^3$. If $g: \mathbb{R} \to \mathbb{R}$ is the function which takes each real number to its (unique!) real cube root then it makes sense to say that g reverses the action of f and, indeed, f reverses the action of g, since

$$gf(x) = x = \mathrm{id}_{\mathbb{R}}(x) \text{ and } fg(x) = x = \mathrm{id}_{\mathbb{R}}(x) \text{ for every } x \in \mathbb{R}.$$

$$\| \quad \| \qquad\qquad \| \quad \|$$

$$g(x^3) = (x^3)^{1/3} \qquad f(x^{1/3}) = (x^{1/3})^3$$

Definition Suppose that $f: X \to Y$ is a function: a function $g: Y \to X$ which goes back from Y to X and is such that the composition gf is id_X and the composition fg is id_Y is called an **inverse** function for (or of) f.

So, an inverse of f (if it exists!) reverses the effect of f. We show first that if an inverse for a function exists, then it is unique.

Theorem 2.2.2 *If a function* $f: X \to Y$ *has an inverse, then this inverse is unique.*

Proof To see this, suppose that each of g and h is an inverse for f. Thus

$$fg = \mathrm{id}_Y = fh \quad \text{and} \quad gf = \mathrm{id}_X = hf.$$

Now consider the composition $(gf)h = (\mathrm{id}_X)h = h$ (cf. Example 4 on p. 94). By Theorem 2.2.1, this is equal to $g(fh) = g(\mathrm{id}_Y) = g$, so h and g are equal. $\square$

Notation The inverse of f, if it exists, is usually denoted by f^{-1}. It should be emphasised that this is inverse with respect to composition, *not* with respect to multiplication. Thus the inverse of the function $f(x) = x + 1$, which adds 1, is the function $g(x) = x - 1$, which subtracts 1, not the function $h(x) = 1/(x + 1)$.

Example 2 on p. 93 shows that the inverse of the function $4x - 3$ exists. On the other hand, the function $f: \mathbb{Z} \to \mathbb{N}$ given by $f(x) = x^2$ cannot have an inverse. That this is so can be seen in two ways. Either note that since f is not onto, there are natural numbers on which an inverse of f could not be defined (what would '$f^{-1}(3)$' be?). Alternatively, since f is not 1-1, its action cannot be reversed ('$f^{-1}(4)$' would have to be both -2 and 2 but then 'f^{-1}' would not be a well defined function).

The following result gives the precise criterion for a function to have an inverse. Although it is straightforward, the proof of this result may seem a little abstract. You should not be unduly disturbed if you do find it so: the purpose of the various parts of the proof will become clearer as you become more familiar with notions such as surjective and injective.

Theorem 2.2.3 *A function $f: X \to Y$ has an inverse if and only if f is a bijection.*

Proof For the first part of the proof, we suppose that f has an inverse and show that f is both injective and surjective. So let f^{-1} denote the inverse of f.
   Suppose that $f(x_1) = f(x_2)$. Apply f^{-1} to both sides to obtain

$$f^{-1}(f(x_1)) = f^{-1}(f(x_2)).$$

Thus

$$f^{-1}f(x_1) = f^{-1}f(x_2).$$

Since $f^{-1}f$ is the identity function on X, we deduce that $x_1 = x_2$ and hence that f is injective.
   To show that f is surjective, take any y in Y. The composite function ff^{-1} is the identity on Y so $y = f(f^{-1}(y))$. Thus y is of the form $f(x)$ where x is $f^{-1}(y) \in X$ and so f is indeed surjective.
   Now we suppose, for the converse, that f is bijective and we define f^{-1} by

$$f^{-1}(y) = x \quad \text{if and only if} \quad f(x) = y.$$

The fact that f is injective means that f^{-1} is well defined (there cannot be more than one x associated to any given y) and the fact that f is surjective means that f^{-1} is defined on all of Y. It follows from the definition that $f^{-1}f$ is the identity on X and that ff^{-1} is the identity on Y. $\square$

The above theorem may be regarded as an 'algebraic' characterisation of bijections. For related characterisations of injections and surjections, see Example 3 on p. 185.

Example   Consider the (four) functions from $\{0, 1\}$ to itself, using the notation of Fig. 2.13. The theorem above tells us that, of these, i and f have inverses. In fact, each is its own inverse: i is the identity function $\mathrm{id}_{\{0,1\}}$; $f^2 = i$.
   For (many) more examples of bijections, refer forward to Section 4.1.

Corollary 2.2.4   *Let $f: X \to Y$ and $g: Y \to Z$ be bijections. Then*

(i) *gf is a bijection from X to Z, with inverse $f^{-1}g^{-1}$, that is*
   $(gf)^{-1} = f^{-1}g^{-1}$,
(ii) *$f^{-1}: Y \to X$ is a bijection, with inverse f, that is $(f^{-1})^{-1} = f$.*
   Also
(iii) *id_X is a bijection (and is its own inverse!).*

Proof   (i) By 2.2.3, there exist inverses $f^{-1}: Y \to X$ and $g^{-1}: Z \to Y$ for f and g. Then the composite function $(f^{-1}g^{-1})(gf)$ equals $f^{-1}(g^{-1}g)f = f^{-1}\mathrm{id}_Y f = f^{-1}f = \mathrm{id}_X$. Similarly $(gf)(f^{-1}g^{-1}) = \mathrm{id}_Z$. So the function $gf: X \to Z$ has an inverse $f^{-1}g^{-1}$ so, by 2.2.3, is a bijection.
   (ii) We have $f^{-1}f = \mathrm{id}_X$ and $ff^{-1} = \mathrm{id}_Y$ since f^{-1} is the inverse of f. Hence the inverse of f^{-1} is f and, in particular (by 2.2.3), f^{-1} is a bijection.
   (iii) It is immediate from the definition that the identity function is injective and surjective.   □

This corollary will be of importance when we discuss permutations in Section 4.1.

Finally in this section, we discuss the cardinality of a (finite) set.
   Suppose that we have two sets X and Y which have a finite number of elements n and m, respectively. If there is an injective map from X to Y then, since distinct elements of X are mapped to distinct elements of Y, there must be at least n different elements in Y. Thus $n \le m$. If there is a surjective map from X to Y, there must exist, for each element of Y, at least one element of X to map to it, and so (since each element of X has just one image in Y) there must be at least as many elements in X as in Y: $n \ge m$. Putting together these observations, we deduce that if there is a bijection from X to Y then X and Y have the same number of elements. This observation forms the basis for the following definition (which is due to Cantor).

Definition   We say that sets X and Y **have the same cardinality** (i.e. have the same 'number' of elements) and write $|X| = |Y|$ if there is a bijection from X to Y. If X is a non-empty set with a finite number of elements then there is a bijection from X to a set of the form $\{1, 2, \ldots, n\}$ for some integer n; we write $|X| = n$ and say that X has n elements. We also set $|\emptyset| = 0$.

In the above definition, we did not require the sets X and Y to be finite. So we have defined what it means for two, possibly infinite, sets to have the *same* number of elements, without having had to define what we mean by an infinite number (in fact, the above idea was used by Cantor as the basis of his definition of 'infinite numbers').

If you are tempted to think that one would never in practice use a bijection to show that two sets have the same number of elements, then consider the following example (with a little thought, you should be able to come up with further examples).

Example   Suppose that a hall contains a large number of people and a large number of chairs. Someone claims that there are precisely the same number of people as chairs. How may this claim be tested? One way is to try (!) to count the number of people and then count the number of chairs, and see if the totals are equal. But there is an easier and more direct way to check this: simply ask everyone to sit down in a chair (one person to one chair!). If there are no people left over and no chairs left over then the function which associates to each person the chair on which they are sitting is a bijection from the set of people to the set of chairs in the hall, and so we conclude that there are indeed the same number of chairs as people. This method has tested the claim without counting either the number of people or the number of chairs.

We can return to explain more clearly a point which arose in the proof of Theorem 1.6.6. There we considered two relatively prime integers a and b, and wished to show that $\phi(ab) = \phi(a)\phi(b)$. The proof given consisted in defining a function f from the set G_{ab} to the set $G_a \times G_b$ by setting $f([t]_{ab}) = ([t]_a, [t]_b)$. It was then shown that f is injective and surjective, so bijective, and hence the result $\phi(ab) = \phi(a)\phi(b)$ followed, since $\phi(n)$ is the number of elements in G_n.

The next result shows how to compute the cardinality of the union of two sets with no intersection.

Theorem 2.2.5   *Let X and Y be finite sets which are disjoint (that is, $X \cap Y = \emptyset$). Then*

$$|X \cup Y| = |X| + |Y|.$$

Proof We include a proof of this (fairly obvious) fact so as to illustrate how one may use the definition of cardinality in proofs.

Suppose that X has n elements: so there is a bijection f from X to the set $\{1, 2, \ldots, n\}$. If Y has m elements then there is a bijection g from Y to the set $\{1, 2, \ldots, m\}$. Define a map h from $X \cup Y$ to the set $\{1, 2, \ldots, n + m\}$ as follows:

$$h(x) = \begin{cases} f(x) & \text{if } x \in X, \\ n + g(x) & \text{if } x \in Y. \end{cases}$$

Since there is no element x in both X and Y, there is no conflict in this two-clause definition of $h(x)$. The images of the elements of X are the integers in the range $\{1, 2, \ldots, n\}$ and the images of the elements of Y are those in the range $\{n + 1, \ldots, n + m\}$. It is easy to check that h is surjective (since both f and g are surjective) and that, since both f and g are injective, h is injective. Thus h is bijective as required. $\square$

Corollary 2.2.6 *Let X and Y be finite sets. Then*

$$|X| + |Y| = |X \cap Y| + |X \cup Y|.$$

Proof The sets $X \cap Y$ and $X \backslash (X \cap Y)$ are disjoint and their union is X. So, by Theorem 2.2.5,

$$|X \backslash (X \cap Y)| + |X \cap Y| = |X|$$

and hence

$$|X \backslash (X \cap Y)| = |X| - |X \cap Y|.$$

Now consider $X \backslash (X \cap Y)$ and Y: these sets are disjoint since if $x \in X \backslash (X \cap Y)$ then x is not a member of $X \cap Y$ and hence is not a member of Y. The union of $X \backslash (X \cap Y)$ and Y is

$$Y \cup (X \cap Y^c) = (Y \cup X) \cap (Y \cup Y^c) = Y \cup X = X \cup Y.$$

So, applying Theorem 2.2.5 again gives

$$|X \cup Y| = |X \backslash (X \cap Y)| + |Y|$$
$$= |X| - |X \cap Y| + |Y| \quad \text{(by the above).}$$

Rearranging gives the required result. $\square$

Example A group of 50 people is tested for the presence of certain genes. Gene X confers the ability to yodel; gene Y endows its bearer with great skill

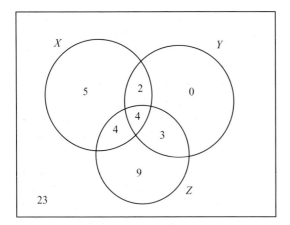

Fig. 2.17

at Monopoly; gene Z produces an allergy to television commercials. It is found that of this group, fifteen have gene X, nine have gene Y and twenty have gene Z. Of these, six have both genes X and Y, eight have genes X and Z and seven have genes Y and Z. Four people have all three genes. How many of this group lack all three of these genes? How many non-yodelling bad Monopoly players are there?

If we draw a Venn diagram as shown in Fig. 2.17, with X being the set of people with gene X and so on, then we may fill in the number of people in each 'minimal region', and so deduce the answers. For example, we are told that the centre region, which represents $X \cap Y \cap Z$, contains four elements. We are also told that the cardinality of $X \cap Y$ is 6. So it must be that $(X \cap Y) \cap Z^c$ has $6 - 4 = 2$ elements. And so on (all the time using 2.2.5 implicitly).

The first question asks for the number of elements of $X^c \cap Y^c \cap Z^c$: that is 23. The second question asks for the cardinality of the set $X^c \cap Y^c = (X \cup Y)^c$: that is 32.

What a mathematician nowadays understands by the term 'function' is very different from what mathematicians of previous centuries understood by the term. Indeed, the way we may regard an expression such as $f(x) = x^2 + 1$ from a purely algebraic point of view would have been foreign to a mathematician of even the eighteenth century, to whom an algebraic expression of this sort would have had very strong geometric overtones. Mathematicians of those times considered that a function must be (implicitly or explicitly) given by a 'rule' of some sort which involves only well understood algebraic operations (addition, division, extraction of roots and so on) together with 'transcendental' functions (such as sine and exponential). Nevertheless, the development of the calculus, independently by Leibniz and Newton, towards the end of the seventeenth century, raised a host of problems about the nature of functions and their behaviour. Resolution of these problems over the following two centuries necessitated a thorough examination of the foundations of analysis and this was one of the main forces involved in changing the face of mathematics during the nineteenth century to something resembling its present-day form.

The work of Euler was probably the most influential in separating the algebraic notion of a function from its geometric background. As for extending the notion of 'function' beyond what is given explicitly or implicitly by a single 'rule', the main impetus here was the development of what is now called Fourier Analysis. On a methods course, you will probably meet/have met the fact that many (physically defined) functions (waveforms, for example) can be represented as infinite sums of simple terms involving sine and cosine.

The idea of representing certain functions as infinite sums of simple functions, in particular, representation by power series ('infinitely long polynomials'), was well established by the late seventeenth century. The general method was stated by Brook Taylor in his *Methodus Incrementorum* of 1715 and 1717 (hence the term 'Taylor series'), although there were many precursors.

The physical problem whose analysis forced mathematicians to re-examine their ideas concerning functions was the problem of describing the motion of a vibrating string which is given an initial configuration and then released (considered by Johann Bernoulli, then d'Alembert, Euler and Daniel Bernoulli) and, somewhat later, Fourier's investigations on the propagation of heat. The analysis of these problems involved representing functions by trigonometric series.

What was new was the generality of those functions which can be represented by trigonometric series throughout their domain. In particular, such functions need not be given by a single 'rule' and they may have discontinuities (breaks) and 'spikes' – hardly in accordance with what most mathematicians of the day would have meant by a function.

A great deal of controversy was generated and this can be largely ascribed to the fact that the idea of 'function' was not at all rigorously defined (so different mathematicians had different ideas as to what was admissible as a function) and, indeed, was too restrictive.

Even for continuous functions ('functions without breaks') it is 1837 before one finds a definition, given by Dirichlet, of continuous function which casts aside the old restrictions. It is also worth noting that it is Dirichlet who in 1829 presents functions of the following sort: $f(x) = 0$ if x is rational; $f(x) = 1$ if x is irrational. By any standards this is a rather peculiar function (it is discontinuous everywhere): nevertheless it is a function.

The ramifications of all this have been of great significance in the development of mathematics: the reader is referred to [Grattan-Guinness], [Manheim]
or one of the more general histories for more on the topic. What we take from
all this is the point that it is probably unwise to try to restrict the methods by
which a function may be defined. In particular, we have seen above that the
modern definition of a function avoids all reference to how a certain function
may be specified, but rather concentrates on the most basic conditions that a
function must satisfy.

The examples that we have given of functions illustrate that our definition
of 'function' is very 'free' and may allow in all kinds of functions which we
do not want to consider. But in that case we may simply restrict attention to the
kinds of function which are relevant for our particular purpose, whether they
be continuous, differentiable, computable, given by a polynomial, or whatever.

Exercises 2.2

1. Describe all the functions from the set $X = \{0, 1, 2\}$ to the set $Y = \{0, 5\}$.
2. Decide which of the following functions are injective, which are surjective
   and which are bijective:
   (i) $f: \mathbb{Z} \to \mathbb{Z}$ defined by $f(x) = x - 1$;
   (ii) $f: \mathbb{R} \to \mathbb{R}^+$ defined by $f(x) = |x|$ (where $\mathbb{R}^+$ denotes the set of non-
         negative real numbers; here $|x|$ is defined to be x if $x \geq 0$ and to be $-x$
         if $x < 0$)
   (iii) $f: \mathbb{R} \to \mathbb{R}$ defined by $f(x) = |x|$;
   (iv) $f: \mathbb{R} \times \mathbb{R} \to \mathbb{R}$ defined by $f(x, y) = x$;
   (v) $f: \mathbb{Z} \to \mathbb{Z}$ defined by $f(x) = 2x$.
3. Draw the graphs of (a) the identity function on $\mathbb{R}$, (b) the constant
   function on $\mathbb{R}$ with value 1.
4. Let $f: \mathbb{R} \to \mathbb{R}$ and $g: \mathbb{R} \to \mathbb{R}$ be defined by
   $f(x) = x + 1, g(x) = x^2 - 2$.
   Find fg, gf, $f^2(= ff)$ and g^2.
5. Find bijections
   (i)   from the set of positive real numbers $\mathbb{R}^+$ to the set $\mathbb{R}$,
   (ii) from the open interval $(-\pi/2, \pi/2) = \{x \in \mathbb{R}: -\pi/2 < x < \pi/2\}$
         to the set $\mathbb{R}$,
   (iii) from the set of natural numbers $\mathbb{N}$ to the set $\mathbb{Z}$ of integers.
6. Describe all the bijections from the set $X = \{0, 1, 2\}$ to itself.
7. Find the inverses of the following functions $f: \mathbb{R} \to \mathbb{R}$:
   (i) $f(x) = (4 - x)/3$;
   (ii) $f(x) = x^3 - 3x^2 + 3x - 1$.

8. Let A, B and C be sets with finite numbers of elements. Show that
$$|A \cup B \cup C| = |A| + |B| + |C| - |A \cap B| - |A \cap C| - |B \cap C|$$
$$+ |A \cap B \cap C|.$$

9. Show that the following data are inconsistent: 'Of a group of 50 students, 23 take mathematics, 14 take chemistry and 17 take physics. 5 take mathematics and physics, 3 take mathematics and chemistry and 7 take chemistry and physics. Twelve students take none of mathematics, chemistry and physics'. [Hint: see the last example of the section.]

10. Let X be a set and let A be one of its subsets. The **characteristic function** of A is the function $\chi_A \colon X \to \{0, 1\}$ defined by
$$\chi_A(x) = \begin{cases} 1 & \text{if } x \in A, \\ 0 & \text{if } x \notin A. \end{cases}$$

    (a) Show that if A and B are subsets of X then they have the same characteristic function if and only if they are equal.

    (b) Show that every function from X to $\{0, 1\}$ is the characteristic function of some subset of X.

    The notation Y^Z is sometimes used for the set of all functions from the set Z to the set Y. Parts (a) and (b) above show that the map which takes a set to its characteristic function is a bijection from the set $P(X)$ of all subsets of X to the set $\{0, 1\}^X$. Since the notation '2' is sometimes used for the set $\{0, 1\}$, this explains why the notation 2^X, instead of $P(X)$, is sometimes used for the set of all subsets of X.

11. Let X be a finite set. Show that $P(X)$ has $2^{|X|}$ elements. That is, in the notation mentioned at the end of the previous example, show that
$$|2^X| = 2^{|X|}.$$
    [Hint: to give a rigorous proof, induct on the number of elements of X.]

2.3 Relations

Consider the function $f \colon \mathbb{R} \to \mathbb{R}$ given by $f(x) = x^2$. Since this function is not injective (or surjective), it does not have an inverse. So we cannot say that associating to a number its square roots defines a function. On the other hand, there certainly is a relationship between a number and its square roots (if it has any). The mathematical definition of a relation allows us to encompass this more general situation.

Definition Let X, Y be sets. A **relation** R from X to Y is simply a subset of the Cartesian product: $R \subseteq X \times Y$. As an alternative to writing $(x, y) \in R$ we also write xRy. We say that x is **related** (in the sense of R) to y if $(x,y) \in R$, that is, if xRy. If $X = Y$ we then talk of a relation **on** X.

This definition may seem very abstract and to be rather a long way from what we might normally term a relation. For, in the above definition, the relation R may be any subset of $X \times Y$: we do not insist on a 'material' connection between those elements x, y such that $(x,y) \in R$. We do find, however, that any normal use of the term relation may be covered by this definition. And, as with the case of functions, the advantage of making this wide, abstract definition is that we do not limit ourselves to a notion of 'relation' which might, with hindsight, be seen as overly restrictive.

The following examples give some idea of the variety and ubiquity of relations.

Example 1 Let $\mathbb{N}$ be the set of natural numbers and consider the relation '$\leq$' on $\mathbb{N}$. This is defined by the condition: $x \leq y$ if and only if x is less than or equal to y ($x, y \in \mathbb{N}$). Alternatively, it may be defined arithmetically by $x \leq y$ if and only if $y - x \in \mathbb{N}$. However one chooses to define it, it is a relation in the sense of the above definition: let $X = \mathbb{N} = Y$ and take the subset $R = \{(x, y): x \leq y\}$ of $\mathbb{N} \times \mathbb{N}$. Then $(x,y) \in R$ if and only if $x \leq y$.

One may note that relations are often specified, not by directly defining a subset of $X \times Y$, but rather, as in this example, by specifying the condition which must be satisfied for elements x and y to be related.

Notation of the sort 'xRy' is more common than '$(x,y) \in R$': the relations of 'less than or equal to' and 'equals' are usually written $x \leq y$ and $x = y$.

Example 2 Any function $f: X \to Y$ determines a relation: namely its graph $\text{Gr}(f) = \{(x, y): x \in X \text{ and } y = f(x)\} \subseteq X \times Y$ (as introduced in Section 2.2). Thus, we may define the associated relation R either by saying that R is the set $\text{Gr}(f)$ or by setting xRy if and only if $y = f(x)$. Thus a function $f: X \to Y$, when regarded as a subset of $X \times Y$, is just a special sort of relation (namely, a relation which satisfies the condition: every $x \in X$ is related to exactly one element of Y).

Example 3 We can define the relation R on the set of real numbers to be the set of all pairs (x,y) with $y^2 = x$. Thus xRy means 'y is a square root of x'. As we mentioned in the introduction to this section, this is not a function, but it is a relation in the sense that we have defined.

Example 4 Let X be the set of integers and let R be the relation: xRy if and only if $x - y$ is divisible by 3.

Thus $1R4$ and $1R7$ but not $2R3$.

Example 5 Let $X = \{1, 2, \ldots, 11, 12\}$ and let D be the relation 'divides' – so xDy if and only if x divides y. As a subset of $X \times X$,

$$D = \{(m,n): m \text{ divides } n\}.$$

Thus, for example $(4,8) \in D$ but $(4,10) \notin D$. (As an exercise in this notation, list D as a subset of $X \times X$.)

Example 6 Let C be the set of all countries. Define the relation B on C by cBd if and only if the countries c,d have a common border.

Here are some rather more 'abstract' relations.

Example 7 Let X and Y be any sets. Then the empty set $\emptyset$, regarded as a subset of $X \times Y$, is a relation from X to Y (the 'empty relation', characterised by the condition that no element of X is related to any element of Y). Another relation is $X \times Y$ itself – this relation is characterised by the fact that every element of X is related to every element of Y.

Example 8 Let X be any set. Then the relation $R = \{(x,x): x \in X\}$ is the 'identity relation' on X: that is, xRx if and only if $x = x$. In other words, this is the relation 'equals'.

Example 9 Given a relation R from a set X to a set Y, we may define the dual, or complementary, relation to be $R^c = (X \times Y) \backslash R$. Thus $xR^c y$ holds if and only if xRy does not hold. For instance, the dual, R^c, of the identity relation on a set is the relation of being unequal: $xR^c y$ if and only if $x \neq y$.

Also, we may define the 'reverse' relation, R^{rev}, of R to be the relation from Y to X which is defined to be $R^{\text{rev}} = \{(y,x): (x, y) \in R\}$.

If $f: X \to Y$ is a function, then the reverse relation from Y to X is the subset $\{(f(x), x): x \in X\}$ of $Y \times X$. This relation will be a function (namely f^{-1}) if and only if f is a bijection.

Observe that the complement and reverse are quite different. Take, for instance, X to be the set of all people (who are alive or have lived). Define the relation R by xRy if and only if x is an ancestor of y. Then the dual R^c of R is the relation defined by $xR^c y$ if and only if x is not an ancestor of y. Whereas the reverse relation R^{rev} is defined by $xR^{\text{rev}} y$ if and only if x is a descendant of y.

Definitions Let R be a relation on a set X. We say that R is

(1) **reflexive** if xRx for all x in X,
(2) **symmetric** if, for all $x, y \in X$, xRy implies yRx,

(3) **weakly antisymmetric** if, for all $x,y \in X$, whenever xRy and yRx hold one
has $x = y$,

(3′) **antisymmetric** if, for all $x,y \in X$, if xRy holds then yRx does not,

(4) **transitive** if, for all $x,y,z \in X$, if xRy and yRz hold then so does xRz.

We reconsider some of our examples in the light of these definitions.

Example 1   The relation '$\leq$' is reflexive ($x \leq x$), not symmetric (e.g. $4 \leq 6$
but not $6 \leq 4$), weakly antisymmetric ($x \leq y$ and $y \leq x$ implies $x = y$), transitive
($x \leq y$ and $y \leq z$ imply $x \leq z$).

Example 2   This example is not a relation on X unless $X = Y$: rather a relation
between X and Y, so (note) the above definitions do not apply.

Example 3   This relation is not reflexive (x^2 is not equal to x in general), not
symmetric, and not transitive ($4R16$ and $2R4$, but not $2R16$). Let us look at weak
antisymmetry in more detail. Suppose xRy and yRx: so $x^2 = y$ and $y^2 = x$. Thus
$x = x^4$ and, since x is real, x is either 0 or 1. Then y is also 0, respectively 1,
since $y = x^2$. It follows that the relation is weakly antisymmetric.

Example 4   The relation is reflexive, symmetric and transitive but not (weakly)
antisymmetric. First, consider reflexivity. Since 0 is divisible by 3 and $0 = x -
x$, we see that, for every $x \in X$, $x R x$. For symmetry, note that if xRy then $x - y$ is
divisible by 3 and so $y - x$ is divisible by 3; that is, yRx. For transitivity, suppose
xRy and yRz, so $x - y$ is divisible by 3, as is $y - z$. Then $(x - y) + (y - z) =
x - z$ is divisible by 3 and so xRz holds. Regarding antisymmetry: note that
$3R6$ and $6R3$ yet 3 and 6 are not equal.

Example 5   This relation is reflexive, not symmetric (for example, 2 divides
4 but 4 does not divide 2), weakly antisymmetric and transitive.

As an exercise, you should examine Examples 6 to 8 in the light of these
definitions.

In dealing with conditions such as those above, one should be careful over logic.
For instance, a relation R on X is symmetric if and only if for every x and y in X,
xRy implies yRx (exercise: is the empty relation $R = \emptyset \subseteq X \times X$ symmetric?).
So, in order to show that a relation R is not symmetric, it is enough to find *one*
pair of elements a,b such that aRb holds but bRa does not. For another example,
to show that a relation is transitive, it must be shown that for *every* triple a,b,c,
if aRb and bRc hold then so does aRc: it is not enough to check it for just some
values of a, b and c.

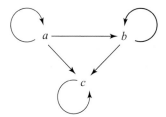

Fig. 2.18

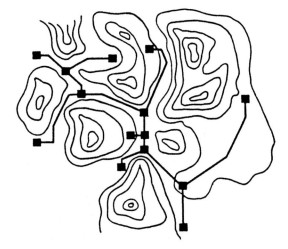

Fig. 2.19 Thin lines, contours; thick lines, railways; small squares, towns.

For more exercises in logic, see Exercises 2.3.2 and 2.3.3 at the end of this section as well as all those in Chapter 3.

Definition A useful pictorial way to represent a relation R on X is by its **digraph** (or **directed graph**) $\Gamma(R)$. To obtain this, we use the elements of X as the vertices of the graph $\Gamma(R)$ and join two of these vertices, x and y, by a directed edge (a directed arrow from x to y) whenever xRy.

Example Let X be the set $\{a,b,c\}$ and let R be the relation specified by

$$aRa,\ bRb,\ cRc,\ aRb,\ aRc,\ bRc.$$

The digraph of R is as shown in Fig. 2.18.

Example Fig. 2.19 is a map showing a number of towns in mountain valleys, and the railway network that connects them. Define the relation R on the set

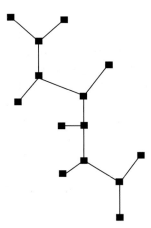

Fig. 2.20

of towns by aRb if and only if a and b are next to each other on the railway line.

The relation is symmetric so, if the directed graph of the relation has a directed edge going from a to b, then it also has one going from b to a. So we make the convention that an edge without any arrow stands for such a pair of directed edges. With this convention, the graph of the relation is as shown in Fig. 2.20.

A relation on a set may be specified by giving its digraph: the set X is recovered as the set of vertices of the digraph, and the pair (x,y) is in the relation if and only if there is a directed edge going from the vertex x to the vertex y.

Yet another way to specify a relation R on a set X is to give its **adjacency matrix**. This is a matrix with rows and columns indexed by the elements of X (listed in an arbitrary but fixed order). Each entry of the matrix is either 0 or 1. The entry at the intersection of the row indexed by x and the column indexed by y is 1 if xRy is true, and is 0 if xRy is false. For convenience, we present examples of adjacency matrices in tabular form.

Example   Let $X = \{a, b, c\}$ and $R = \{(a, a),\ (b, b),\ (c, c),\ (a, b),\ (a, c),\ (b, c)\}$: so R is the relation with the digraph in Fig. 2.18. Its adjacency matrix is as shown:

	a	b	c
a	1	1	1
b	0	1	1
c	0	0	1

It is possible to interpret some of the properties of relations in terms of their adjacency matrices. Thus a relation is reflexive if the entries down the main diagonal are all 1, it is symmetric if the matrix is symmetric (that is, if the entry at position (x,y) is equal to that at (y,x)) and it is weakly antisymmetric if the entries at (x,y) and (y,x) are never both 1 unless $x = y$. The transitivity of R can also be characterised in terms of the adjacency matrix but, since this is considerably more complicated, we omit this (see, for example [Kalmanson, p. 330]). We can immediately see that the above relation is reflexive, weakly antisymmetric and (if we check case by case) we can see that it is transitive.

Example For the set $\{1, 2, \ldots, 11, 12\}$ and the relation D (xDy if and only if x divides y) the adjacency matrix is

	1	2	3	4	5	6	7	8	9	10	11	12
1	1	1	1	1	1	1	1	1	1	1	1	1
2	0	1	0	1	0	1	0	1	0	1	0	1
3	0	0	1	0	0	1	0	0	1	0	0	1
4	0	0	0	1	0	0	0	1	0	0	0	1
5	0	0	0	0	1	0	0	0	0	1	0	0
6	0	0	0	0	0	1	0	0	0	0	0	1
7	0	0	0	0	0	0	1	0	0	0	0	0
8	0	0	0	0	0	0	0	1	0	0	0	0
9	0	0	0	0	0	0	0	0	1	0	0	0
10	0	0	0	0	0	0	0	0	0	1	0	0
11	0	0	0	0	0	0	0	0	0	0	1	0
12	0	0	0	0	0	0	0	0	0	0	0	1

Certain general types of relations, characterised by combinations of properties such as symmetry, reflexivity, ... frequently arise in mathematics and, indeed, in many spheres. We consider what are probably the two most important: partial orderings and equivalence relations.

Definition A relation R on a set X is a **partial order(ing)** if R is reflexive, weakly antisymmetric and transitive. Thus, for all $x \in X$ one has xRx; for all $x,y \in X$ if xRy and yRx then $x = y$; for all $x,y,z \in X$, if xRy and yRz then xRz.

Example 1 Define a relation R on the set of real numbers by xRy if and only if $x \leq y$. This relation is one of the most familiar examples of a partial order in mathematics. In many examples of partial orders which arise in practice, there will be some sense in which the relation 'xRy' can be read as 'x is smaller than or equal to y'.

Example 2   Both examples discussed above in connection with adjacency matrices are partial orders.

Example 3   Let A be the set $\{a,b,c\}$ and define X to be the set of all subsets of A: so X has the 8 elements

$$\emptyset, \{a\}, \{b\}, \{c\}, \{a,b\}, \{a,c\}, \{b,c\}, A.$$

Define a relation R on X by $(U,V) \in R$ if and only if U is a subset of V. Then R is a partially ordered set whose adjacency matrix is as shown:

	Ø	{a}	{b}	{c}	{a,b}	{a,c}	{b,c}	A
Ø	1	1	1	1	1	1	1	1
{a}	0	1	0	0	1	1	0	1
{b}	0	0	1	0	1	0	1	1
{c}	0	0	0	1	0	1	1	1
{a,b}	0	0	0	0	1	0	0	1
{a,c}	0	0	0	0	0	1	0	1
{b,c}	0	0	0	0	0	0	1	1
A	0	0	0	0	0	0	0	1

One may define a similar partial order on the set of all subsets of any set. Thus, if X is the set of all subsets of a set A, the relation R on X defined by

$$(B, C) \in R \text{ if and only if } B \text{ is a subset of } C$$

is a partial order.

Example 4   The relation D on the set $\mathbb{Z}$ of integers which is given by xDy if and only if x divides y is another example of a partial order.

We may also define a **strict partial order** to be a set X with a relation R on it which is antisymmetric and transitive. For instance, the relation '$<$' on $\mathbb{N}$ given by $x < y$ if and only if $y - x$ is positive is a strict partial order. Another example, defined on the set of all people (past and present), is given by aRb if and only if a is an ancestor of b. (It is a strict partial order since a person cannot be an ancestor of himself or herself.)

It is straightforward to show that if R is a partial order on X then the relation S on X defined by xSy if and only if xRy and $x \neq y$ is a strict partial order. Conversely, if S is a strict partial order on the set X then the relation R defined by xRy if and only if xSy or $x = y$ is a partial order on X. The notation $(P, \leq)$ and term **partially ordered set** (or **poset** for short) are often used for a set P equipped with a partial order $\leq$.

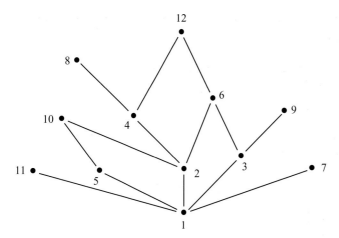

Fig. 2.21

There is a graphical way of representing partially ordered sets which have a finite number of elements: by use of Hasse diagrams.

Definition   Let R be a strict partial order on a set X. If x, y are elements of X then y is an **immediate successor** of x (and x is an **immediate predecessor** of y) if xRy and if there is no z in X with xRz and zRy. Roughly, y is an immediate successor of x if y is 'greater than' x and if there is no element strictly between x and y.

In the case that R is a partial order (as opposed to a strict partial order), we modify the definition in the obvious way, saying that y is an immediate successor of x if xRy and $x \neq y$ and if, whenever $z \in X$ is such that xRz and zRy, then either $z = x$ or $z = y$.

The **Hasse diagram** of the (strict) partial order R on the set X is obtained as follows. Place one point on the plane for each element of X. The points must be placed in such a way that a line may be drawn going in a general upwards direction from each element x in X to each of its immediate successors. Draw in these lines.

Example 1   Let $X = \{1, 2, \ldots, 11, 12\}$ and let R be given by xRy if and only if x divides y. Note that 6 is an immediate successor of 2, but 12 is not, since $2R6$ and $6R12$. The Hasse diagram is as shown in Fig. 2.21.

Example 2   Let $X = \{1, 2, \ldots, 11, 12\}$ and let R be the usual ordering '$\leq$'. The Hasse diagram is then Fig. 2.22.

Example 3   Let X be the set of all subsets of $\{0, 1, 2\}$ and let R be the relation 'is a subset of': Fig. 2.23.

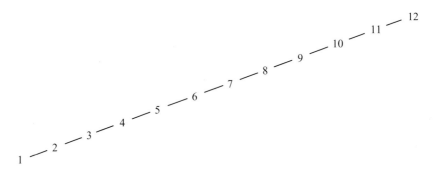

Fig. 2.22

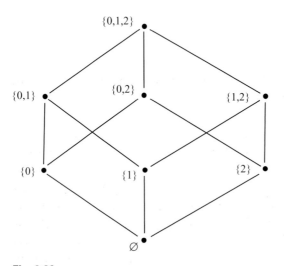

Fig. 2.23

Example 4 Let X be the set of integers $\{1, 2, 3, 6, 9, 18\}$ and let R be given by xRy if and only if x divides y. The Hasse diagram is as shown in Fig. 2.24.

We now come to our second special type of relation.

Definition A relation R on the set X is an **equivalence relation** if R is reflexive, symmetric and transitive.

Example 1 For any set X, the identity relation R on X, given by xRy, if and only if $x = y$, is an equivalence relation.

Example 2 Let X be the set of all integers and fix an integer $n \geq 2$. Define a relation E on X by aEb if and only if $a - b$ is divisible by n. Thus E is the relation

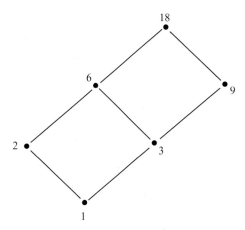

Fig. 2.24

of congruence modulo n. It is quickly checked that E satisfies the conditions for being an equivalence relation.

Example 3   The notion of logical equivalence (see Section 3.1 below) is, as its name suggests, an equivalence relation on the set of propositional terms.

Example 4   (For the reader who has met some linear algebra.) Let X be the set of $n \times n$ matrices with real entries. Matrices A and B in X are defined to be 'similar' if there are invertible matrices P and Q such that $B = P^{-1}AQ$. It is straightforward to check that similarity of matrices is an equivalence relation (i.e. if we define the relation S on X by $(A,B) \in S$ if and only if A is similar to B, then S is an equivalence relation). Matrices $A, B \in X$ are said to be 'equivalent' if there is an invertible matrix P such that $B = P^{-1}AP$. It is easy to verify that the relation E of equivalence in this sense is also an equivalence relation.

Example 5   Let $f: X \to Y$ be a function. Define a relation F on X by $x F x_1$ if and only if $f(x) = f(x_1)$. Then F is an equivalence relation.

Definition   Let X be any set. By a partition of X we mean a particular way of dividing up the set X into 'blocks'. More precisely: a **partition** of X is a collection $\{X_i: i \in I\}$ of non-empty subsets of X which is

disjoint, in the sense that $X_i \cap X_j = \emptyset$ if i is different from j, and
covering, in the sense that each x in X belongs to one (and by disjointness, only one) X_i.

Example   A group of people is divided up into separate teams to work on various projects: this 'division' determines a partition of the set of people involved, with each team constituting one member of the partition (so the 'X_i' are the various teams).

We now have on the one hand partitions, on the other hand equivalence relations. We will see that these amount to the same thing (this is not to say that the intuitive ideas coincide but rather that when the ideas are formalised mathematically we obtain 'equivalent' notions).

It may be worthwhile to warn the reader that the following proof may appear more 'abstract' than any in this book so far (mainly because the objects which the theorem refers to – equivalence relations and partitions – probably seem less concrete than numbers or even sets and functions). If you feel that the theorem and its proof do not make much sense to you, try picking a *particular* equivalence relation or partition and then going through the proof, working out what the details of the proof mean in the context of the example that you have chosen.

Theorem 2.3.1   *Let X be any set.*

(i) *Suppose that $\{X_i: i \in I\}$ is a partition of X. Then the relation R on X which is defined by xRy if and only if x and y belong to the same member of the partition is an equivalence relation.*

(ii) *If, conversely, E is an equivalence relation on X then E determines the partition whose 'blocks' X_i are the **equivalence classes** $[x]_E = \{y \in X: yEx\}$ of members x of X.*

Proof   (i) Suppose that we are given a partition $\{X_i: i \in I\}$ of X. The relation R as defined is clearly reflexive. Also R is symmetric since if x and y are both in X_i then so are y and x. Finally R is transitive since if x and y are both in X_i (say) and if y and z are both in X_j (say) then, by disjointness, $i = j$ and so x and z lie in the same member of the partition.

(ii) Now suppose that E is an equivalence relation on X. Define $[x] = [x]_E$ to be the subset $\{y \in X: yEx\}$, containing x. It is claimed that the distinct sets of this kind form a partition of X.

First note that if $b \in [a]$ then $[b] = [a]$. For if $c \in [a]$ then we have cEa. Since $b \in [a]$ we also have bEa and so, by symmetry, aEb. Transitivity then implies cEb and so $c \in [b]$. Thus we have shown $[a] \subseteq [b]$. For the converse, suppose $d \in [b]$, so dEb. Since also bEa, transitivity yields dEa: that is, $d \in [a]$, as required.

Disjointness now follows quickly. Suppose that $[a]$ and $[b]$ have an element c (say) in common. Then by the above $[c] = [a]$ and also $[c] = [b]$. Hence $[a] = [b]$, as required.

Finally, the sets are covering since each element a is in some set of the form $[x]$: namely $[a]$. Thus we do have a partition of X. □

As one example of this essential equivalence between partitions and equivalence relations (via equivalence classes) consider the notion of congruence modulo n. The equivalence classes determined by the relation aRb if and only if $n|a - b$ are the (n) congruence classes modulo n. Conversely, given the partition of $\mathbb{Z}$, $\{nk: k \in \mathbb{Z}\}, \{nk + 1: k \in \mathbb{Z}\}, \ldots, \{nk + (n - 1): k \in \mathbb{Z}\}$, the corresponding equivalence relation is R.

For another example, take X to be the set of all the points on the real plane $\mathbb{R} \times \mathbb{R}$, apart from the origin $(0,0)$: define an equivalence relation on X by setting x equivalent to y if and only if $(0,0)$, x and y all lie on a straight line. By Theorem 2.3.1 this equivalence relation determines a partition of the plane into a disjoint covering family of subsets. These subsets are the equivalence classes of points and they are simply the straight lines (minus the origin) which pass through the origin.

De Morgan and C.S. Peirce studied relations in the abstract in the latter part of the nineteenth century.

Exercises 2.3

1. For each of the following relations R on the set X decide whether R is reflexive, symmetric, (weakly) antisymmetric or transitive:
   (a) $X = \mathbb{Z}$, $\qquad$ aRb if and only if $a \le b + 1$;
   (b) $X = \mathbb{Z}$, $\qquad$ aRb if and only if $a + b$ is even;
   (c) $X = \mathbb{P}$, $\qquad$ aRb if and only if a and b are coprime:
   (d) $X = \mathbb{Z}$, $\qquad$ aRb if and only if $a + b$ is divisible by 3;
   (e) $X = \mathbb{R}$, $\qquad$ aRb if and only if $a^2 \le b^2$;
   (f) $X = \mathbb{R} \times \mathbb{R}$, $\quad$ $(a, b)R(c, d)$ if and only if $a = c$;
   (g) $X = \mathbb{N} \times \mathbb{N}$, $\quad$ $(a, b)R(c, d)$ if and only if either $a < c$ or $(a = c$ and $b \le d)$.

2. (a) Show that any relation which is both symmetric and antisymmetric must be the empty relation.
   (b) Show that if a relation is antisymmetric then it is weakly antisymmetric.
   (c) Give an example of a non-empty relation which is symmetric and weakly antisymmetric (!).
   (d) Show that if a relation is symmetric then so is its complement.
   (e) Show that if a relation is transitive then so is its reverse.
   (f) Give an example of a non-empty relation which is symmetric and transitive but which is not reflexive.

3. What is wrong with the following argument?

   '*Theorem*' Every transitive symmetric relation is reflexive.

   '*Proof*' Let R be a relation on the set X and suppose that R is transitive and symmetric. Let $x \in X$. From xRy we have, by symmetry, yRx, and so, by transitivity, we deduce xRx. Thus R is reflexive, as required.

   There is something wrong with the argument, since the 'Theorem' is false – if you look at the solution for Exercise 2.3.2 (f), you will find an example of a transitive symmetric relation which is not reflexive!

4. Let X be the set of all European countries (choose your own definitions of 'European' and 'country'): alternatively let X be the set of all states of the USA. Define the relation B on X by xBy if and only if x and y have a common border (let us make the convention that x does not have a common border with itself). Draw the digraph of this relation. (Since the relation is symmetric, it would be reasonable to use an edge without an arrow to stand for a pair of directed edges between two vertices.)

5. Let X be the set $\{a, b, c, d, e\}$ and R be the relation whose adjacency matrix is shown. Prove that R is a partial order on X and draw its Hasse diagram.

	a	b	c	d	e
a	1	0	1	1	1
b	0	1	0	1	1
c	0	0	1	1	1
d	0	0	0	1	1
e	0	0	0	0	1

6. Let X be the set $\{1, 2, 3, 4\}$ and let

   $$R = \{(1, 1), (1, 2), (1, 3), (1, 4), (2, 2), (2, 4), (3, 3), (3, 4), (4, 4)\}.$$

   Write down the adjacency matrix for R. Show that R is a partial order and draw its Hasse diagram.

7. Let X be the set $\{1, 2, 3, 4\}$ and let

   $$R = \{(1, 1), (1, 2), (2, 1), (2, 2), (3, 3), (3, 4), (4, 3), (4, 4)\}.$$

   Show that R is an equivalence relation and write down its equivalence classes.

8. Let $S = \{1, 2, 3, 4\}$ and let X be the set $S \times S$. Define a relation R on X by

   $$(a, b)R(c, d) \text{ if and only if } a + b = c + d.$$

   Show that R is an equivalence relation and list the equivalence classes.

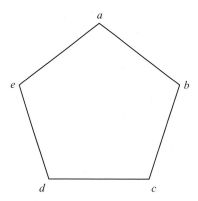

Fig. 2.25

9. Let a pentagon have vertices denoted a to e as shown in Fig. 2.25. Define a relation R on the set of vertices, by aRb if and only if a and b do not lie on the same edge of the pentagon. Decide whether or not R is transitive and draw the digraph of R.

10. Let X be the set of $n \times n$ matrices with real entries and let S and E be respectively the (equivalence) relations of similarity and equivalence as defined in Example 4 on p. 113. Show that the partition of X corresponding (in the sense of 2.3.1) to E **refines** that corresponding to S, in the sense that for all matrices A and B, $(A, B) \in E$ implies $(A, B) \in S$ (and hence every S-equivalence class is a disjoint union of E-equivalence classes).

2.4   Finite state machines

'Calculating machines' have a longer history than is often realised. The first mechanical digital calculator was built by Blaise Pascal sometime between 1642 and 1644. Pascal's machine was limited to addition and subtraction but, in 1673, Leibniz built a machine which could also multiply and divide (it is interesting to note, in connection with this, Leibniz' advocacy of the binary system of numeration and his dream of a 'logical calculus'). Calculating machines based on Leibniz' design were in general use until they were very recently supplanted by electronic calculators.

Charles Babbage (1792–1871) designed mechanical calculators which would carry out a sequence of computations. He proposed to the Royal Astronomical Society in 1822 that he build a giant 'Difference Engine' – a machine which would compute and even set in type mathematical tables (the current hand-calculated ones were infested with errors). He had already built a small machine of this kind. The construction began well but foundered over financial and related difficulties, and the 'Difference Engine' was never completed.

Babbage went on to conceive a much more sophisticated calculating machine which would have a 'memory' and would be programmable by punched cards (such cards were already used to control weaving looms). But again financial difficulties and Babbage's seeming inability ever to stop tinkering and to finish a project, brought the construction to a standstill.

Babbage's work slipped into obscurity but advances were made, such as the development by Hollerith, towards the end of the century, of punched card systems for handling large masses of data.

Large analogue (as opposed to digital) computers – called 'differential analysers' – were built in the USA in the 1930s, and prototype digital computers were built by various scientists in the USA and UK in the late 1930s and early 1940s. By that stage, electrical and electronic components rather than mechanical ones were being used.

In a paper of 1937, Alan Turing described a theoretical computing machine which would be able to compute according to any rule or set of rules fed to it – a programmable computer. Such a theoretical machine is now called a Turing machine and could, if given the appropriate instructions and enough time and space, perform any computation which might be described as algorithmic (i.e. proceeding according to some rule or set of rules).

What was probably the first working electronic computer, named COLOSSUS, became operational in 1943 at Bletchley – a top-secret code-breaking centre in England. This machine was built by Turing and others who were engaged in deciphering German secret messages.

The first general-purpose electronic computer – ENIAC – became operational in 1944–5. It was built by a team at the Moore School (attached to the University of Pennsylvania) in Philadelphia.

The first stored-program digital computer, the 'Baby', was built by Tom Kilburn and Freddie Williams at the University of Manchester and became operational in 1948.

Of course, present-day computers are immensely more powerful and faster than those original ones, but all can be seen as realisations of Turing's idea.

A Turing machine can, in principle, model any computation. In this section, we will consider a restricted class of Turing machines. Although these do not have the flexibility of Turing machines (there are certain computations which they cannot perform), they are relatively easy to construct in practice and the class of computations which they can perform has certain properties which are interesting from the point of view of formal language theory.

There are several, related, types of finite state machines. We first consider the most basic of these.

Definition   A **finite state machine** M is a triple (S, A, μ) where S is the set of **states** of M and includes a distinguished **initial state** 0, A is an **alphabet** – its elements are called **letters** – and μ is a **transition function** $\mu \colon S \times A \to S$. Thus μ assigns a state to every pair of the form (s, a) where s is a state and a is a letter of A.

The picture to bear in mind is that of a machine with an input tape which it can read: the entries on the tape are letters of the alphabet A. It will operate in a discrete manner (as does any digital computer). When it is started up, its internal configuration or state is 0. It reads the first letter on the tape. Depending on what that letter is, its internal state may change, becoming i say. It then reads the next letter on the tape: that letter, together with its current internal state, determines what its new state is to be, ... and so on. Thus, at a given time, the machine is in a certain internal state; it reads a letter; it then moves to another (or remains in the same) state and begins to read the next letter on the tape. The transition function $\mu \colon S \times A \to S$ has the following interpretation: if the machine is in a state s and it reads the letter a on the tape then its internal state becomes $\mu((s, a))$.

We will usually denote the states of M by integers and 0 will always denote the initial state. The members of A will usually be denoted by letters of the Roman alphabet. We will make the convention that the machine begins to read the tape at its left-most end and always moves from left to right. Let us consider two examples.

Example 1   Let M be the machine with three states $\{0, 1, 2\}$, alphabet $A = \{a, b\}$ and μ given by the table shown (the entry at the intersection of the row labelled i and the column labelled x is the value $\mu(i, x)$).

	a	b
0	0	1
1	1	2
2	2	0

Consider the sequence baababa. Initially, the machine is in state 0. On reading the first b, the machine moves into state 1. It remains in 1 after reading the two a's and moves into state 2 after reading the next b. It stays there after reading the a, but moves back to state 0 on reading the next b and then remains in state 0 on reading the final a.

Example 2   Let M be the machine with states $\{0, 1, 2, 3\}$, alphabet $A = \{a,b,c\}$ and with μ given by

	a	b	c
0	1	2	1
1	0	3	2
2	2	1	2
3	3	3	3

We consider what this machine does on reading the sequence abcab: starting in state 0, reading a takes the machine into state 1, reading b gives state 3, reading c leaves it in 3, as will all further transitions. Thus the machine reads abcab and moves from initial state 0 to final state 3.

An alternative way to give the transition function μ is to use the **state diagram**. This is a directed graph with labelled edges. The vertices (usually denoted by numbers in circles) of this graph are the elements of S. Two vertices i and j are joined by an arrow with label a (for instance) if when the machine is in state i and reads letter a it moves to state j. Thus the state diagrams of our two examples are as shown in Fig. 2.26.

We now discuss a type of finite state machine that is very important for applications. Such machines are components of cash machines, lifts and, indeed, they are used in computers to recognise, for instance, key words of a programming language and to respond appropriately. They are also important in theoretical computer science and in the theory of formal languages.

Definition   A finite state **automaton** is a finite state machine $M = (S, A, \mu)$ together with a subset F of S, known as the set of **acceptance states** of M.

We may regard automata as being intended to *recognise* certain sequences of letters (words) in the alphabet A (a password or PIN, for example). Formally, we have the following definition.

Definition   Let $M = (S, A, \mu)$ with set F of acceptance states be a finite state automation. A sequence σ of letters in the alphabet A is **accepted by** M if, after reading the sequence σ, the automaton is in state s for some s in F.

Example 1   We return to the first example above and let F consist of the state 2. Thus a sequence of a's and b's is accepted by the automaton if it is in the state

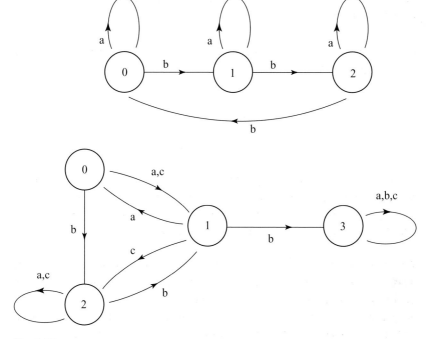

Fig. 2.26

2 after the sequence has been read. If a sequence consisting entirely of a's is read, the automaton stays in state 0, and so this sequence is not accepted. If our sequence has one b in it, the automaton arrives at state 1. For a sequence with two b's, such as aabaaba, the automaton is sent into state 1 when it encounters the first b and, on encountering the second, moves to state 2 where it remains: so such a sequence is accepted. Continuing in this way, we can see that sequences with three or four b's are not accepted and that, in general, a sequence is accepted if and only if the number of b's in the sequence is congruent to 2 modulo 3.

If the set F of acceptance states were changed to be {1} then the set of sequences accepted would be those in which the number of b's is congruent to 1 modulo 3. Similarly, if F were {0}, the sequences accepted would be precisely those with k b's, where k is divisible by 3.

Example 2 Let M be the automaton with states {0, 1, 2, 3} and alphabet {a,b}, for which F is {1} and μ is given by the state diagram shown in Fig. 2.27. Consider the words accepted by this automaton. The a's it reads move the machine back and forth between states 0 and 1 and between states 2 and 3. Movement between states 0 and 3 and between 1 and 2 is governed by the bs

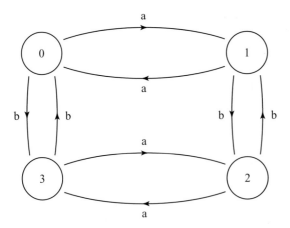

Fig. 2.27

read. If a word has an even number of as, the automaton must be in one of the states 0 or 3 after reading the word. For a word with an odd number of as, the automaton will be in either state 1 or state 2. Similarly, after reading a word with an even number of b's, the automaton is in one of the states 0 or 1 and it is in state 2 or 3 after reading a word with an odd number of bs. Since $F = \{1\}$, the accepted words are those with an odd number of as and an even number of bs.

Another special type of finite state machine arises by modifying the machine to produce output. This is done by considering A as the input alphabet, adding an output alphabet B and giving an output table $v: S \times A \to B$ which is a rule assigning an output symbol $v(i,a)$ to each pair consisting of a state i and an input a. If the machine is in state i and reads the letter a then it outputs $v(i,a)$ before moving to state $\mu(i,a)$.

Example 1   We return to the example where M is the machine with states $\{0, 1, 2, 3\}$, alphabet $A = \{a, b, c\}$ and where μ is given by

	a	b	c
0	1	2	1
1	0	3	2
2	2	1	2
3	3	3	3

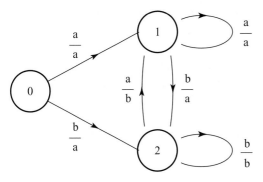

Fig. 2.28

Thinking of A as the input alphabet, we let B be the set $\{\alpha, \beta\}$ and define v by the table

	a	b	c
0	α	α	β
1	α	β	α
2	β	β	α
3	α	α	α

Thus on reading the sequence acbaab, the machine would go through the sequence of states 0,1,2,1,0,1,3 and would output $\alpha\alpha\beta\alpha\alpha\beta$.

Example 2   As another example of a machine with output, we consider the unit delay machine M. This has states $\{0, 1, 2\}$, input and output alphabets $\{a,b\}$ and the functions μ,v given by the tables as shown:

	μ		v	
	Next state		Output	
	a	b	a	b
0	1	2	a	a
1	1	2	a	a
2	1	2	b	b

The tables for μ and v have been combined in an obvious way. The state diagram of this machine is given in Fig. 2.28.

Each arrow in this diagram is labelled by two letters of A. The upper letter is the input required to move in the direction of the arrow and the lower is the corresponding output. Thus if the sequence abbaba is read, the machine goes through states 0,1,2,2,1,2 and then ends in state 1 and it outputs the sequence aabbab. The output starts with the letter a and then repeats the input sequence up to its penultimate letter. A little thought will show that this is what the machine does to any sequence.

There is a strong connection between formal languages and the 'machines' which we have discussed above. A formal language consists of certain words, defined over a given alphabet. It is specified by a 'grammar' – a set of rules which determine how words of the language may be built from other words of the language. A measure of complexity of the language is obtained by asking what is the simplest kind of machine that will accept precisely the words of that language. There is a classification of these languages, depending on the kind of grammatical rules. It turns out that the simplest languages are precisely those that are accepted by a finite state machine. The most general languages are those that are accepted by a Turing machine. Between these extremes, we have the types of language accepted by other types of automata. For more on this topic, see [Salomaa] for instance.

Exercises 2.4

1. Draw state diagrams for the machines shown:
   (a) The machine M with $S = \{0, 1, 2\}$, $A = \{a,b\}$ and μ given by

	a	b
0	0	1
1	1	2
2	2	0

   (b) The machine M with $S = \{0, 1, 2\}$, $A = \{a,b,c\}$ and μ given by

	a	b	c
0	1	0	2
1	0	0	1
2	2	0	2

2. Construct the tables of transition functions for the finite state machines whose state diagrams are shown in Fig. 2.29.
3. Let M be the finite automaton as specified. Determine the words accepted by M:
   (a) the machine in 2.4.1(a) above with $F = \{1\}$;
   (b) the machine in 2.4.2(b) above with $F = \{1\}$;
   (c) the machine in 2.4.2(b) above with $F = \{2\}$;
   (d) the machine in 2.4.2(c) above with $F = \{3\}$.
4. Let M be the automaton with $F = \{1, 3\}$ and state diagram as shown in Fig. 2.30.

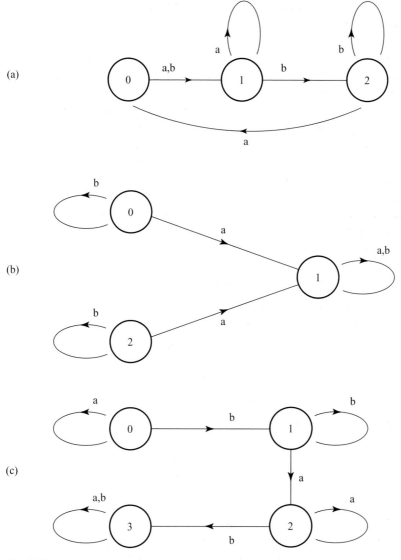

Fig. 2.29

Determine the words accepted by *M* and design a finite automaton with just two states which accepts the same words as those accepted by *M*.

5. Design finite state machines to meet the following specifications.
   (i) An automaton with alphabet {a,b} that will only accept sequences composed entirely of the letter a.

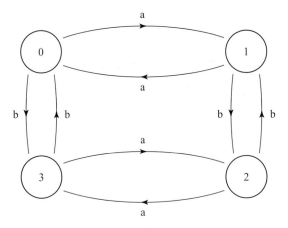

Fig. 2.30

(ii) A finite state machine with input alphabet {a,b} and output alphabet
{α,β} that will output a sequence whose last term is β exactly when a
word with an even number of b's is read.

(iii) A finite state machine which will read a word in {a, b, c} and output
each occurrence of a and b but will replace every second occurrence
of c by a.

6. A cautious millionaire has a home safe. Design an automaton, to be
attached to the safe, which will read four-digit decimal numbers but will
only accept the millionaire's personal number (which is 1357).

Summary of Chapter 2

We introduced sets and the operations of intersection, union and complement on
sets and we gave a list of basic identities involving these operations. The product
operation on sets was introduced. In the second section we gave an abstract, very
general, definition of function, discussed composition of functions and special
types of functions, in particular bijections, then used this to define what it means
for sets to have the same cardinality. In Section 2.3 we considered relations and
various properties which these may have. Partial orders and their Hasse diagrams
were introduced, as were equivalence relations and the partitions which they
induce. In the last section, we introduced finite state machines, both without
and with output.

3 Logic and mathematical argument

Mathematics and mathematical reasoning is precise. This is in contrast to most discourse where deductions and chains of reasoning have gaps and imprecision. A chain of reasoning in mathematics should be compelling, in the sense that anyone who can follow the argument (which might, of course, be a very difficult task), should feel obliged to accept its correctness. Of course, in practice mathematical arguments can contain errors. But again, once an error has been pointed out, it should not be a matter of opinion whether or not there is an error. Over the centuries, indeed millennia, people have puzzled over what makes a chain of reasoning watertight. Especially over the past 150 years the logic of mathematics has been thoroughly investigated. A rather remarkable outcome of this mathematical investigation of logic is that it is, in principle, possible to formalise mathematical reasoning to the point where any purported mathematical argument could be checked by a computer. By this, we mean that the chain of reasoning can be checked (the correctness or applicability of the conclusions rather depends on the assumptions made at the outset).

It is interesting that, in practice, mathematicians do not always or even often produce proofs that are in a form which are easily computer checkable. Simple lemmas and computations are susceptible to such formal checking but, typically, proofs are quite (or very) complex and are written in a way which makes them comprehensible to human readers who have the necessary factual background and who have developed sufficient intuitive understanding that they do not need all the details to be written down. Of course this leaves open the possibility of error not being detected and it is by no means unusual for long and complex proofs in research papers to contain errors or 'gaps' which need to be filled (the proof of 'Fermat's Last Theorem' illustrates this). But, in finding and developing proofs, mathematicians are guided more by their intuitions and understanding than by precise logic and, as a result, typically these errors are corrected and gaps are filled easily enough once they have been noticed. Nevertheless sometimes proofs contain not gaps, but vast chasms, which can take much time and effort to fill and which, very occasionally, simply cannot be filled: sometimes proofs do have to be withdrawn. It is also the case that errors can go unnoticed for some time. But, once a result has been proved, mathematicians will use it to draw further conclusions and this process usually points up any significant errors.

We spend the first section describing the most basic part of the logic that underlies mathematical reasoning: the logic of simple propositions. This is the logic of 'and', 'or', 'not' and 'implies': the logic of *combining* and *manipulating* statements. In order to *construct* mathematical statements, we need more, including the quantifiers 'for all' and 'there exists': these are discussed in the second section. In the final section, we review various of the proof strategies that we use in this book.

3.1 Propositional logic

Propositional logic enables us to handle the elementary logical connections between statements (or 'propositions'). By a **proposition** we mean an assertion which has a definite **truth value**: true (denoted by 't') or false (denoted by 'f'). For instance the following are propositions.

The sum of the first n positive integers is $n(n + 1)/2$.
2 is an odd integer.
2 is a prime number.
Every even number greater than 2 is the sum of two prime numbers.

Each of these statements is either true or false (in the case of the last no-one knows which), but not both! On the other hand the following are not propositions.

Look here!
Is every even number greater than 2 the sum of two prime numbers?
n is a prime number.

Of course, in the last example, if the context were such that 'n' denoted a *particular* natural number, then the sentence would be a proposition but, in the absence of such a context, the statement is neither true nor false since we have left open what 'n' denotes. Notice also that a question is not a proposition, since a question cannot itself be true or false.

Even these few examples may well have raised in the reader's mind a number of questions of a sort which we do not deal with here: for a more extensive discussion of formal logic and its relation to natural language, the reader should consult, for instance, [Hodges].

We use letters such as p, q and r to stand for propositions. For example, p might be the proposition '2 is an odd number' and q might be the statement '2 is a prime number'.

Each proposition p has a **negation**, which is itself a proposition, denoted by $\neg p$ and read 'not p'. For instance, the negation of the statement '2 is an odd number' is '2 is not an odd number'. The proposition p is true exactly if $\neg p$ is false. The relationship between a proposition p and its negation $\neg p$ may be expressed in a 'truth table'.

p	$\neg p$
t	f
f	t

This table says (read along the rows) that if the proposition p is true then $\neg p$ is false, and if p is false then $\neg p$ is true.

Given two propositions p and q, we can form new propositions from them: their **disjunction** $p \vee q$, and their **conjuction** $p \wedge q$.

The proposition $p \vee q$, read as 'p or q', is true exactly if *at least one* of p, q is true, while $p \wedge q$, read as 'p and q', is true exactly if *both* p and q are true. Our statement means that as well as explaining the truth values taken by the propositions $p \vee q$ and $p \wedge q$, we have (by implication) made the standard English words 'or' and 'and' into functions on truth values, with two truth values as input and one as output. Note that, as is usual in mathematics, we use 'or' in the *inclusive* sense that $p \vee q$ is true if either or both p and q are true. For example the proposition '2 is a prime number or 2 is an even number' is true. (When Boole introduced his calculus of propositions (see [Boole]), he actually used 'or' in the *exclusive* sense – 'p or q but not both' – but the inclusive sense turns out to be the more convenient in mathematics.)

As an example of disjunction and conjunction of propositions, if p is the proposition '2 is an odd number' and q is the statement '2 is a prime number' then $p \vee q$ is the statement '2 is an odd number or 2 is a prime number' (which is true) and $p \wedge q$ is the statement '2 is an odd number and 2 is a prime number' (which is false).

In general we do not know the truth values of p and q but we do know that each is either true or false: so there are four possible ways to assign truth values to p, q. Namely:

both p and q are true (in which case $p \vee q$ is true);
p is true and q is false (in which case $p \vee q$ is true);
p is false and q is true (in which case $p \vee q$ is true);
both p and q are false (in which case $p \vee q$ is false).

These four possibilities give the four rows (under the heading line) of the first table below. The reader should satisfy himself or herself that the rows of the second truth table correctly express the relationship between the truth value of $p \wedge q$ and the truth values of its constituents p and q.

p	q	$p \vee q$		p	q	$p \wedge q$
t	t	t		t	t	t
t	f	t		t	f	f
f	t	t		f	t	f
f	f	f		f	f	f

Note that forming the proposition $p \wedge q$ may be thought of as building up a compound proposition from two simpler ones: we say that $p \wedge q$ is a (**propositional**) **term in** p and q. We define 'term in' to be a transitive relation. Thus, for example, if p itself is of the form $r \wedge s$ and if q is of the form $s \vee (\neg t)$, where r, s and t are propositions, then $p \wedge q$, that is $(r \wedge s) \wedge (s \vee (\neg t))$, as well as being a term in p and q, is also a term in r, s and t. The expressions '**Boolean combination of**' and '**Boolean expression in**' are also used instead of 'term in'.

In order to avoid ambiguity when reading propositions, we make the convention that $\neg$ has priority over $\wedge$ and $\vee$. So $\neg p \wedge q$ will mean $(\neg p) \wedge q$ rather than $\neg(p \wedge q)$.

Suppose now that the proposition p is a term in (that is, built up, using $\vee$, $\wedge$ and $\neg$, from) the propositions $q_1, \ldots, q_n$. Then we may draw up a **truth table** for p which shows how the truth value (true 't' or false 'f') of p depends on the truth values of $q_1, q_2, \ldots, q_n$. The first n columns of the truth table are labelled by $q_1, q_2, \ldots, q_n$, and the last column is labelled by p: we may insert additional columns to ease the actual computations (as in the table just below, also see other examples below). There will be, apart from the heading row, 2^n rows of the table, corresponding to the 2^n different ways of assigning truth values to $q_1, q_2, \ldots, q_n$.

Example   Let p be the proposition $(q \wedge r) \vee (r \wedge \neg s)$. This is a term in the three propositions q, r and s, so the truth table for p will have $2^3 = 8$ rows.

q	r	s	$q \wedge r$	$r \wedge \neg s$	p
t	t	t	t	f	t
t	t	f	t	t	t
t	f	t	f	f	f
t	f	f	f	f	f
f	t	t	f	f	f
f	t	f	f	t	t
f	f	t	f	f	f
f	f	f	f	f	f

Given two propositions p and q, we write $p \to q$ for the proposition which is read as 'p implies q'. It is also read as 'if p then q'. This proposition is defined to have truth value 't' except when p is true and q is false, when it is 'f'. Again, in doing this, we have implicitly made the ordinary English sentence construction 'if ... then ...' into a function on truth values.

Our definition implies that if p is false, then $p \to q$ is true, no matter whether q is true or false. In particular, the truth of $p \to q$ does not imply that there is any real connection between p and q. Thus, for example, because '$2 = 3$' is false, the following proposition is true: '$2 = 3$ implies 6 is a prime number'. This, no doubt, seems peculiar, but it turns out to be the only sensible way to assign a truth value 't' or 'f' to such statements. Also, if we rephrase the proposition as 'if $2 = 3$ then 6 is a prime number', perhaps it seems a little less strange that this is a true proposition: all it says is that *if* $2 = 3$ then 6 is prime so, since $2 = 3$ is false, the proposition is true 'by default'.

The truth table for '$p \to q$' follows.

p	q	$p \to q$
t	t	t
t	f	f
f	t	t
f	f	t

Notice that this is the 'same' truth table as that for $\neg p \vee q$ (which is shown below; we have added a column for $\neg p$ as a computational aid). That is, given any assignment of truth values to p and q, the propositions $p \to q$ and $\neg p \vee q$ have the same truth value. Indeed, quite commonly '$p \to q$' is introduced simply as a shorthand for '$\neg p \vee q$'.

p	q	$\neg p$	$\neg p \vee q$
t	t	f	t
t	f	f	f
f	t	t	t
f	f	t	t

There are many ways in the English language of asserting (the truth of) an implication $p \to q$: here are some of them. You should take the time to think why these really are all equivalent to $p \to q$.

'p is a sufficient condition for q'
'if p then q'
'p only if q'
'q if p'

'q is a necessary condition for p' (to hold) (because if p holds then necessarily
  so does q)
(and various other phrases such as 'q follows from p', 'since p holds so does
  q' etc.)

Other terms used in mathematics which can be explained using propositional
logic include the following.

An assertion $p \to q$ is an **implication**. Its **converse** is the implication
$q \to p$. These are different: take p to be 'n is a multiple of 6' (think of n
as fixed by a context) and q to be 'n is a multiple of 2'. Then certainly $p \to q$
('if n is a multiple of 6 then n is a multiple of 2') is true but the converse, $q \to$
p ('if n is a multiple of 2 then n is a multiple of 6') is not.

On the other hand, the **contrapositive** of the implication $p \to q$ is the implication $\neg q \to \neg p$ and this *is* logically equivalent to $p \to q$ (compute the truth
tables to check this). In our example $\neg q \to \neg p$ reads 'if n is not a multiple of 2
then n is not a multiple of 6', which *is* equivalent to $p \to q$ ('if n is a multiple
of 6 then n is a multiple of 2').

As an exercise in the use of truth tables, the reader might like to check that
for any propositions p and q, the propositions $\neg(\neg p \wedge \neg q)$ and $p \vee q$ have the
same truth tables. In terminology that we will define later in this section, we can
therefore say that 'p or q' is logically equivalent to 'not (not p and not q)' (think
about this to see if you agree that these come to the same thing). It follows that,
if we wished, we could define 'or' in terms of 'and' and 'not', replacing $p \vee q$
by $\neg(\neg p \wedge \neg q)$.

We write $p \leftrightarrow q$ for the statement '$(p \to q) \wedge (q \to p)$'. This is read as '$p$ is
equivalent to q' or 'p if and only if q' (and 'if and only if' is often abbreviated
in mathematics to 'iff'). Since we have a conjunction sign between the two
implications, it follows that $p \leftrightarrow q$ is only true when *both* $p \to q$ and $q \to p$
are true. Another way to say this is that $p \leftrightarrow q$ is only true when $p \to q$ and its
converse are true. As we have already noted it is quite possible for a statement
to be true but its converse to be false. We also express $p \leftrightarrow q$ by saying that
'p is a necessary and sufficient condition for q'. The statement $p \leftrightarrow q$ has the
following truth table

p	q	$p \leftrightarrow q$
t	t	t
t	f	f
f	t	f
f	f	t

The computation for this is shown below.

p	q	$p \to q$	$q \to p$	$p \leftrightarrow q$
t	t	t	t	t
t	f	f	t	f
f	t	t	f	f
f	f	t	t	t

So $p \leftrightarrow q$ is true exactly when p and q have the same truth values. Therefore both '$2 = 2 \leftrightarrow 7$ is a prime number' and '$2 = 3 \leftrightarrow 6$ is a prime number' are true.

A Boolean term is said to be a **tautology** if it is true no matter what truth assignments are given to its component propositions. Two Boolean terms, built from the same propositions, which have the same truth tables (that is, which take the same truth value for each assignment of truth values to their component propositions) are said to be **logically equivalent**. We can tell whether or not a Boolean term is a tautology by calculating its truth table: it is a tautology if and only if every row of its truth table ends with 't'. Also two terms are logically equivalent if and only if they have 'the same' truth tables (corresponding rows of their truth tables end with the same truth value).

Example To decide whether either of the Boolean expressions $(p \to q) \to (q \to p)$ or $\neg(p \land q) \leftrightarrow (\neg p \lor \neg q)$ is a tautology, we calculate their truth tables as follows.

p	q	$p \to q$	$q \to p$	$(p \to q) \to (q \to p)$
t	t	t	t	t
t	f	f	t	t
f	t	t	f	f
f	f	t	t	t

p	q	$\neg(p \land q)$	$\neg p \lor \neg q$	$\neg(p \land q) \leftrightarrow (\neg p \lor \neg q)$
t	t	f	f	t
t	f	t	t	t
f	t	t	t	t
f	f	t	t	t

Therefore the first Boolean term is not a tautology, but the second is a tautology.

Although somewhat tedious, this method enables us to determine whether a Boolean term is a tautology. The second term above is a tautology of the form $r \leftrightarrow s$ (with $r = \neg(p \land q)$ and $s = \neg p \lor \neg q$): tautologies of this form are referred to as **logical identities** since such a term being a tautology means that the terms r and s always take the same truth values as each other and so are logically equivalent. For example $\neg(\neg p) \leftrightarrow p$ is a tautology and hence is a logical identity, reflecting the fact that $\neg(\neg p)$ and p are logically equivalent.

Further examples of logical identities are the following, in which T denotes any proposition which is always true (any tautology, such as $p \leftrightarrow p$) and F one which is always false (any **contradiction**, such as $\neg p \leftrightarrow p$):

Theorem 3.1.1   *The following are logical identities:*

$(p \wedge p) \leftrightarrow p$ and	
$(p \vee p) \leftrightarrow p$	*idempotence;*
$(p \wedge \neg p) \leftrightarrow F$	*consistency;*
$(p \vee \neg p) \leftrightarrow T$	*law of the excluded middle;*
$(p \wedge q) \leftrightarrow (q \wedge p)$ and	
$(p \vee q) \leftrightarrow (q \vee p)$	*commutativity;*
$p \wedge (q \wedge r) \leftrightarrow (p \wedge q) \wedge r$ and	
$p \vee (q \vee r) \leftrightarrow (p \vee q) \vee r$	*associativity;*
$\neg(p \wedge q) \leftrightarrow \neg p \vee \neg q$ and	
$\neg(p \vee q) \leftrightarrow \neg p \wedge \neg q$	*De Morgan laws;*
$p \wedge (q \vee r) \leftrightarrow (p \wedge q) \vee (p \wedge r)$ and	
$p \vee (q \wedge r) \leftrightarrow (p \vee q) \wedge (p \vee r)$	*distributivity;*
$\neg(\neg p) \leftrightarrow p$	*double negative;*
$(p \rightarrow q) \leftrightarrow (\neg q \rightarrow \neg p)$	*contrapositive law;*
$p \wedge T \leftrightarrow p$ and	
$p \vee T \leftrightarrow T$	*properties of T;*
$p \wedge F \leftrightarrow F$ and	
$p \vee F \leftrightarrow p$	*properties of F;*
$p \wedge (p \vee q) \leftrightarrow p$ and	
$p \vee (p \wedge q) \leftrightarrow p$	*absorption laws.*

Each of the above identities may be established using truth tables; you should try some as exercises. But to prove them all this way would be inefficient, for some of them may be deduced easily from others. For instance, by the law of the excluded middle, $p \vee T$ is equivalent to $p \vee (p \vee \neg p)$ which, by associativity, is equivalent to $(p \vee p) \vee \neg p$ which, by idempotence, is equivalent to $p \vee \neg p$ then, by another appeal to the law of the excluded middle, this is equivalent to T. Thus, the logical identity $p \vee T \leftrightarrow T$ follows from certain of the others.

Note that in Theorem 3.1.1 p, q and r may themselves be compound propositions: so a particular case of De Morgan's Law $\neg(p \wedge q) \leftrightarrow (\neg p \vee \neg q)$ is

$$\neg((p \wedge \neg r) \wedge q) \leftrightarrow (\neg(p \wedge \neg r) \vee \neg q).$$

The above logical identities are 'rules of logic' which are used all the time in mathematical reasoning. For instance, in a mathematical argument we might replace the statement 'It is not true that both $x \geq 0$ and $y \geq 1$' by the statement 'Either $x < 0$ or $y < 1$'. This is an example of the first of De Morgan's Laws above. We recognise (with a bit of thought) that the above two statements are logically equivalent.

For another example, if x is a real variable then we can replace the condition '$x > -1$ and $x^2 \geq 0$' by the condition '$x > -1$' because we know that, for real numbers, $x^2 \geq 0$ is always true. This is the same 'law of thought' or 'rule of logic' as the first property of T above.

Example As an example of the process of deducing new identities from a (small) basic set, we will show that $((p \wedge q) \rightarrow r) \leftrightarrow (p \rightarrow (q \rightarrow r))$ is a tautology, that is, we will show that $(p \wedge q) \rightarrow r$ and $p \rightarrow (q \rightarrow r)$ are logically equivalent. To do this, we will show that both terms are logically equivalent to the Boolean term $\neg p \vee (\neg q \vee r)$.

The following terms are equivalent:

$$(p \wedge q) \rightarrow r \qquad \neg(p \wedge q) \vee r \qquad (\neg p \vee \neg q) \vee r;$$

the first pair since $(a \rightarrow b) \leftrightarrow (\neg a \vee b)$ is a tautology and the second pair since $\neg(a \wedge b) \leftrightarrow (\neg a \vee \neg b)$ is a tautology.

Also the following terms are logically equivalent:

$$p \rightarrow (q \rightarrow r) \qquad p \rightarrow (\neg q \vee r) \qquad \neg p \vee (\neg q \vee r);$$

the first pair since $(a \rightarrow b) \leftrightarrow (\neg a \vee b)$ is a tautology and the second pair for the same reason.

Therefore the required identity follows since, by associativity, $(\neg p \vee \neg q) \vee r$ is logically equivalent to $\neg p \vee (\neg q \vee r)$.

The above list of logical identities may well seem familiar: if the reader has not already done so, then he or she should compare this list with that given as Theorem 2.1.1. Surely the similarity between these lists is no coincidence! Indeed it is not, and we will explain this in two ways.

The first is that the rules of logic which were used to establish Theorem 2.1.1 are simply the rules appearing in the above list. More precisely, the properties of '∩', '∪' and complementation '^{c}' are precisely analogous to those of '∧', '∨' and '¬' (look even at the words used in defining the set-theoretic operations).

As illustration, suppose that we are given sets X and Y. Let p be the proposition 'x is an element of X' and let q be 'x is an element of Y'. Then the statement 'x is an element of $X \cap Y$' means that x is in X and x is in Y, so is represented by the proposition $p \wedge q$. Similarly, 'x is an element of $X \cup Y$' is represented by $p \vee q$ and 'x is not an element of X' by $\neg p$. If X is a subset of Y, we have that if x is in X then x is in Y. We therefore represent $X \subseteq Y$ by $p \rightarrow q$. Also, equality between sets, $X = Y$, is represented by logical equivalence: $p \leftrightarrow q$. Using this, we may translate any logical identity into a theorem about sets (and vice versa). For instance, $p \wedge q \leftrightarrow q \wedge p$ in Theorem 3.1.1 translates into $X \cap Y = Y \cap X$ in Theorem 2.1.1 (and $X \cap X^c = \emptyset$ in Theorem 2.1.1 translates to $p \wedge \neg p \leftrightarrow F$ in Theorem 3.1.1 since $\emptyset$ and F correspond, as do U and T).

Another way to regard the similarity is to say that in each case we have an example of a Boolean algebra, as will be defined in Section 4.4, and that the properties expressed in 2.1.1 and 3.1.1 are just special cases of the defining properties of Boolean algebras.

The notion that the laws of reasoning might be amenable to an algebraic treatment, the idea of a 'logical calculus', seems to have appeared first in the work of Leibniz, a many-talented individual who, along with Newton, was one of the inventors of the integral and differential calculus. Leibniz' ideas on a logical calculus were not taken very seriously at the time and, indeed, for some time afterwards. It was only with Augustus De Morgan and, especially, George Boole, around the middle of the nineteenth century, that an algebraic treatment of logic was formalised. Boole noted that the 'logical operations', usually expressed using words such as 'and', 'or' and 'not', obey certain algebraic laws. He extracted those laws and came up with what is now termed 'Boolean algebra'.

Exercises 3.1

1. Here are a few examples of English-language propositions for you to write down in terms of simpler (constituent) propositions, as defined below.
   Let p be 'It is raining on Venus'.
   Let q be 'The Margrave of Brandenburg carries his umbrella'.
   Let r be 'The umbrella will dissolve'.
   Let s be 'X loves Y'.
   Let t be 'Y loves Z'.
   (a) Write down propositions in terms of p, q, r, s and t for the following.
       (i) 'If it is raining on Venus and the Margrave of Brandenburg carries his umbrella then the umbrella will dissolve'.
       (ii) 'If Y does not love Z and if it is raining on Venus then either X loves Y or the Margrave of Brandenburg carries his umbrella but not both'.

(b) Render into reasonable English each of the propositions expressed by
     the following:
     (i) $(p \wedge q) \vee r$;   (ii) $p \wedge (q \vee r)$;
     (iii) $\neg p \rightarrow (s \wedge (r \rightarrow \neg t))$;   (iv) $\neg(\neg s \vee \neg t) \rightarrow p$.

2. Write down the truth tables for each of the following Boolean terms and so
   decide which are tautologies and which are contradictions:
   (i) $p \wedge (\neg q \vee p)$;   (ii) $(p \wedge q) \vee r$;
   (iii) $p \wedge \neg p$;   (iv) $p \vee \neg p$;
   (v) $(p \vee q) \rightarrow p$;   (vi) $(p \wedge q) \rightarrow p$.

3. Which among the following Boolean terms are logically equivalent to each
   other?
   $$p \wedge (p \rightarrow q), q, (p \wedge q) \leftrightarrow p, p \rightarrow q, p \wedge q.$$

4. Use the properties listed in Theorem 3.1.1 to establish the following:
   (i) $(\neg p \leftrightarrow q) \leftrightarrow ((\neg q) \leftrightarrow p)$;
   (ii) $((p \rightarrow \neg q) \wedge (p \rightarrow \neg r)) \leftrightarrow (\neg(p \wedge (q \vee r)))$;
   (iii) $(p \rightarrow (q \vee r)) \leftrightarrow (\neg q \rightarrow (\neg p \vee r))$.

5. Suppose that X is a subset of Y. Let p be the proposition 'x is an element of
   X' and let q be the proposition 'x is an element of Y'. Write down a
   propositional term which represents the statement 'x is an element of
   $Y \backslash X$'. Hence or otherwise, establish the following identities for sets:
   (i) $A \cap B = A \backslash B^c$;
   (ii) $A \cup (B \backslash A) = A \cup B$;
   (iii) $A \backslash (B \cup C) = (A \backslash B) \cap (A \backslash C)$;
   (iv) $A \backslash (B \cap C) = (A \backslash B) \cup (A \backslash C)$.

3.2   Quantifiers

The logic of propositions that we discussed in the previous section captures only
a small part of mathematical reasoning. It is merely a way of handling the logic
of already formed propositions: it says nothing about how those propositions can
be formed in the first place. If you look at various of the mathematical statements
that we make in this book you will see that they typically involve functions (like
that which squares a number or that which adds together any two numbers),
relations (like that of one number being less than another or that of congruence
modulo n) and constants (such as 0 and 1). They also involve quantifiers, the
use of which is signalled by phrases such as 'for all', 'for every', 'there exists',
'we can find'. Mathematical logics much richer than the propositional logic of
the previous section and containing all the above ingredients can be constructed
and are rich enough for the expression of essentially all mathematics. We do

not discuss these here (for more see [Enderton, *Logic*], for example) but we do give a brief discussion of quantifiers, the use of which pervades mathematical reasoning.

Consider the following two statements:

(i) for every real number a there is a real number b such that $a \leq b$;
(ii) there is a real number b such that for every real number a we have $a \leq b$.

Think about these statements: we hope that you agree that the first is true (given a we can take $b = a + 1$ for example) and that the second is false (since it says that b is the largest real number and there is no such thing). We can introduce quantifiers as a mathematical shorthand which allows us to write assertions such as those above in a very compact (and unambiguous) way.

The **universal quantifier** is usually written as $\forall$ and read as 'for all' (also 'for every', 'for any', etc.): so the first phrase of (i) above can be abbreviated as '$\forall a$'. The use of an upside down A here is to remind one of its connection with the word 'all'. The word universal refers to the universe over which the variable 'x' varies. Unless we state otherwise, in our examples we will always take this to be the 'universe' of real numbers. The **existential quantifier** is usually written as $\exists$ and read as 'there exists' (also 'there is', 'we have' etc.). The use of a backwards E reminds one of the word 'exist'. Now the first phrase of (ii) above can be abbreviated as '$\exists b$'.

As further examples, $\forall x (x^2 \geq 0)$ is read as 'for all x, $x^2 \geq 0$' and $\exists x(x < 0)$ is read as 'there is an x with $x < 0$'. Things get more interesting when we combine quantifiers. For instance, $\forall x \forall y((x > 2 \wedge y > 2) \rightarrow (x + y < xy))$ reads as 'for all x and for all y if $x > 2$ and $y > 2$ then $x + y < xy$', which can be shortened to 'for all $x > 2$ and $y > 2$ we have $x + y < xy$'.

We can abbreviate the statements (i) and (ii) above as follows:

(i) $\forall a \exists b(a \leq b)$;
(ii) $\exists b \forall a(a \leq b)$.

The point to take from this example is that 'for all' and 'there exists' do not commute! The only formal difference between these statements is that the quantifiers have been interchanged and we noted that one is true whereas the other is false, so they are certainly not logically equivalent statements. (Quantifiers of the same kind do commute: $\forall x \forall y \ldots$ is equivalent to $\forall y \forall x \ldots$ and similarly for $\exists$.) Students can and do make errors of logic, and unjustified interchange of quantifiers is a common one! Using the formal symbols $\forall$ and $\exists$ can clarify the logic of a statement or argument.

We give two examples of valid and frequently used deductions which may be made when dealing with quantifiers and then we give an example of the kind of mistake which can be made when dealing with them.

First, negation interchanges $\forall$ and $\exists$ when it 'moves past' one of these quantifiers. That is, for any statement p (involving the variable x, so we will sometimes write $p(x)$ for emphasis), $\neg\forall x p$ is equivalent to $\exists x \neg p$ and $\neg\exists x p$ is equivalent to $\forall x \neg p$.

To illustrate the first, if $p(x)$ is '$x > 0$' then $\neg\forall x p(x)$ reads as 'not (for all x, $x > 0$)' or, more naturally, 'it is not the case that every x is greater than 0.' The formula $\exists x \neg p(x)$ reads as 'there exists x such that not $(x > 0)$' or, more naturally, 'there is some x which is not greater than 0'. This is logically equivalent to the first statement and is an example of the general rule that $\neg\forall x p(x)$ is equivalent to $\exists x \neg p(x)$.

The equivalence of the formulae $\neg\exists x p$ and $\forall x \neg p$ is the equivalence of the statements 'there is no x which satisfies p' and 'every x satisfies not-p' (that is, 'every x fails to satisfy p'). This actually follows completely formally from the first rule. To see this, apply the first rule with $\neg p$ in place of p to obtain that $\neg\forall x \neg p$ is equivalent to $\exists x \neg\neg p$; then use that $\neg\neg p$ is equivalent to p to deduce that $\neg\forall x \neg p$ is equivalent to $\exists x p$. It follows (negate each statement) that $\neg\neg\forall x \neg p$ is equivalent to $\neg\exists x p$ and hence that $\forall x \neg p$ is equivalent to $\neg\exists x p$.

For a second example of a correct deduction, first we notice a simple fact: from $\forall x(r(x) \rightarrow p(x))$ we may deduce the statement $\forall x(r(x)) \rightarrow \forall x(p(x))$. To see this, notice that if we know that it is always the case (for all x) that the statement $r(x)$ implies the statement $p(x)$ then, if we know that $\forall x(r(x))$ is true ($r(x)$ holds for every x) then $\forall x(p(x))$ holds ($p(x)$ holds for every value of x).

Now, assuming $\forall x(r(x) \rightarrow p(x))$ holds we have, since $\neg p(x) \rightarrow \neg r(x)$ is logically equivalent to $r(x) \rightarrow p(x)$ (being the contrapositive) that $\forall x(\neg p(x) \rightarrow \neg r(x))$ holds and hence, from the fact above, that $\forall x(\neg p(x)) \rightarrow \forall x(\neg r(x))$ holds. That is, from $\forall x(r(x) \rightarrow p(x))$ we may deduce $\forall x(\neg p(x)) \rightarrow \forall x(\neg r(x))$.

Here is an example of a non-implication: if we assume the truth of $\forall x(r(x) \rightarrow p(x) \vee q(x))$ then it does *not* follow that either $\forall x(r(x) \rightarrow p(x))$ or $\forall x(r(x) \rightarrow q(x))$ is true, that is, $\forall x(r(x) \rightarrow p(x) \vee q(x))$ does not imply $(\forall x(r(x) \rightarrow p(x))) \vee (\forall x(r(x) \rightarrow p(x)))$. To show this, we can take $r(x)$ to be $x^2 > 0$, $p(x)$ to be $x > 0$ and $q(x)$ to be $x < 0$. Then certainly $\forall x(r(x) \rightarrow p(x) \vee q(x))$ is true (it says that if the square of an element is strictly greater than 0 then either the element is strictly greater than 0 or the element is strictly less than 0). But neither $\forall x(r(x) \rightarrow p(x))$ nor $\forall x(r(x) \rightarrow q(x))$ is true (for instance, the first says that if $x^2 > 0$ then $x > 0$, which is false). In this example it is quite easy to see that the statements are not equivalent but it is not unusual to see students make a mistake in logic based on this.

There are a number of further rules of deduction involving quantifiers which are used continually in mathematical argument and we refer to [Enderton, *Logic*], for example, for a full list. There is a full list, in the sense that one may write down a (small) number of shapes of rules of deduction from which all other rules of deduction follow. In principle, a computer can be programmed to use these rules in order to generate all valid mathematical deductions. The existence of such a 'generating set' of rules of deduction is related to Gödel's celebrated Completeness Theorem in mathematical logic. To state that theorem properly we would have to expand (considerably) on what we mean by 'valid' above and we refer to books on mathematical logic for this.

Another theorem from logic says that there exists nothing like truth tables for statements with quantifiers. We have seen that the method of computing truth tables allows us to decide, given any propositional term (and given enough time), whether that statement is a tautology or not. So we can, in principle, check any implication between propositional terms. There is no such method for statements involving quantifiers. We are not saying that no such method has been found: rather that it has been proved that there can be no such method!

Of course, the correctness or otherwise of many implications has been or can be established but there is no general method or collection of methods which will apply in all cases. That is, given two mathematical statements p and q there is no general method which we can apply in order to decide whether the implication $p \rightarrow q$ is correct: in particular, there is no general method which will either provide us with a deduction which starts with p as assumption and ends with q as conclusion or tell us that no such deduction exists.

It does follow, from what we said above, that one could programme a computer to start generating all correct mathematical implications: if an implication is correct it will eventually be output by the computer, and every implication output by the computer will be correct. But if you have a particular implication that you want to check (say, do the axioms for a group prove that such-and-such a formula holds in every group?) then, if it is correct, the computer will eventually output it (but you have no idea when) whereas, if it is false, you will never discover this just by waiting for the computer. If it is false it will never be output but you cannot discover this fact just by waiting to see whether it appears.

In the following exercises, you will be asked to determine whether certain statements involving quantifiers are true or false. This will not involve you in waiting for an arbitrarily long time! Rather, the sentences will, like those discussed in the text, have a fairly clear truth value which can often be determined by rewriting the statement in English rather than in symbols. One of the main objects of the exercises is for you to become accustomed to 'translating' statements from symbols into words since this is such a common part of mathematical argument.

Exercises 3.2

1. Let $W(x)$ be the statement 'x likes whisky', let $S(x)$ be 'x is Scottish'.
   (a) Give English-language readings of the following statements with quantifiers.
      (i) $\forall x(S(x) \to W(x))$
      (ii) $\forall x(W(x) \to S(x))$
      (iii) $\exists x(S(x) \wedge \neg W(x))$
      (iv) $\neg\forall x(S(x) \to W(x))$
      (v) $\neg\forall x(S(x) \wedge W(x))$
      (vi) $\exists x\exists y(x \neq y \wedge W(x) \wedge W(y))$.
   (b) Write down formal statements which have the following meanings.
      (i)   There is someone who is not Scottish and likes whisky.
      (ii)   If there is someone who likes whisky then there is someone who is Scottish and likes whisky.
      (iii) Everyone who is not Scottish does not like whisky.
      (iv)   There are at least two people who are not Scottish and who like whisky.
2. Decide which of the following are correct.
   (i) $(\forall x(r \to p)) \wedge (\forall x(p \to q))$ implies $\forall x(r \to q)$.
   (ii) $(\exists x(r \wedge p)) \wedge (\exists x(p \wedge q))$ implies $\exists x(r \wedge q)$.
   (iii) $\exists x\forall y(x < y)$ implies $\forall y\exists x(x < y)$.
   (iv) $\forall y\exists x(x < y)$ implies $\exists x\forall y(x < y)$.

3.3   Some proof strategies

In this section we gather together some of what we might call 'methods of proof' or 'proof strategies' that are used in the book. We do not claim that this list is complete or is in any way a classification of strategies. Our aim is simply to point out that, although the details of a proof depend on the specific situation, there are certain forms of argument that are used time and again in mathematical proofs. Longer proofs will use more than one of these strategies, possibly many times.

Certainly we are not giving recipes for constructing proofs: but the comments below might help you in understanding and even producing proofs since they make explicit some of their building blocks. After each 'strategy' we give some references to places where these are used in the text. We have left it for you to identify, within each argument that we reference, exactly where the strategy is used.

You should also look out for these strategies being used when you read proofs in this, and other, books.

Argument by contradiction   We want to prove a statement so we prove that its negation leads to a contradiction. It is the law of the excluded middle (see the list in 3.1.1) which is the basis for the validity of this way of arguing. Examples: 1.1.1, 1.1.2, Example on p. 58, 4.1.3, analysis of groups of small order p. 225.

Argument by cases   Sometimes it is easier or even necessary to treat different possibilites by different arguments, so we split into cases but it is necessary to make sure that all possibilities are covered! Examples: 1.3.2, 4.1.2, 4.2.7, 5.1.3 (i).

Sometimes this is combined with argument by contradiction: we split the range of all possibilities into two or more cases and show that all but one leads to a contradiction, so that case must hold. Example: the proof of 1.2.2.

Argument by contrapositive   The idea here is that $p \to q$ is logically equivalent to $\neg q \to \neg p$. It may sometimes be easier to prove $\neg q \to \neg p$ than $p \to q$. Example: deductions with quantifiers on p. 139.

Choosing the least   This method can be used when dealing with situations involving or indexed by positive integers. For example, we choose the least positive integer (or natural number), in some set, satisfying some condition and we want to establish some property of this integer. We show that if it did not have this property then we could produce a smaller integer in the set, contradicting the fact that our first choice was already supposed to be the smallest in the set. Examples: 1.1.1, 1.1.2, 4.2.3, 5.4.3.

Showing equality indirectly   If they are sets, we can show $X = Y$ by showing that $X \subseteq Y$ and $Y \subseteq X$. If they are integers we can show $a = b$ by showing that $a \leq b$ and $b \leq a$. If they are positive integers we can show $a = b$ by showing that each divides the other. Examples: p. 81, 1.1.4, 5.4.3.

'Doing the same to both sides'   This can be used for an equation or inequality for instance. Examples: 1.1.6, 1.3.3, 1.4.4, 5.1.1.

Mathematical induction   Used to prove 'obvious' properties – they may seem obvious from what has been proved up to that point but, still, they should be proved (for instance extending a result from '$n = 2$' to general n). Examples: 1.3.2, 4.2.1 (iv).

Used where we do something to simplify (say by 'removing one term') and so reduce to the case covered by the induction hypothesis. Examples: 1.3.2, 1.3.3.

Used where we start with the induction hypothesis and 'add the next term to each side'. Example on p. 17.

Showing that a construction terminates   If at each stage the construction produces, say, a natural number, and these numbers are strictly decreasing at each new stage then (by the well-ordering principle) the construction must stop. Examples: 1.1.5, 1.3.3 (though it is not explicitly written that way).

Use of key results   Certain results are used time and time again in proofs of other results. Sometimes they are major theorems but sometimes they are just very useful lemmas. Examples: 1.1.3 used in 1.1.6, 1.4.3, 1.5.2; 5.2.4 used in 5.2.6 and Section 5.3; as well as more obvious examples like Fermat's and Euler's Theorems (1.6.3 and 1.6.7) and Lagrange's Theorem (5.2.3).

Use of definitions   Many mathematical exercises are of the form 'prove that every set (or integer or . . .) which satisfies property A also satisfies property B'. Before being able to attempt such an exercise, it is essential to have a clear idea of what properties A and B are! This may involve going back to an earlier chapter (or previous lecture notes) to find precise definitions.

Sometimes, one may be asked to explain why something is not true, in other words to give a 'disproof'. This can be done by giving a **counterexample**. To do this we give explicit values of the variables and show that, with these values, the result does not hold. For example if we are trying to show that some statement about integers is not true, we might be able to express our condition algebraically and arrive at something like 'if the condition is true then $ad - bc = a + d$'. If we are considering our variables as integers, we should now give explicit values (such as $a = 1$, $b = 2$, $c = 0$ and $d = 0$) to show that $ad - bc$ need not equal $a + d$ and so deduce that the original statement does not hold.

We emphasise that constructing a proof is quite different (and a lot harder!) than reading a proof. Of course, standard proofs may well combine many of the techniques above. In order to get some idea of which techniques might apply in any given case, it is best to try to write proofs as soon as possible in your mathematical studies. Some proofs are straightforward to find in the sense that, if you understand the definitions, understand what is being assumed and can see where you are heading then it is rather obvious what steps to take in reaching that goal. But usually one needs some insight to guide one's efforts in finding a proof. Some proofs are based on a clever idea (for instance 1.3.4). Others, though they might not be so difficult to understand once found, require deep understanding of 'what is going on'. Fermat's, Euler's and Lagrange's Theorems surely come under this heading. A professional mathematician would not be likely to describe these as being, in the present-day context, deep theorems, simply because the ideas are now so familiar (to mathematicians). But in the contexts in which they were first proved, they required deep understanding of structure behind what is obvious and, indeed, they were instrumental in shaping some of the major concepts of mathematics which we can use so easily now.

To conclude this section, we will give some examples in the style of our end-of-section exercises, and consider what proof strategies might apply.

Example 1   Prove that, for any positive integer n the last digit of n (n written in base 10) is the same as the last digit of n^5.

As with many problems, the initial difficulty is in finding a mathematical formulation of the problem. In this case, we want to show that n and n^5 have the same last digit. This will happen if and only if $n^5 - n$ ends in a zero. Another way to say that a number ends in a zero is to say that that the number is divisible by 10. Thus, we can rephrase our original problem as: prove that, for any positive integer n, 10 divides $n^5 - n$.

As a general rule, the appearance of the words 'for any positive integer' suggests trying to use mathematical induction. We try this first: with the statement that 10 divides $1^5 - 1$ (the base case) being clear. So now suppose, for the inductive hypothesis, that 10 divides $n^5 - n$. We then consider $(n + 1)^5 - (n + 1)$. In order to proceed by induction, therefore, we need to be able to do something with $(n + 1)^5$. An expansion of this expression requires the binomial theorem (1.2.1). This gives (after working out the binomial coefficients)

$$(n + 1)^5 - (n + 1) = n^5 - n + 5n^4 + 10n^3 + 10n^2 + 5n + 1 - 1$$
$$= (n^5 - n) + 5(n^4 + n) + 10\,(n^3 + n^2).$$

It is now almost clear that 10 divides the right-hand side. We know 10 divides $(n^5 - n)$ and (of course) $10\,(n^3 + n^2)$, so the proof will be complete once we show that 10 divides $5(n^3 + n^3)$. Clearly 5 divides this number so, by 1.1.6 (ii) we are left with the problem of showing why 2 divides $n^3 + n^2 = n^2(n + 1)$. This is clear because (considering cases) either n is even so 2 divides n or n is odd (in which case 2 divides $n + 1$). This completes the proof by induction. However, that was a somewhat complicated proof of its type, so we just pause before proceeding to our next example to see if we could find alternatives to this proof.

As we saw, we considered $n^5 - n$. We could start by factorising this to get $n(n^4 - 1)$. By Fermat's Theorem, 1.6.3, $n^4 - 1$ is divisible by 5 (when n is not divisible by 5). Again if n is even, then 2 divides n, but if n is odd then 2 divides $n^4 - 1$ since then n^4 will be odd. Thus, using Fermat (a 'key result'), we see that 10 divides $n^5 - n$ except, possibly when five divides n. If 10 divides n, then n ends in a zero, and 10 divides n^5, so now we are left only with the case when 5 divides n, but 10 does not (so n is odd). In that case 5 divides n and 2 divides $n^4 - 1$, so 10 divides $n^5 - n$.

We have now seen two proofs of this fact, the second being a combination of cases and key results. There are many more proofs yet of this fact. The reader might try to find another using congruence classes modulo 10 and so a division into 10 cases. This illustrates the fact that it is worth thinking carefully about the possible strategies.

Example 2   As a second example, prove that if n^2 is odd then n is odd.

Again several proofs are possible, but before discussing these, the reader should perhaps pause and decide which seems the best to try.

In fact, the most straightforward one would be to consider the contrapositive statement: if n is even, then n^2 is even. It is clear that this holds, since if n is even 2 divides n and so 2 divides n^2 (in fact 4 would divide n^2 in that case). Thus if n^2 is odd then n is odd.

Example 3   Show that $\sqrt{2}$ cannot be written as a rational number (a quotient of two integers). It is worth checking our list of proof strategies to decide which one to try. The best possibility seems to be proof by contradiction. Accordingly, we suppose that $\sqrt{2}$ can be written in the form a/b for integers a/b, where we can suppose that all the common factors of a and b have been cancelled to give a reduced fraction. Then, squaring both sides would give $2 = a^2/b^2$ or $2a^2 = b^2$. This would mean that 2 divides b^2 and so b cannot be odd (otherwise b^2 would be odd). Since 2 therefore divides b, 4 divides b^2, that is $b^2 = 4c$ for some integer c. Then $2a^2 = 4c$ so $a^2 = 2c$ and a^2 is even so a is even. Thus 2 divides both a and b so the fraction was not reduced. This contradiction shows that $\sqrt{2}$ is not rational.

Exercises 3.3

In each of the following cases think about and discuss which proof strategies are likely to be helpful and write down at least one proof.

1. For any integer n, $n^2 + n + 1$ is not an even number.
2. If a, b are integers, then $a + b$ is odd precisely if one of a, b is odd.
3. If a, b are integers with $a + b$ an even integer, then $a - b$ is an even integer. Give a counterexample to show that if $a + b$ is even then ab need not be even.

Summary of Chapter 3

In this chapter, we discussed some ideas from elementary mathematical logic. The first section was concerned with propositions (statements which have a truth value) and ways to combine them using negation, conjunction, disjunction and implication. We also discussed a standard way to decide the truth values of a Boolean expression built from propositions and these operations, using truth tables. The second section introduced the idea of quantifiers: the universal quantifier ($\forall$), and the existential quantifier ($\exists$). Since these symbols are often used in mathematical arguments, our aim is for the reader to become familiar with them and to be able to use them freely. Finally, in Section 3.3 we considered some strategies of proof and illustrated them by examples.

4 Examples of groups

The mathematical concept of a group unifies many apparently disparate ideas. It is an abstraction of essential mathematical content from particular situations. Abstract group theory is the study of this essential content. There are several advantages to working at this level of generality. First, any result obtained at this level may be applied to many different situations, and so the result does not have to be worked out or rediscovered in each particular context. Futhermore, it is often easier to discover facts when working at this abstract level since one has shorn away details which, though perhaps pertinent at some level of analysis, are irrelevant to the broad picture.

Of course, to work effectively in the abstract one has to develop some intuition at this level. Although some people can develop this intuition by working only with abstract concepts, most people need to combine such work with the detailed study of particular examples, in order to build up an effective understanding.

That is why we have deferred the formal definition of a group until the third section of this fourth chapter. For you will see that you have already encountered examples of groups in Chapter 1, so, when you come to the definition of a group in Section 4.3, you will be able to interpret the various definitions and theorems which follow that in terms of the examples that you know. In Sections 4.1 and 4.2 we consider permutations: these provide further examples of groups and they have significantly different properties from the arithmetical groups of Chapter 1. A key section of this book is Section 4.3, in which the definition of a group is given. We illustrate this concept by many examples. Finally, in Section 4.4 we give examples of other kinds of algebraic structures.

perm just rearranging in elements aset.

4.1   Permutations

Definition   Let X be a set. A **permutation** of X is a bijection from X to itself (in other words, a 'rearrangement' of the elements of X).

Thus, for example, the identity function, id_X, on any set X is a permutation of X (albeit a rather uninteresting one).

For finite sets X there are two notations available for expressing the action of a permutation of X. These are used in preference to the usual notation for functions.

The first of these, known as two-row notation, was introduced by Cauchy in a paper of 1815. To use this for a permutation π, list the elements of X in some fixed order $a, b, c, \ldots$, then write down a matrix with 2 rows and n columns which has $a, b, c, \ldots$ along the top row and has $\pi(a)$, $\pi(b)$, $\pi(c), \ldots$ along the second row (thus underneath each element x of X appears its image $\pi(x)$):

$$\begin{pmatrix} a & b & c & \cdots \\ \pi(a) & \pi(b) & \pi(c) & \cdots \end{pmatrix}.$$

Example   Suppose that X is the set $\{a, b, c, d\}$ and that π is the permutation on X given by $\pi(a) = d, \pi(b) = c, \pi(c) = a$ and $\pi(d) = b$. Then the two-row notation for π is

$$\begin{pmatrix} a & b & c & d \\ d & c & a & b \end{pmatrix}.$$

If X is a finite set with, say, n elements then there is a bijection from the set of integers $\{1, 2, \ldots, n\}$ to X. If we write x_i for the image of $i \in \{1, \ldots, n\}$, then we may think of such a bijection as being just a way of listing the elements of X as $\{x_1, x_2, \ldots, x_n\}$. When we use two-row notation to express permutations it saves time to write not x_1 but just 1, not x_2 but 2, $\ldots$, and so on. It even makes sense to 'identify' the elements of X with the integers $\{1, \ldots, n\}$. Hence all our discussion of permutations may be placed within the context of permutations of sets of the form $\{1, \ldots, n\}$ (thus we permute not the elements but the labels for the elements).

The fact that the function π is a bijection means that in this two-row notation no integer occurs more than once in the second row (since π is injective) and each integer in the set $\{1, \ldots, n\}$ occurs at least once in the second row (since π is surjective). Thus the second row is indeed a rearrangement, or permutation, of the first row.

Example Take $X = \{1,2,3\}$; there are $3! = 6$ permutations of this set:

$$\begin{pmatrix} 1 & 2 & 3 \\ 1 & 2 & 3 \end{pmatrix}, \begin{pmatrix} 1 & 2 & 3 \\ 1 & 3 & 2 \end{pmatrix}, \begin{pmatrix} 1 & 2 & 3 \\ 2 & 1 & 3 \end{pmatrix}, \begin{pmatrix} 1 & 2 & 3 \\ 3 & 2 & 1 \end{pmatrix}, \begin{pmatrix} 1 & 2 & 3 \\ 2 & 3 & 1 \end{pmatrix}, \begin{pmatrix} 1 & 2 & 3 \\ 3 & 1 & 2 \end{pmatrix}.$$

Since permutations are functions from a set to itself, we may compose them. There are many examples in the following pages.

Definition Let n be a positive integer. Denote by $S(n)$ the set of all permutations of the set $\{1, \ldots, n\}$, equipped with the operation of composition (of functions). $S(n)$ is called the **symmetric group** on n symbols (or elements).

We now consider some properties of this operation of composition of permutations. We will use Greek letters ρ ('rho'), σ ('sigma') and τ ('tau') for permutations as well as π (with which we assume you are familiar).

Theorem 4.1.1 *Let n be a positive integer. Then $S(n)$ satisfies the following conditions:*

(Cl) *if π, σ are members of $S(n)$ then so is the composition $\pi\sigma$;*
(Id) *the identity function* $\mathrm{id} = \mathrm{id}_{\{1,\ldots,n\}}$ *is in $S(n)$;*
(In) *if π is in $S(n)$ then the inverse function π^{-1} is in $S(n)$.*

Also $S(n)$ has $n!$ elements.

Proof The three conditions (Cl), (Id), (In) (short for 'closure', 'identity' and 'inverse') may be rephrased as

(Cl) the composition of any two bijections is a bijection,
(Id) the indentity function is a bijection,
(In) the inverse of a bijection (exists and) is a bijection.

Each of these has already been established in Corollary 2.2.4.

To see that $S(n)$ has $n!$ elements, note that, in terms of the two-row notation, there are n choices for the entry in the second row under 1, for each such choice there are $n - 1$ choices left for the entry under 2 (thus there are $n(n-1)$ choices for the first two entries of the second row), and so on. $\square$

The notation $S(n)$ is sometimes used for the set of permutations (with the operation of composition) of any set with n elements. Of course this will not be the 'same' structure as that we have defined above but it is 'essentially' the same structure (refer to 'isomorphism of groups' in Section 5.3 below).

We will regard the operation of composition of permutations as a kind of 'multiplication'. Suppose that we have two permutations π, σ in $S(n)$ given in the two-row notation; how do we calculate the 'product' $\pi\sigma$? Since this is composition of functions, $\pi\sigma$ means: do σ, then do π. The result may be computed using two-row notation. An easy way to do this is to write (the two-row notation for) π underneath (that for) σ, and then to reorder the columns of π so that they occur in the order given by the second row of σ. This gives us four rows in which the second and third rows are identical. The two-row notation for the composition is obtained by deleting these identical rows, and writing only the first and fourth.

Example   Consider the permutations in $S(5)$

$$\pi = \begin{pmatrix} 1 & 2 & 3 & 4 & 5 \\ 4 & 2 & 1 & 3 & 5 \end{pmatrix} \qquad \sigma = \begin{pmatrix} 1 & 2 & 3 & 4 & 5 \\ 2 & 3 & 4 & 5 & 1 \end{pmatrix}.$$

The four-row array for computing $\pi\sigma$ is

$$\begin{pmatrix} 1 & 2 & 3 & 4 & 5 \\ 2 & 3 & 4 & 5 & 1 \end{pmatrix}$$

$$\begin{pmatrix} 1 & 2 & 3 & 4 & 5 \\ 4 & 2 & 1 & 3 & 5 \end{pmatrix}.$$

Reordering the third and fourth rows together in the order determined by the second row gives

$$\begin{pmatrix} 1 & 2 & 3 & 4 & 5 \\ 2 & 3 & 4 & 5 & 1 \end{pmatrix}$$

$$\begin{pmatrix} 2 & 3 & 4 & 5 & 1 \\ 2 & 1 & 3 & 5 & 4 \end{pmatrix}$$

and so the composition is

$$\begin{pmatrix} 1 & 2 & 3 & 4 & 5 \\ 2 & 1 & 3 & 5 & 4 \end{pmatrix}.$$

This method is a little cumbersome to write down and so it is usually abbreviated as follows. The entry which will come below '1' (say) in the two-row notation for $\pi\sigma$ is found by looking at the entry below '1' in the two-row notation for σ – say that entry is k – and then looking below 'k' in the two-row notation for π: that entry (m say) is the one to place below '1' in the two-row notation for $\pi\sigma$. Proceed in the same way for $2, \ldots, n$.

It should be clear why this works: the first function, σ, takes 1 to k (since 'k' occurs below '1' in the notation for σ), and then the second function, π, takes

k to m – therefore the composition takes 1 to m, and so 'm' is placed below '1' in the notation for $\pi\sigma$.

Example In $S(3)$ we have

$$\begin{pmatrix} 1 & 2 & 3 \\ 3 & 1 & 2 \end{pmatrix} \begin{pmatrix} 1 & 2 & 3 \\ 2 & 1 & 3 \end{pmatrix} = \begin{pmatrix} 1 & 2 & 3 \\ 1 & 3 & 2 \end{pmatrix}.$$

Notice that

non abelian

$$\begin{pmatrix} 1 & 2 & 3 \\ 2 & 1 & 3 \end{pmatrix} \begin{pmatrix} 1 & 2 & 3 \\ 3 & 1 & 2 \end{pmatrix} = \begin{pmatrix} 1 & 2 & 3 \\ 3 & 2 & 1 \end{pmatrix} \neq \begin{pmatrix} 1 & 2 & 3 \\ 1 & 3 & 2 \end{pmatrix}.$$

Hence the operation of composition is **non-commutative** in the sense that $\pi\sigma$ need not equal $\sigma\pi$. Therefore, it is important to remember that we are using the convention that $\pi\sigma$ is the function obtained by applying σ and then applying π.

Example Consider $S(5)$: write id for the identity function, and take

$$\pi = \begin{pmatrix} 1 & 2 & 3 & 4 & 5 \\ 2 & 3 & 1 & 5 & 4 \end{pmatrix}, \sigma = \begin{pmatrix} 1 & 2 & 3 & 4 & 5 \\ 1 & 2 & 4 & 5 & 3 \end{pmatrix}, \tau = \begin{pmatrix} 1 & 2 & 3 & 4 & 5 \\ 2 & 1 & 5 & 4 & 3 \end{pmatrix}$$

then (as the reader should check)

$$\sigma\pi = \begin{pmatrix} 1 & 2 & 3 & 4 & 5 \\ 2 & 4 & 1 & 3 & 5 \end{pmatrix}, \pi\sigma = \begin{pmatrix} 1 & 2 & 3 & 4 & 5 \\ 2 & 3 & 5 & 4 & 1 \end{pmatrix}, \tau^2 = \tau\tau = \begin{pmatrix} 1 & 2 & 3 & 4 & 5 \\ 1 & 2 & 3 & 4 & 5 \end{pmatrix} = \text{id},$$

$$\sigma^2 = \begin{pmatrix} 1 & 2 & 3 & 4 & 5 \\ 1 & 2 & 5 & 3 & 4 \end{pmatrix}, \quad \sigma^3 = \sigma^2\sigma = \begin{pmatrix} 1 & 2 & 3 & 4 & 5 \\ 1 & 2 & 3 & 4 & 5 \end{pmatrix} = \text{id},$$

$$\pi^2 = \begin{pmatrix} 1 & 2 & 3 & 4 & 5 \\ 3 & 1 & 2 & 4 & 5 \end{pmatrix}, \pi^3 = \begin{pmatrix} 1 & 2 & 3 & 4 & 5 \\ 1 & 2 & 3 & 5 & 4 \end{pmatrix}, \pi^4 = \begin{pmatrix} 1 & 2 & 3 & 4 & 5 \\ 2 & 3 & 1 & 4 & 5 \end{pmatrix},$$

$$\pi^5 = \begin{pmatrix} 1 & 2 & 3 & 4 & 5 \\ 3 & 1 & 2 & 5 & 4 \end{pmatrix}, \pi^6 = \begin{pmatrix} 1 & 2 & 3 & 4 & 5 \\ 1 & 2 & 3 & 4 & 5 \end{pmatrix} = \text{id}.$$

It must be stressed that the reader probably will understand little of what follows in this and the next section, unless complete confidence in multiplication of permutations has been acquired. With this in mind, a large selection of calculations is provided in the exercises at the end of this section.

The two-row notation is also very useful when calculating the inverse of a permutation. The inverse is calculated by exchanging the upper and lower rows, and then reordering the columns so that the entries on the upper row occur in the natural order.

how to calculate the inverse

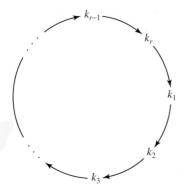

Fig. 4.1

Example   In $S(7)$ the inverse of

$$\begin{pmatrix} 1 & 2 & 3 & 4 & 5 & 6 & 7 \\ 5 & 3 & 7 & 1 & 4 & 2 & 6 \end{pmatrix}$$

is

$$\begin{pmatrix} 5 & 3 & 7 & 1 & 4 & 2 & 6 \\ 1 & 2 & 3 & 4 & 5 & 6 & 7 \end{pmatrix} = \begin{pmatrix} 1 & 2 & 3 & 4 & 5 & 6 & 7 \\ 4 & 6 & 2 & 5 & 1 & 7 & 3 \end{pmatrix}.$$

We now consider the other notation for permutations.

Definition   A permutation $\pi \in S(n)$ is **cyclic** or a **cycle** if the elements $1, \ldots,$ n may be rearranged, as say $k_1, \ldots, k_r, k_{r+1}, \ldots, k_n$ (we allow the possibilities that $r + 1 = 1$ or $r = n$), in such a way that π fixes each of $k_{r+1}, \ldots, k_n$ and 'cycles' the remainder, sending k_1 to k_2 sending k_2 to $k_3, \ldots,$ sending k_{r-1} to k_r and finally sending k_r back to k_1. The integer r (that is, the number of elements in, or the length of, the cycling part) is called the **length** of π. (The algebraic significance of this integer will be explained later.) We say that the length of the identity permutation is 1. A cycle of length 2 is called a **transposition**. A cycle of length r is termed an r-**cycle**.

There is a special notation for cycles: write down, between parentheses, the integers which are moved by the cycle, in the order in which they are moved. Thus the cycle above could be denoted by $(k_1 k_2 \ldots k_r)$. See Fig. 4.1.

The point of the cycle at which to start may be chosen arbitrarily, so for any given cycle there will be a number of ways (equal to its length) of writing such a notation for it. For example, if π is the member of $S(5)$ which sends 1 to 3,

3 to 4, 4 to 1, and fixes 2 and 5 (so π is a cycle of length 3), then π may be written using this notation as (1 3 4), or as (3 4 1), or as (4 1 3).

Example

$$\begin{pmatrix} 1 & 2 & 3 & 4 & 5 \\ 2 & 3 & 1 & 4 & 5 \end{pmatrix}, \begin{pmatrix} 1 & 2 \\ 2 & 1 \end{pmatrix}, \begin{pmatrix} 1 & 2 & 3 & 4 \\ 4 & 1 & 3 & 2 \end{pmatrix}, \begin{pmatrix} 1 & 2 & 3 & 4 & 5 & 6 & 7 \\ 4 & 7 & 6 & 2 & 1 & 5 & 3 \end{pmatrix}$$

are cycles of lengths 3, 2, 3 and 7 respectively, but

$$\begin{pmatrix} 1 & 2 & 3 & 4 & 5 \\ 2 & 1 & 4 & 5 & 3 \end{pmatrix}, \begin{pmatrix} 1 & 2 & 3 & 4 \\ 2 & 1 & 4 & 3 \end{pmatrix}, \begin{pmatrix} 1 & 2 & 3 & 4 & 5 & 6 & 7 \\ 2 & 5 & 1 & 7 & 3 & 4 & 6 \end{pmatrix}$$

are not cycles. Notations for the four cycles listed are (1 2 3), (1 2) (a transposition), (1 4 2) and (1 4 2 7 3 6 5).

Note that in the definition of cyclic permutation the case $r + 1 = 1$ corresponds to the permutation which does not move anything – in other words to the identity permutation id (which is therefore a cycle: its cycle notation would be empty, so we continue to write it as 'id'). The case $r = n$ is the case of a cycle which moves every element (the fourth, but not the first, cycle in the above example is of this kind).

Definition Let π and σ be elements of $S(n)$. Then π and σ are **disjoint** if every integer in $\{1, \ldots, n\}$ which is moved by π is fixed by σ and every integer moved by σ is fixed by π (we say that π **moves** $k \in \{1, \ldots, n\}$ if $\pi(k) \neq k$, otherwise π **fixes** k).

Theorem 4.1.2 *If π and σ are disjoint permutations in $S(n)$, then π and σ commute, that is, $\pi\sigma = \sigma\pi$.*

Proof For any permutation ρ in $S(n)$, let Mov(ρ) be the set of integers in $\{1, 2, \ldots, n\}$ which are moved by ρ. More formally

$$\text{Mov}(\rho) = \{m : 1 \leq m \leq n \quad \text{and} \quad \rho(m) \neq m\}.$$

To say that π and σ are disjoint is just to say that the intersection of Mov(π) with Mov(σ) is empty.

We have the following possibilities for $m \in \{1, \ldots, n\}$:

$m \in \text{Mov}(\pi)$;

$m \in \text{Mov}(\sigma)$;

m is in neither Mov(π) nor Mov(σ).

In the first case m is sent to $\pi(m)$ by both $\pi\sigma$ and $\sigma\pi$. For we have $\pi\sigma(m) = \pi(\sigma(m)) = \pi(m)$ (since m is moved by π it is fixed by σ): on the other hand we have $\sigma\pi(m) = \sigma(\pi(m)) = \pi(m)$. The last equality follows since $\pi(m)$ is moved by π (so is fixed by σ): for otherwise we would have $\pi(\pi(m)) = \pi(m)$ and so, since π is 1-1, $\pi(m) = m$, a contradiction.

The other two cases are dealt with by similar arguments and we leave these to the reader. Thus $\pi\sigma = \sigma\pi$, since they have the same effect on the elements of $\{1, \ldots, n\}$. □

You might find it useful to go through the above proof with particular choices of disjoint π and σ.

Remark As we have already noted, the conclusion of 4.1.2 can fail for non-disjoint cycles. For another example, consider, in $S(3)$, $(1\ 2)(1\ 3) = (1\ 3\ 2)$, whereas $(1\ 3)(1\ 2) = (1\ 2\ 3) \neq (1\ 3\ 2)$. *warn they are applied*

The next result says that any permutation may be written as a product of disjoint cycles. But first we present an example illustrating how to 'decompose' a permutation in this way.

Example Let π be the following permutation in $S(14)$.

$$\begin{pmatrix} 1 & 2 & 3 & 4 & 5 & 6 & 7 & 8 & 9 & 10 & 11 & 12 & 13 & 14 \\ 4 & 9 & 10 & 7 & 5 & 2 & 6 & 13 & 1 & 3 & 11 & 12 & 14 & 8 \end{pmatrix}.$$

We begin by considering the repeated action of π on 1: π sends 1 to 4, which in turn is sent to 7, which is sent to 6, to 2, to 9, and then back to 1. So we find the 'circuit' to which 1 belongs, and write down the cycle in $S(14)$ that corresponds to this 'circuit', namely $(1\ 4\ 7\ 6\ 2\ 9)$. Now we look for the first integer in $\{1, \ldots, 14\}$ that is not moved by this cycle: that is 3. The 'circuit' to which 3 belongs takes 3 to 10, which in turn goes back to 3. The cycle of $S(14)$ corresponding to this is $(3\ 10)$; note that this cycle is disjoint from $(1\ 4\ 7\ 6\ 2\ 9)$. The next integer which has not yet been encountered is 5: π fixes 5, so we do not need to write down a cycle for 5. The next integer not yet treated is 8: the cycle corresponding to the repeated action of π on 8 is $(8\ 13\ 14)$; note that this is disjoint from each of the other two cycles found. Finally π fixes both 11 and 12. Thus we obtain an expression of π as a product of disjoint cycles:

$$\pi = (1\ 4\ 7\ 6\ 2\ 9)(3\ 10)(8\ 13\ 14).$$

By Theorem 4.1.2, we may rearrange the order of the cycles occurring, say as

$$(3\ 10)(1\ 4\ 7\ 6\ 2\ 9)(8\ 13\ 14)$$

but the actual cycles which occur are uniquely determined by π. See Fig. 4.2.

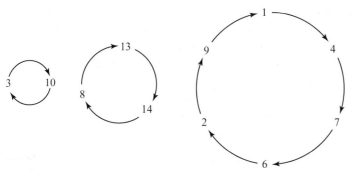

Fig. 4.2

Theorem 4.1.3 *Let π be an element of $S(n)$. Then π may be expressed as a product of disjoint cycles. This **cycle decomposition** of π is unique up to rearrangement of the cycles involved.*

Proof First look for the smallest integer which is not fixed by π: suppose that this is k. Apply π successively to k: let k_1 be k, k_2 be $\pi(k_1)$, k_3 be $\pi(k_2)$ and so on. Since the set $\{1,2,\ldots, n\}$ is finite, we will obtain repetitions after sufficiently many steps. Let r be the smallest integer such that k_r equals k_s for some s strictly less than r. If s were greater than 1 then we could write

$$\pi(k_{s-1}) = k_s = k_r = \pi(k_{r-1}).$$

Since π is injective, we deduce $k_{s-1} = k_{r-1}$, contrary to the minimality of r. It follows that s is 1, and so $(k_1 k_2 \ldots k_r)$ is an r-cycle.

   Repeat the process for the smallest integer not fixed by π and not in the set $(k_1, k_2, \ldots, k_r)$ of integers already encountered. Continuing in this way, we obtain an expression of π as a product of disjoint cycles. From the construction it follows that the decomposition will be unique up to rearrangement of the cycles. □

You should note that the proof just given simply formalises the procedure used in the preceding example.

Two further examples of cycle notation

$$\begin{pmatrix} 1 & 2 & 3 & 4 & 5 & 6 & 7 & 8 & 9 & 10 & 11 & 12 \\ 7 & 2 & 10 & 12 & 5 & 4 & 8 & 1 & 6 & 3 & 9 & 11 \end{pmatrix} = (1\ 7\ 8)(3\ 10)(4\ 12\ 11\ 9\ 6),$$

$$\begin{pmatrix} 1 & 2 & 3 & 4 & 5 & 6 & 7 & 8 & 9 & 10 \\ 4 & 7 & 8 & 6 & 2 & 10 & 3 & 9 & 5 & 1 \end{pmatrix} = (1\ 4\ 6\ 10)(2\ 7\ 3\ 8\ 9\ 5).$$

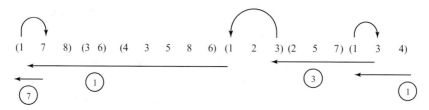

Fig. 4.3

In order to multiply together two permutations which are written using cycle notation, one can write down their two-row notations, multiply, and then write down the cycle notation for the result. But this is a cumbersome process, and the multiplication is best done directly. The basic manipulation involved is what we will call a switch. Suppose we are given a product, π, of cycles and we want to compute its cycle decomposition. We visualise the effect of π on an integer i moving from right to left, encountering the various cycles, possibly being switched to a new value at each encounter. To switch i, seek the first occurrence of i to the left of its present position. This lies in a cycle of π, and i is now switched to the number, k say, to which this cycle takes i. Now think of k continuing to move to the left, and repeat this switching process until the left-hand end is reached. The number, m say, which finally emerges at the left-hand end is $\pi(i)$.

The multiplication is carried out by repeating these switches, starting the process with each integer in $\{1, 2, \ldots, n\}$ in sequence if the result is to be written in two-row notation, or in the order determined as the process continues otherwise. The method is illustrated by an example.

Example 1 Compute the cycle decomposition of the product π:

$$(1\ 7\ 8)(3\ 6)(4\ 3\ 5\ 8\ 6)(1\ 2\ 3)(2\ 5\ 7)(1\ 3\ 4).$$

Start with the integer 1 at the right-hand end of the above product. The first cycle encountered involves 1, switching it to 3. The number 3 continues to move to the left, and is switched back to 1 by the third cycle from the right since this cycle takes 3 to 1. Now 1 continues to move to the left, and is switched to 7 by the last cycle encountered. Therefore the product sends 1 to 7. See Fig. 4.3.

If we want to write the result in cycle notation, then it is most convenient to repeat the process next starting with the integer $7 = \pi(1)$: 7 is switched to 2; 2 to 3; then 3 goes to 5. Therefore 7 is sent to 5.

Continuing in this way ($5 = \pi(7)$ is treated next), we obtain an 8-cycle (1 7 5 8 3 6 4 2). Now we look for the first integer which has not yet been

$(23)(123) = (13)$

'fed into' the right-hand end: in this case there are none, so the answer is just the above 8-cycle.

(12)

Example 2 In order to give further examples of multiplication of cycles, and to illustrate Theorem 4.1.1, we present the complete multiplication table for $S(3)$. The entry at the intersection of the row labelled σ and the column labelled τ is $\sigma\tau$.

$(23)(132)$

$(123)(123) = (132)$

	id	(123)	(132)	(12)	(13)	(23)
id	id	(123)	(132)	(12)	(13)	(23)
(123)	(123)	(132)	id	(13)	(23)	(12)
(132)	(132)	id	(123)	(23)	(12)	(13)
(12)	(12)	(23)	(13)	id	(132)	(123)
(13)	(13)	(12)	(23)	(123)	id	(132)
(23)	(23)	(13)	(12)	(132)	(123)	id

$(132)(132) = (123)$

Example 3 The following permutations in $S(4)$ have a multiplication table as shown:

id; $(1\ 3\ 4\ 2); (1\ 4)(2\ 3); (1\ 2\ 4\ 3); (2\ 3); (1\ 4); (1\ 2)(3\ 4); (1\ 3)(2\ 4)$.

	id	(1342)	(14)(23)	(1243)	(23)	(14)	(12)(34)	(13)(24)
id	id	(1342)	(14)(23)	(1243)	(23)	(14)	(12)(34)	(13)(24)
(1342)	(1342)	(14)(23)	(1243)	id	(13)(24)	(12)(34)	(23)	(14)
(14)(23)	(14)(23)	(1243)	id	(1342)	(14)	(23)	(13)(24)	(12)(34)
(1243)	(1243)	id	(1342)	(14)(23)	(12)(34)	(13)(24)	(14)	(23)
(23)	(23)	(12)(34)	(14)	(13)(24)	id	(14)(23)	(1342)	(1243)
(14)	(14)	(13)(24)	(23)	(12)(34)	(14)(23)	id	(1243)	(1342)
(12)(34)	(12)(34)	(14)	(13)(24)	(23)	(1243)	(1342)	id	(14)(23)
(13)(24)	(13)(24)	(23)	(12)(34)	(14)	(1342)	(1243)	(14)(23)	id

We finish the section by describing how to write down the inverse of a cycle: one simply reverses the order of the terms which appear (and then, if one wishes to, rewrites the resulting cycle with the smallest integer first). For example,

$(1\ 2\ 3\ 4\ 5)^{-1} = (5\ 4\ 3\ 2\ 1) = (1\ 5\ 4\ 3\ 2)$ — don't have to rewrite

It follows that if a permutation is written as a product of disjoint (hence commuting) cycles then the inverse is found by applying this process to each of its component cycles. If a permutation is written as a product of not necessarily disjoint cycles then the order of the components must also be reversed, because $(\pi\sigma)^{-1} = \sigma^{-1}\pi^{-1}$ (by 2.2.4(i)).

Permutations were important in the development of group theory, in that permutation groups of the roots of a polynomial were a key feature of Galois' work on solvability of polynomial equations by radicals. They also figure in the work of Lagrange, Cauchy and others, as actions on polynomials (see the proof of Theorem 4.2.8 below). For more on this, see the notes at the end of Section 4.3.

The reader is strongly advised to attempt the exercises that follow before continuing to the next section.

Exercises 4.1

Let π_1, π_2, π_3, π_4 and π_5 be the following permutations:

$$\pi_1 = \begin{pmatrix} 1 & 2 & 3 & 4 & 5 & 6 & 7 & 8 & 9 \\ 3 & 2 & 1 & 6 & 5 & 4 & 9 & 8 & 7 \end{pmatrix},$$

$$\pi_2 = \begin{pmatrix} 1 & 2 & 3 & 4 & 5 & 6 & 7 & 8 & 9 \\ 9 & 8 & 7 & 6 & 5 & 4 & 3 & 2 & 1 \end{pmatrix},$$

$$\pi_3 = \begin{pmatrix} 1 & 2 & 3 & 4 & 5 & 6 & 7 & 8 & 9 \\ 4 & 5 & 6 & 7 & 8 & 9 & 1 & 2 & 3 \end{pmatrix},$$

$$\pi_4 = \begin{pmatrix} 1 & 2 & 3 & 4 & 5 & 6 & 7 & 8 & 9 & 10 & 11 & 12 \\ 9 & 1 & 2 & 3 & 4 & 5 & 6 & 7 & 8 & 11 & 12 & 10 \end{pmatrix},$$

$$\pi_5 = \begin{pmatrix} 1 & 2 & 3 & 4 & 5 & 6 & 7 & 8 & 9 & 10 & 11 & 12 \\ 12 & 7 & 2 & 8 & 4 & 6 & 3 & 9 & 5 & 1 & 11 & 10 \end{pmatrix}.$$

1. Calculate the following products:

   $\pi_1\pi_2$, $\pi_2\pi_3$, $\pi_3\pi_1$, $\pi_3\pi_2$, $\pi_2\pi_1\pi_3$, $\pi_2\pi_2\pi_2$, $\pi_4\pi_5$, $\pi_5\pi_4$, $\pi_1\pi_3$, $\pi_2\pi_2$, $\pi_2\pi_1$, $\pi_3\pi_3$, $\pi_2\pi_1\pi_2$, $\pi_2\pi_3\pi_2$, $\pi_4\pi_4$, $\pi_5\pi_5$.

2. Find the inverses of π_1, π_2, π_3, π_4 and π_5.

3. Write each permutation in Example 4.1.1 as a product of disjoint cycles.

4. Compute the following products, writing each as a product of disjoint cycles:

   (i) (1 2 3 4 5)(1 3 6 8)(6 5 4 3)(1 3 6 8);

   (ii) (1 12 10)(2 7 3)(4 6 9 5)(1 3)(4 6)(7 9);

   (iii) (1 4 7)(2 5 8)(3 6 9)(1 2 3 4 5 6 7 8 9)(10 11).

5. Write down the complete multiplication table for the following set of permutations in $S(4)$:

   id, (1 2 3 4), (1 3)(2 4), (1 4 3 2), (1 3), (2 4), (1 2)(3 4) and (1 4)(2 3).

6. The study of the symmetric group $S(52)$ has engaged the attention of many sharp minds. As an aid to their investigations, devotees of this pursuit make use of a practical device which provides a concrete realisation of $S(52)$. This device is technically termed a 'deck of playing cards'. We now give

some exercises based on the permutations of these objects. Since it is a time-consuming task even to write down a typical permutation of 52 objects, we will work with a restricted deck which contains only 10 cards – say the ace to ten of spades (denoted $1, \ldots, 10$) for definiteness.

Permutations of the deck are termed 'shuffles' and 'cuts': let us regard these as elements of $S(10)$.

Define s to be the 'interleaving' shuffle which hides the top card:

$$s = \begin{pmatrix} 1 & 2 & 3 & 4 & 5 & 6 & 7 & 8 & 9 & 10 \\ 2 & 4 & 6 & 8 & 10 & 1 & 3 & 5 & 7 & 9 \end{pmatrix}.$$

Let t be the interleaving shuffle which leaves the top card unchanged:

$$t = \begin{pmatrix} 1 & 2 & 3 & 4 & 5 & 6 & 7 & 8 & 9 & 10 \\ 1 & 3 & 5 & 7 & 9 & 2 & 4 & 6 & 8 & 10 \end{pmatrix}.$$

Finally, let c be the cut:

$$c = \begin{pmatrix} 1 & 2 & 3 & 4 & 5 & 6 & 7 & 8 & 9 & 10 \\ 6 & 7 & 8 & 9 & 10 & 1 & 2 & 3 & 4 & 5 \end{pmatrix}.$$

Show that cutting the deck according to c and then applying the shuffle s has the same effect as the single shuffle t.

Write s, t, c, cs and scs using cycle notation.

For each of these basic and combined shuffles s, t, c, cs and scs, how many times must the shuffle be repeated before the cards are returned to their original positions?

4.2 The order and sign of a permutation

Definition Let π be a permutation. The positive **powers**, π^n, of π are defined inductively by setting $\pi^1 = \pi$ and $\pi^{k+1} = \pi \cdot \pi^k$ (k a positive integer). We also define the negative powers: $\pi^{-k} = (\pi^{-1})^k$ where k is a positive integer, and finally set $\pi^0 = \mathrm{id}$.

The following index laws for powers are obtained using mathematical induction.

Theorem 4.2.1 *Let π be a permutation and let r, s be positive integers. Then*

(i) $\pi^r \pi^s = \pi^{r+s}$, ~ exp rules .

(ii) $(\pi^r)^s = \pi^{rs}$,

(iii) $\pi^{-r} = (\pi^r)^{-1}$,

(iv) *if π, σ are permutations such that $\pi\sigma = \sigma\pi$ then $(\pi\sigma)^r = \pi^r\sigma^r$.*

[handwritten annotations: Pf by induction. $\pi^r \pi^s = \pi^{r+s}$ [if they commute] $\pi^r \sigma^r = (\pi\sigma)^r$]

Proof (i) The proof is by induction on r. If $r = 1$, then $\pi \cdot \pi^s = \pi^{s+1}$ by definition. Now suppose that

$$\pi^r \pi^s = \pi^{r+s}.$$

Then

$$\pi^{r+1}\pi^s = (\pi \cdot \pi^r)\pi^s$$
$$= \pi(\pi^r \pi^s)$$
$$= \pi(\pi^{r+s})$$
$$= \pi^{r+s+1}$$

as required.

The proofs of (ii) and (iii) are also achieved using mathematical induction and are left as exercises (for (iii) use the fact (2.2.4(i)) that $(fg)^{-1} = g^{-1}f^{-1}$ if f and g are bijections from a set to itself).

The fourth part also is proved by induction. We actually need a slightly stronger statement: that $(\pi\sigma)^k = \pi^k\sigma^k$ and $\sigma\pi^k = \pi^k\sigma$ (we use the second equation within the proof). By assumption the result is true for $k = 1$. So suppose inductively that $(\pi\sigma)^k = \pi^k\sigma^k$ and $\sigma\pi^k = \pi^k\sigma$. Then

$$(\pi\sigma)^{k+1} = \pi\sigma(\pi\sigma)^k \quad \text{(by definition)}$$
$$= \pi\sigma\pi^k\sigma^k \quad \text{(by induction)}$$
$$= \pi\pi^k\sigma\sigma^k \quad \text{(also by induction)}$$
$$= \pi^{k+1}\sigma^{k+1} \quad \text{(by definition)}$$

Also

$$\sigma\pi^{k+1} = \sigma\pi\pi^k = \pi\sigma\pi^k \quad \text{(by assumption)}$$
$$= \pi\pi^k\sigma \quad \text{(by induction)}$$
$$= \pi^{k+1}\sigma \quad \text{(by definition)}.$$

So we have proved both parts of the induction hypothesis for $k + 1$ and the result therefore follows by induction. $\square$

Theorem 4.2.2 Let π be an element of $S(n)$. Then there is an integer m, greater than or equal to 1, such that $\pi^m = \text{id}$.

Proof Consider the successive powers of π: π; π^2; π^3; ... Each of these powers is a bijection from $\{1, \ldots, n\}$ to itself. Since there are only finitely many such functions (4.1.1) there must be repetitions within the list: say $\pi^r = \pi^s$ with

$r < s$. Since π^{-1} exists, we may multiply each side by π^{-r} to obtain (using 4.2.1(iii)) id $= \pi^{s-r}$. So m may be taken to be $s - r$. □

Definition The **order** of a permutation π, $o(\pi)$, is the least positive integer n such that π^n is the identity permutation. Note that the order of id is 1 and id is the only permutation of order 1.

Example The order of any transposition is 2.

Example The successive powers of the cycle (3 4 2 5) are (3 4 2 5), (3 2)(4 5), (3 5 2 4), id. Thus the order of (3 4 2 5) is 4.

Example The successive powers of the permutation (1 3)(2 5 4) are (1 3)(2 5 4), (2 4 5), (1 3), (2 5 4), (1 3)(2 4 5), id, (1 3)(2 5 4), and so on. In particular, the order of (1 3)(2 5 4) is six.

Theorem 4.2.3 *Let π be a permutation of order n. Then $\pi^r = \pi^s$ if and only if r is congruent to s modulo n.*

Proof From the proof of 4.2.2 it follows that if $\pi^r = \pi^s$ then $\pi^{s-r} = $ id. If, conversely, $\pi^{s-r} = $ id then, multiplying each side by π^r and using 4.2.1, we obtain $\pi^s = \pi^r$. We will therefore have proved the result if we show that $\pi^k = $ id $= (\pi^0)$ precisely if k is congruent to 0 modulo n, that is, precisely if k is divisible by n.

To see this, observe first that if k is a multiple of n, say $k = nt$, then, using 4.2.1(ii),

$$\pi^k = \pi^{nt} = (\pi^n)^t = (\text{id})^t = \text{id}.$$

Suppose conversely that $\pi^k = $ id. Apply the division algorithm (1.1.1) to write k in the form $nq + r$ with $0 \le r < n$. Then, again using 4.2.1, we have

$$\text{id} = \pi^k = \pi^{nq+r} = \pi^{nq}\pi^r = (\pi^n)^q \pi^r = (\text{id})^q \pi^r = \text{id} \cdot \pi^r = \pi^r.$$

The definition of n (as giving the least positive power of π equal to id) now forces r to be zero: that is, n divides k. □

How can we quickly find the order of a permutation? For cycles the order turns out to be just the length of the cycle.

Theorem 4.2.4 *Let π be a cycle in $S(n)$. Then $o(\pi)$ is the length of the cycle π.*

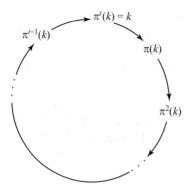

Fig. 4.4

Proof Think of the elements which are moved by π arranged in a circle, so that π^n just has the effect of moving each element n steps forward in the circuit (Fig. 4.4). From this picture it should be clear that if t is the length of π then the least positive integer n for which π^n equals the identity is t.

We can argue more formally as follows. If $\pi = \mathrm{id}$ then the result is clear; so we may suppose that there is $k \in \{1,\dots, n\}$ with $\pi(k) \neq k$. Since π is a cycle, the set of integers moved by π is precisely $\mathrm{Mov}(\pi) = \{k, \pi(k), \pi^2(k), \dots, \pi^{t-1}(k)\}$, where t is the length of π. There are no repetitions in the above list, so the order of π is at least t.

On the other hand, $\pi^t(k) = k$, and hence for every value of r

$$\pi^t(\pi^r(k)) = \pi^{t+r}(k) = \pi^{r+t}(k) = \pi^r(\pi^t(k)) = \pi^r(k).$$

Therefore π^t fixes every element of the set $\mathrm{Mov}(\pi)$. Since π fixes all other elements of $\{1, \dots, n\}$ so does π^t. Thus, $\pi^t = \mathrm{id}$.

Therefore the least positive power of π equal to the identity permutation is the tth, so $o(\pi) = t$, as required. $\square$

Next we consider those permutations that are products of two disjoint permutations.

Lemma 4.2.5 *If π, σ are disjoint permutations in $S(n)$ then the order of $\pi\sigma$ is the least common multiple, $\mathrm{lcm}(o(\pi), o(\sigma))$, of the orders of π and σ.*

Proof Suppose that $o(\pi) = r$ and $o(\sigma) = s$, and let $d = ra$, $d = sb$ where $d = \mathrm{lcm}(r,s)$. Then certainly we have

$$(\pi\sigma)^d = \pi^d\sigma^d = \pi^{ra}\sigma^{sb} = (\pi^r)^a(\sigma^s)^b = \mathrm{id}$$

(handwritten annotations: "ord = length", "lcm (disjoint", "naked disjoint cycles.")

(the first equality by Theorem 4.2.1 since π and σ commute). So it remains to show that d is the least positive integer for which $\pi\sigma$ raised to that power is the identity permutation.

So suppose that $(\pi\sigma)^e = \text{id}$. Since π and σ commute it follows, by Theorem 4.2.1, that $\pi^e\sigma^e = \text{id}$. Let $k \in \{1, \ldots, n\}$. If k is moved by π then it is fixed by σ, and hence by σ^e: so $k = \text{id}(k) = \pi^e\sigma^e(k) = \pi^e(k)$. On the other hand if k is fixed by π then certainly it is fixed by π^e. Therefore $\pi^e = \text{id}$. Since $\pi^e\sigma^e = \text{id}$, it then follows that also $\sigma^e = \text{id}$. So by Theorem 4.2.3 it follows that r divides e and s divides e: hence d divides e (by definition of least common multiple), as required. $\square$

Example The permutation

$$c = \begin{pmatrix} 1\ 2\ 3\ 4\ 5\ 6\ 7 \\ 5\ 7\ 3\ 1\ 4\ 6\ 2 \end{pmatrix}$$

may be written as the product $(1\ 5\ 4)(2\ 7)$, of disjoint permutations. Therefore the order of π is the lcm of 3 and 2: that is $o(\pi) = 6$. (It is an instructive exercise, which illustrates the proof of 4.2.5, to compute the powers of π, and their cycle decompositions, up to the sixth.)

Example The permutation $\pi = (1\ 6)(3\ 7\ 4\ 2)$ is already expressed as the product of disjoint cycles, one of length 4 and the other of length 2. The order of π is therefore the lcm of 4 and 2 (which is 4): $o(\pi) = 4$. Note in particular that the order is not in this case the product of the orders of the separate cycles. You should compute the powers of π, and their cycle decompositions (up to the fourth), to see why this is so.

Theorem 4.2.6 *Let π be an element of $S(n)$, and suppose that $\pi = \tau_1\tau_2 \ldots \tau_k$ is a decomposition of π as a product of disjoint cycles. Then the order of π is the least common multiple of the lengths of the cycles $\tau_1, \ldots, \tau_k$.*

Proof The proof is by induction on k. When k is 1, the result holds by Theorem 4.2.4. Now suppose, inductively, that the theorem is true if π can be written as a product of $k - 1$ disjoint cycles. If π is a permutation which is of the form

$$\tau_1\tau_2 \ldots \tau_k,$$

with the τ_j disjoint, then apply the induction hypothesis to the product $\tau_1\tau_2 \ldots \tau_{k-1}$ to deduce that

$$o(\tau_1\tau_2 \ldots \tau_{k-1}) = \text{lcm}(o(\tau_1), \ldots, o(\tau_{k-1})),$$

and then apply Lemma 4.2.5 to the product $(\tau_1 \tau_2 \ldots \tau_{k-1})\tau_k$ to obtain the result (since $\mathrm{lcm}(\mathrm{lcm}(o(\tau_1), \ldots, o(\tau_{k-1})), o(\tau_k)) = \mathrm{lcm}(o(\tau_1), \ldots, o(\tau_k)))$. $\square$

In passing, we say a little about the shape of a permutation. By the 'shape' of a permutation π we mean the sequence of integers (in non-descending order) giving the lengths of the disjoint cyclic components of π. Thus if π has shape $(2,2,5)$ then π is a product of three disjoint cycles, two of length 2 and one of length 5; the permutation $(1\ 3\ 4)(2\ 5\ 8\ 6)$ has shape $(3,4)$. We say that permutations π and σ are **conjugate** if there exists some permutation τ such that $\sigma = \tau^{-1}\pi\tau$. Then it may be shown that two permutations have the same shape if and only if they are conjugate. This is proved for transpositions – permutations of shape (2) – below (see the proof of Theorem 4.2.9(iv)) but, since we do not need the general result, we simply refer the reader to [Ledermann, Proposition 21] for a proof of the general result, which is due to Cauchy.

Finally in this section, we consider the sign of a permutation. There are a number of (equivalent) ways to define this notion: here we take the following route.

Definition Let $n \geq 2$ be an integer. Define the polynomial $\Delta = \Delta(x_1, \ldots x_n)$ in the indeterminates $x_1, \ldots, x_n$ to be

$$\Delta(x_1, \ldots, x_n) = \Pi\{(x_i - x_j): i, j \in \{1, \ldots, n\}, i < j\}$$

the product of all terms of the form $(x_i - x_j)$ where $i < j$. For instance:

$$\Delta(x_1, x_2) = (x_1 - x_2);$$
$$\Delta(x_1, x_2, x_3) = (x_1 - x_2)(x_1 - x_3)(x_2 - x_3);$$
$$\Delta(x_1, x_2, x_3, x_4) = (x_1 - x_2)(x_1 - x_3)(x_1 - x_4)(x_2 - x_3)(x_2 - x_4)(x_3 - x_4).$$

One may, of course, multiply out the terms but there is no need to do so: it will be most convenient to handle such polynomials in this factorised form.

Now let $n \geq 2$ and let $\pi \in S(n)$. We define a new polynomial, denoted $\pi\Delta$, from $\Delta = \Delta(x_1, \ldots, x_n)$ and π by the following rule: wherever Δ has a factor $x_i - x_j$, $\pi\Delta$ has the factor $x_{\pi(i)} - x_{\pi(j)}$. It is important to observe that $\pi\Delta$ is as Δ but with x_i replaced throughout by $x_{\pi(i)}$ for each i. More formally, we define $\pi\Delta$ by

$$\pi\Delta(x_1, \ldots, x_n) = \Pi\{(x_{\pi(i)} - x_{\pi(j)}): i, j \in \{1, \ldots, n\}, i < j\}.$$

For example, suppose that $n = 3$ and that π is the transposition $(2\ 3)$. Then $\pi\Delta$ is obtained from Δ by replacing x_2 by x_3 and x_3 by x_2:

$$\pi\Delta(x_1, x_2, x_3) = (x_1 - x_3)(x_1 - x_2)(x_3 - x_2).$$

Now we note that this is just $-\Delta(x_1, x_2, x_3)$. To see this, just interchange the first two factors of $\pi \Delta$ and also write $(x_3 - x_2)$ as $-(x_2 - x_3)$.

The general case is similar: the only effect of applying a permutation to Δ in this way is to interchange the order of the factors and to replace some factors $x_i - x_j$ by $x_j - x_i$. We state this as our next result.

Lemma 4.2.7 *Let $\pi \in S(n)$ and let $\Delta(x_1, \ldots, x_n)$ be the polynomial as defined above. Then either $\pi \Delta = \Delta$ or $\pi \Delta = -\Delta$.*

Proof Consider a single factor $x_i - x_j$ of Δ. Since π is a bijection there are unique values k and l in $\{1, \ldots, n\}$ such that $\pi(k) = i$ and $\pi(l) = j$: also $k \neq l$ since $i \neq j$. There are two possibilities.

If $k < l$ then the factor $x_k - x_l$ occurs in Δ, and it is transformed in $\pi \Delta$ into $x_i - x_j$.

If $k > l$ then the factor $x_l - x_k$ occurs in Δ, and it is transformed in $\pi \Delta$ into $x_j - x_i = -(x_i - x_j)$.

Thus for every factor $x_i - x_j$ of Δ, either it or minus it occurs as a factor of $\pi \Delta$. Clearly (by the construction of $\pi \Delta$) Δ and $\pi \Delta$ have the same number of factors. It follows therefore (on collecting all the minus signs together) that $\pi \Delta$ is either Δ or $-\Delta$. □

We advise you to work through the above proof with some particular example(s) of π is $S(3)$ and $S(4)$.

Definition Let $\pi \in S(n)$. Define the **sign** of π, $\mathrm{sgn}(\pi)$, to be 1 or -1 according as $\pi \Delta = \Delta$ or $-\Delta$. Thus $\pi \Delta = \mathrm{sgn}(\pi) \cdot \Delta$. If $\mathrm{sgn}(\pi)$ is 1 then π is said to be an **even** permutation: if $\mathrm{sgn}(\pi) = -1$ then π is an **odd** permutation.

Theorem 4.2.8 *Let $\pi, \sigma \in S(n)$. Then $\mathrm{sgn}(\sigma \pi) = \mathrm{sgn}(\sigma) \cdot \mathrm{sgn}(\pi)$.*

Proof We compute, in two slightly different ways, the effect of applying the composite permutation $\sigma \pi$ to $\Delta = \Delta(x_1, \ldots, x_n)$. First we apply π to Δ, to get $\pi \Delta$: the effect is to replace, for each i, x_i by $x_{\pi(i)}$ throughout. Without rearranging, we immediately apply the permutation σ: this results in each $x_{\pi(i)}$ being replaced throughout by $x_{\sigma(\pi(i))} = x_{\sigma \pi(i)}$. The net result is that for each i, x_i has been replaced throughout by $x_{\sigma \pi(i)}$. So the resulting polynomial is, by definition, $(\sigma \pi)\Delta$ and, by definition, $(\sigma \pi)\Delta = \mathrm{sgn}(\sigma \pi) \cdot \Delta$.

Now we also have, by definition, that $\pi \Delta = \mathrm{sgn}(\pi) \cdot \Delta$. So, when we apply σ to $\pi \Delta$, we are just applying σ to $\mathrm{sgn}(\pi) \cdot \Delta$ (which is either Δ or $-\Delta$). The result of that is therefore equal to $\mathrm{sgn}(\pi) \cdot \sigma \Delta$, which equals $\mathrm{sgn}(\pi) \cdot \mathrm{sgn}(\sigma) \cdot \Delta$.

So the net result of applying $\sigma\pi$ to Δ may be expressed in two ways, as $\text{sgn}(\sigma\pi)\cdot\Delta$ and as $\text{sgn}(\pi)\text{sgn}(\sigma)\cdot\Delta$. Equating these expressions, we obtain that the polynomials $\text{sgn}(\sigma\pi)\cdot\Delta$ and $\text{sgn}(\pi)\cdot\text{sgn}(\sigma)\cdot\Delta$ are identical. Hence it must be that $\text{sgn}(\sigma\pi)=\text{sgn}(\pi)\cdot\text{sgn}(\sigma)$, as required. $\square$

You may observe that what we are using in the proof above is an 'action' of the symmetric group $S(n)$ on the set of polynomials $p(x_1,\ldots,x_n)$ in the variables $x_1,\ldots,x_n$. Given $\pi\in S(n)$ and $p(x_1,\ldots,x_n)$, we define the polynomial πp to be as p but with each x_i replaced by $x_{\pi(i)}$. What we used was, in essence, that if $\pi,\sigma\in S(n)$ then $(\sigma\pi)p=\sigma(\pi(p))$.

There are other routes to defining the sign of a permutation (see, for example, [Fraleigh, Chapter 5] and [MacLane and Birkhoff, Chapter III, Section 6]).

Theorem 4.2.9 *Let π and σ be in $S(n)$. Then*

(i) $\text{sgn}(\text{id})=1$,
(ii) $\text{sgn}(\pi)=\text{sgn}(\pi^{-1})$,
(iii) $\text{sgn}(\pi^{-1}\sigma\pi)=\text{sgn}(\sigma)$,
(iv) *if τ is a transposition then $\text{sgn}(\tau)=-1$.*

Proof (i) This is immediate from the definition of sign.
(ii) By Theorem 4.2.8 we have

$$\text{sgn}(\pi^{-1})\text{sgn}(\pi)=\text{sgn}(\pi^{-1}\pi)=\text{sgn}(\text{id})=1$$

using (i). So either both π^{-1} and π are even or both are odd, as required.
(iii) This is immediate by Theorem 4.2.8 and (ii).
(iv) The proof proceeds by showing this for increasingly more general transpositions.

First notice that the result is obviously true for $\tau=(1\ 2)$ since the only factor of Δ whose sign is changed by interchanging 1 and 2 is x_1-x_2.

Secondly, note that any transposition involving '1' is a conjugate of $(1\ 2)$:

$$(1\ k)=(2\ k)(1\ 2)(2\ k)=(2\ k)^{-1}(1\ 2)(2\ k).$$

So by (iii) $\text{sgn}(1\ k)=\text{sgn}(1\ 2)=-1$.

Finally we notice that every transposition is conjugate to one involving '1':

$$(m\ k)=(1\ k)(1\ m)(1\ k)=(1\ k)^{-1}(1\ m)(1\ k).$$

So, by another application of (iii), we obtain $\text{sgn}(m\ k)=-1$, as required. $\square$

Note, by the way, that we have illustrated the remark after 4.2.6 by showing that every two transpositions are conjugate.

Example   For any positive integer n, let $A(n)$ denote the set of all even permutations (permutatations with sign $+1$) in $S(n)$. We notice, using Theorems 4.2.8 and 4.2.9, that the product of any two elements of $A(n)$ is in $A(n)$, that id is in $A(n)$ and that the inverse of any element of $A(n)$ is in $A(n)$. Also, provided $n \geq 2$, $(1\ 2)$ is in $S(n)$ but is not in $A(n)$, and so not every permutation is even.

Since $(1\ 2)$ is odd, multiplying an even permutation by $(1\ 2)$ gives an odd permutation, and multiplying an odd permutation by $(1\ 2)$ gives an even permutation. The map f from the set of even permutations to the set of odd permutations defined by $f(\pi) = (1\ 2)\pi$ is a bijection, so it follows that half the elements of $S(n)$ are even and the other half are odd. Hence $A(n)$ has $n!/2$ elements. You can think of the map f more concretely by imagining the elements of $A(n)$ written out in a row; then, beneath each such element π, write its image $(1\ 2)\pi$. It is easy to show that the second row contains no repetitions and contains all odd permutations, so it is clear that $A(n)$ contains exactly half of the $n!$ elements of $S(n)$.

We finish the section by showing that every permutation may be written (in many ways) as a product of transpositions (not disjoint in general of course!). It will follow by Theorems 4.2.8 and 4.2.9 that a permutation is even or odd according as the number of transpositions in such a product is even or odd (hence the terminology).

$$\text{sgn}\,\pi = (-1)^{\text{length } n - 1}$$

Theorem 4.2.10   *Every cycle is a product of transpositions. If π is a cycle then* $\text{sgn}(\pi) = (-1)^{\text{length}(\pi)-1}$.

Proof   To see that a cycle $(x_1\ x_2 \ldots x_k)$ can be written as a product of transpositions, we just check:

$$(x_1\ x_2 \ldots x_k) = (x_1\ x_k) \ldots (x_1\ x_3)(x_1\ x_2).$$

There are $k - 1 = \text{length}(\pi) - 1$ terms on the right-hand sign each with sign -1, by Theorem 4.2.9(iv). By Theorem 4.2.8 it follows that

$$\text{sgn}(\pi) = \text{sgn}((x_1\ x_k) \cdots (x_1\ x_3)(x_1\ x_2)) = (-1)^{\text{length}(\pi)-1}. \quad \square$$

Next we extend this result to arbitrary permutations.

Theorem 4.2.11 *Suppose $n \geq 2$. Every permutation in $S(n)$ is a product of transpositions. Although there are many ways of writing a given permutation π as a product of transpositions, the number of terms occurring will always be either even or odd according as π is even or odd.*

Proof It is immediate from Theorems 4.1.3 and 4.2.10 that every permutation may be written as a product of transpositions. Suppose that we write π as a product of transpositions. Then, by the multiplicative property of sign (Theorem 4.2.8) and Theorem 4.2.9(iv), we have that $\text{sgn}(\pi)$ is -1 raised to the number of terms in the decomposition. Thus the statement follows. $\square$

Exercises 4.2

1. Determine the order and sign of each of the following permutations:
   (i) $(1\ 2\ 3\ 4\ 5)(8\ 7\ 6)(10\ 11)$;
   (ii) $(1\ 3\ 5\ 7\ 9\ 11)(2\ 4\ 6\ 8\ 10)$;
   (iii) $(1\ 2)(3\ 4)(5\ 6\ 7\ 8)(9\ 10)$;
   (iv) $(1\ 2\ 3\ 4\ 5\ 6\ 7\ 8)(1\ 8\ 7\ 6\ 5\ 4\ 3\ 2)$.

2. Give an example of two cycles of lengths r and s respectively whose product does not have order $\text{lcm}(r,s)$.

3. Give an example of a permutation of order 2 which is not a transposition.

4. Show that if π and σ are permutations such that $(\pi\sigma)^2 = \pi^2\sigma^2$ then $\pi\sigma = \sigma\pi$.

5. Find permutations π, σ such that $(\pi\sigma)^2 \neq \pi^2\sigma^2$.

6. Compute the orders of the permutations

$$(2\ 1\ 4\ 6\ 3), (1\ 2)(3\ 4\ 5) \text{ and } (1\ 2)(3\ 4).$$

7. Compute the orders of the following products of *non-disjoint* cycles:

$$(1\ 2\ 3)(2\ 3\ 4); (1\ 2\ 3)(3\ 2\ 4); (1\ 2\ 3)(3\ 4\ 5).$$

8. Complete the proof of Theorem 4.2.1.

9. List the elements of $A(4)$ and give the order of each of them.

10. Show that every element of $S(n)$ $(n \geq 2)$ is a product of transpositions of the form $(k\ k+1)$.
    [Hint: $(k\ k+2) = (k\ k+1)(k+1\ k+2)(k\ k+1)$.]

11. What is the highest possible order of an element in
    (i) $S(8)$, (ii) $S(12)$, (iii) $S(15)$?
    You may be interested to learn that there is no formula known for the highest order of an element of $S(n)$.

Start

1	2	3	4
5	6	7	8
9	10	11	12
13	14	15	

End?

15	14	13	12
11	10	9	8
7	6	5	4
3	2	1	

Fig. 4.5

12. Refer back to Exercise 4.1.6 for notation and terminology. Compute the orders and signs of $s, t, c, cs,$ and scs. You should find that the order of $t = cs$ is 6.

    Suppose that someone shuffles the cards according to the interleaving s, having attempted to make the cut c but, in making the cut, failed to pick up the bottom card, so that the first permutation actually performed was

    $$x = \begin{pmatrix} 1 & 2 & 3 & 4 & 5 & 6 & 7 & 8 & 9 & 10 \\ 5 & 6 & 7 & 8 & 9 & 1 & 2 & 3 & 4 & 10 \end{pmatrix}.$$

    Believing that the composite permutation sc $(=t)$ has been made, and having read the part of this section on orders, this person repeats the shuffle sc five more times, is somewhat surprised to discover that the cards have not returned to their original order, but then continues to make sc shuffles, hoping that the cards will eventually return to their original order. Show that this will not happen.

    [Hint: use what you have learned about the sign of a permutation.]

13. A well known children's puzzle has 15 numbered pieces arranged inside a square as shown (Fig. 4.5). A move is made by sliding a piece into the empty position. Consider the empty position as occupied by the number 16, so that every move is a transposition involving 16. Show that the order of the pieces can never be reversed. [Hint: show that if a product of transpositions each involving the number 16 moves 16 to an even-numbered position on the 4×4 board then the number of transpositions must be even. Also consider the sign of the permutation which takes the 'start' board to the 'end?' board.]

4.3 Definition and examples of groups

We are now ready to abstract the properties which several of our structures share. We make the following general definition.

Definition   A **group** is a set G, together with an operation $*$, which satisfies the following properties:

(G1) for all elements g and h of G, $g * h$ is an element of G (closure);

(G2) for all elements, g, h and k of G,

$$(g * h) * k = g * (h * k) \quad \text{(associativity)};$$

ensures G is not empty

(G3) there exists an element e of G, called the **identity** (or **unit**) of G, such that for all g in G we have

$$e * g = g * e = g \quad \text{(existence of identity)};$$

(G4) for every g in G there exists an element g^{-1} called the **inverse** of g, such that

$$g * g^{-1} = g^{-1} * g = e \quad \text{(existence of inverse).} \quad comm.$$

Definition   The group $(G, *)$ is said to be **commutative** or **Abelian** (after Niels Henrik Abel (1802–29)) if the operation $*$ satisfies the commutative law, that is, if for all g and h in G we have $g * h = h * g$.

Normally we will use multiplicative notation instead of '$*$', and so write 'gh' instead of '$g * h$': for that reason some books use '1' instead of 'e' for the identity element of G. Occasionally (and especially if the group is Abelian) we will use additive notation, writing '$g + h$' instead of '$g * h$' and $-g$ for g^{-1}, in which case the symbol '0' is normally used for the identity element of G.

Note that the condition (G3) ensures that the set G is non-empty. Associativity means that $(gh)k = g(hk)$, and so we may write ghk without ambiguity. It follows, by induction, that this is true for any product of group elements: different bracketing does not change the value of the resulting element.

Use of the terms '*the* identity' and '*the* inverse' presupposes that the objects named are uniquely defined. We now justify this usage.

Theorem 4.3.1   *Let G be any group. Then there is just one element e of G satisfying the condition for being an identity of G. Also, for each element g in G*

If G is a group there must be an identity + inverse.

there is just one element g^{-1} in G satisfying the condition for being an inverse of g.

Proof Suppose that both e and f satisfy the condition for being an identity element of G. Then we have

$$f = ef = e:$$

the first equality holds since e acts as an identity, the second equality holds since f acts as an identity. So there is just one identity element.

Given g in G, suppose that both h and k satisfy the condition for being an inverse of g. Then we have

$$h = he = h(gk),$$

since k is an inverse for g. But, by associativity, this equals $(hg)k$ and then, since h is an inverse for g, this in turn equals $ek = k$. Thus $h = k$, and the inverse of g is indeed unique. $\square$

Let us now consider some examples of groups.

$$\left(\mathbb{Z}, + \right)$$
$$id = 0 \quad inv \text{ of } n = -n$$

4.3.1 Groups of numbers

Example 1 The integers $(\mathbb{Z}, +)$ with addition as the operation, form a group. The closure and associativity properties are part of the unwritten assumptions we have made about $\mathbb{Z}$. The identity element for addition is 0. The inverse of n is $-n$. This group has an infinite number of elements.

In contrast, the set of natural numbers $\mathbb{N}$ equipped with addition is not a group, since not all its elements have additive inverses (*within* the set $\mathbb{N}$).

Note also that the integers with multiplication as the operation do not form a group since, for instance, 2 does not have a multiplicative inverse within the set $\mathbb{Z}$.

Example 2 The integers modulo n, $(\mathbb{Z}_n, +)$ (i.e. the set of congruence classes modulo n), equipped with addition modulo n (i.e. addition of congruence classes) as the operation, form a group. The identity element is the congruence class $[0]_n$ of 0 modulo n. The inverse of $[k]_n$ is $[-k]_n$. This example was discussed in Chapter 1, and it is an example of a group with a finite number of elements.

Notice again that if multiplication is taken as the operation then the set of congruence classes modulo n is not a group since not all elements have inverses (for example, $[0]_n$ has no multiplicative inverse).

$$id = [0]_n \qquad [k]_n = [-k]_n$$

Handwritten annotations at top of page:

1.4.3 inverse only occures iff gcd(a,n)=1

$[a]_n$ inv. is $[r]$ Examples of groups

$ar + ns = 1$

Example 3   Consider $(G_n, \cdot)$, the set of invertible congruence classes modulo n under multiplication. This is a group. The identity element is $[1]_n$. The inverse of each element of G_n exists by definition of G_n (and is found as in Theorem 1.4.3). The number of elements in this group is $\phi(n)$ (the Euler ϕ-function was defined in Section 1.6).

$\phi(p^n) = p^n - p^{n-1}$

Example 4   Other familiar number systems – the real numbers $\mathbb{R}$, the complex numbers $\mathbb{C}$, the rational numbers $\mathbb{Q}$ – are groups under addition. In each case, in order to obtain a group under the operation of multiplication, we must remove zero from the set.

Example 5   An interesting example of a finite non-Abelian group associated with a number system is provided by the **quaternions** $\mathbb{H}$ discovered by William Rowan Hamilton (1805–65). These can be regarded as 'hyper-complex' numbers, being of the form $a1 + bi + cj + dk$ where a,b,c,d are real numbers. The product of any two of these numbers can be computed by using the following rules for multiplying i, j and k:

$$i^2 = j^2 = k^2 = -1, \quad ij = k, \quad ji = -k, \quad jk = i, \quad kj = -i, \quad ki = j, \quad ik = -j.$$

Let $\mathbb{H}_0 = \{\pm 1, \pm i, \pm j, \pm k\}$. Since $\mathbb{H}_0$ has only finitely many elements (eight), the closure under multiplication and associativity of multiplication are properties which can be checked (although the direct checking of associativity is very tedious). The set $\mathbb{H}_0$, under the multiplication defined above, is a group with 'multiplication table' as shown.

	1	−1	i	−i	j	−j	k	−k
1	1	−1	i	−i	j	−j	k	−k
−1	−1	1	−i	i	−j	j	−k	k
i	i	−i	−1	1	k	−k	−j	j
−i	−i	i	1	−1	−k	k	j	−j
j	j	−j	−k	k	−1	1	i	−i
−j	−j	j	k	−k	1	−1	−i	i
k	k	−k	j	−j	−i	i	−1	1
−k	−k	k	−j	j	i	−i	1	−1

We should perhaps point out the convention for forming a table, such as that above, which enables us to see the effect of an operation '$*$' on a set G. The table has the elements of the set G in some definite order as a heading row, and the elements, in the same order, as a leading column. The entry at the intersection of the row labelled by g and the column labelled by h is $g * h$.

We have by now encountered a number of situations in which tables of the above type have arisen (see Section 1.4 and the end of Section 4.1). One reason

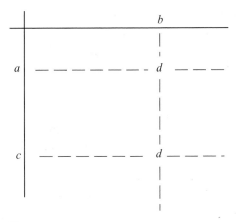

Fig. 4.6

for the use of such a table, which completely describes the operation, is that it helps one to determine whether or not the set under the operation is a group. The closure property of the set G under the operation is reduced to the question of whether or not every entry in the table is in the set G. The existence of an identity element can be determined by seeking an element of G such that the row labelled by that element is the same as the heading row of the table (and similarly for the columns). The inverse of an element g of G can be read off from the table by looking for the identity element in the row containing g and noting the heading for the column in which it occurs: the element heading that column is the inverse of g. For instance, in the example $\mathbb{H}_0$ above, to find the inverse of j we look along the row labelled 'j' until we find the identity element 1, then look to the head of that column: we conclude that $j^{-1} = -j$.

It is also possible to tell from its table whether or not G is Abelian: G will be Abelian exactly if the table is symmetric about its main diagonal (as in Example 6 below, but not as in Example 5 above). The only property which the table fails to give directly is associativity. In Example 5 above, in order to check this from the table, we would need to work out the products $(gh)k$ and $g(hk)$ for each of the eight choices for g, h and k: a total of 1024 calculations! Clearly some other method of checking associativity is usually sought (see Example 4 on p. 176).

When one is constructing such a group table it is useful to bear in mind that every row or column must contain each element of the group exactly once. For if one had an entry occurring twice in, say, the same column (as shown in Fig. 4.6) then one would have $ab = cb$. Multiplying on the right by b^{-1} would give the contradiction $a = c$.

In a paper of 1854, Cayley describes how to construct the group table. He also emphasises that each row and column contains each element exactly once.

Example 6   As an example of using a table to define a group, let G be the set $\{e, a, b, c\}$ with operation given by the table

	e	a	b	c
e	e	a	b	c
a	a	e	c	b
b	b	c	e	a
c	c	b	a	e

[handwritten: $ab=c$ $ba=c$]

Associativity can be checked case by case, and so one may verify that this is an Abelian group with 4 elements.

Example 7   The set, S, of all complex numbers (see the Appendix for these) of the form e^{ir} with r a real number, under multiplication ($e^{ir}e^{is} = e^{i(r+s)}$) is a group. The identity element is $e^{i0} = 1$, and the inverse of the element e^{ir} is $e^{i(-r)}$. This is an infinite Abelian group.

[handwritten: Perm aren't abelian $n \geq 3$.]

4.3.2   Groups of permutations

Example 1   We can now see Theorem 4.1.1 as saying that the set $S(n)$, of permutations of the set $\{1, 2, \ldots, n\}$, is a group under composition of functions. This is a group with $n!$ elements, and it is non-Abelian if $n \geq 3$. Notice that associativity follows by Theorem 2.2.1.

Example 2   We also saw in Section 4.2 (Example on p. 167) that the set $A(n)$ of even permutations is a group under composition of functions. This is a group with $n!/2$ elements (assuming $n \geq 2$), the **alternating group** on n elements. However, the set of all odd permutations in $S(n)$, with composition as the operation, fails to be a group since the product of two odd permutations is not odd, and so the set is not closed under the operation (if $n = 1$ it fails to be a group since it is empty!).

Example 3   Let G consist of the following four elements of $S(4)$:

$$G = \{id, (1\ 2)(3\ 4), (1\ 3)(2\ 4), (1\ 4)(2\ 3)\}.$$

Equipped with the usual product for permutations, the operation on this set is associative (since composition of any permutations is). To check the other group properties, it is easiest to work out the table.

$(12)(34)(13)(24)$

$(14)(23)$

	id	(12)(34)	(13)(24)	(14)(23)
id	id	(12)(34)	(13)(24)	(14)(23)
(12)(34)	(12)(34)	id	(14)(23)	(13)(24)
(13)(24)	(13)(24)	(14)(23)	id	(12)(34)
(14)(23)	(14)(23)	(13)(24)	(12)(34)	id

Notice that this is essentially the same table as that given in Example 6 in Section 4.3.1. For if we write e for id, a for (1 2)(3 4), b for (1 3)(2 4) and c for (1 4)(2 3), then we transform this table into that in the other example. This allows us to conclude that the operation in that example is indeed associative, because the tables for the two operations are identical (up to the relabelling mentioned above); so, since multiplication of permutations is associative, the other operation must also be associative. This is an example of a 'faithful permutation representation' of a group – where a set of permutations is found with the 'same' multiplication table as the group.

4.3.3   Groups of matrices

Example 1   Let GL(2,$\mathbb{R}$) be the set of all invertible 2×2 matrices with real entries, equipped with matrix multiplication as the operation (a matrix is said to be **invertible** if it has an inverse with respect to multiplication).

This operation is associative: even if you have not seen this proved before, you may verify it quite easily for 2×2 matrices: you need to check that.

$$\left(\begin{pmatrix} a & b \\ c & d \end{pmatrix} \begin{pmatrix} e & f \\ g & h \end{pmatrix} \right) \begin{pmatrix} i & j \\ k & l \end{pmatrix} = \begin{pmatrix} a & b \\ c & d \end{pmatrix} \left(\begin{pmatrix} e & f \\ g & h \end{pmatrix} \begin{pmatrix} i & j \\ k & l \end{pmatrix} \right).$$

The identity for the operation is the 2×2 identity matrix

$$\begin{pmatrix} 1 & 0 \\ 0 & 1 \end{pmatrix}$$

and the other conditions are easily seen to be satisfied. This example may be generalised by replacing '2' by 'n' so as to get the **general linear** group, GL(n, $\mathbb{R}$), of all invertible $n \times n$ matrices with real entries.

Example 2   Let G be the set of all upper-triangular 2×2 matrices with both diagonal elements non-zero. Equivalently (as you may check), G is the set of all invertible upper-triangular 2×2 matrices: those of the form

$$\begin{pmatrix} a & b \\ 0 & c \end{pmatrix}$$

where a, b and c are real numbers with both a and c non-zero. Equipped with the operation of matrix multiplication, this set is closed (you should check this), contains the identity matrix, and the inverse of any matrix in G is also in G, as you should verify.

Example 3   Let G be the set of all 2×2 invertible diagonal matrices with real entries: that is, matrices of the form

$$\begin{pmatrix} a & 0 \\ 0 & b \end{pmatrix}$$

where a and b are both non-zero. Then G is a group under matrix multiplication. The verification of this claim is left to the reader.

Example 4   Let X and Y be the matrices

$$X = \begin{pmatrix} 0 & -1 \\ 1 & 0 \end{pmatrix} \qquad Y = \begin{pmatrix} i & 0 \\ 0 & -i \end{pmatrix}$$

where i is a square root of -1. It can be seen that if $\mathbf{I}$ denotes the 2×2 identity matrix then

$$X^2 = Y^2 = -\mathbf{I} \quad \text{and} \quad XY = -YX.$$

Putting $Z = XY$, we deduce, or check, that

$$Z^2 = -\mathbf{I}, \quad YZ = X, \quad ZY = -X, \quad ZX = Y, \quad XZ = -Y.$$

It follows that the eight matrices $\mathbf{I}$, $-\mathbf{I}$, X, $-X$, Y, $-Y$, Z and $-Z$ have the same multiplication table as the quaternion group $\mathbb{H}_0$ (Example 5 on p. 172) since that table was constructed using essentially the same equations.

This is a matrix representation of $\mathbb{H}_0$, and it provides a nice proof of the fact that the operation on $\mathbb{H}_0$ is associative since matrix multiplication is associative. (Notice that the set of matrices of the form $a\mathbf{I} + bX + cY + dZ$ where a, b, c, d are real numbers gives a representation of the quaternions as 2×2 matrices with complex entries.)

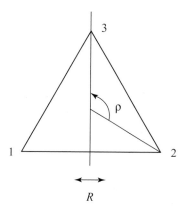

Fig. 4.7

4.3.4   Groups of symmetries of geometric figures

Now we turn to a rather different source of examples. Groups may arise in the form of groups of symmetries of geometric figures. By a symmetry of a geometric figure we mean an orthogonal affine transformation of the plane (or 3-space, if appropriate) which leaves the figure invariant. If the terms just used are unfamiliar, no matter: the meaning of 'symmetry' should become intuitively clear when you look at the following examples. We may say, roughly, that a symmetry of a geometric figure is a rigid movement of it which leaves it looking as it was before the movement was made.

Example 1   Consider an equilateral triangle such as that shown in Fig. 4.7. (The triangle itself is unlabelled, but we assign an arbitrary numbering to the vertices so as to be able to keep track of the movements made.)

There are a number of ways of 'picking up the triangle and then setting it down again' so that it looks the same as when we started, although the vertices may have been moved. In particular, we could rotate it anti-clockwise about its mid-point by an angle of $2\pi/3$: let us denote that operation (or 'symmetry') by ρ. Or we could reflect the triangle in the vertical line shown: let us denote that symmetry by R. Of course there are other symmetries, including that which leaves everything as it was (we denote that by e), but it will turn out that all the other symmetries may be described in terms of ρ and R.

We may define a group operation on this set of symmetries: if σ and τ are symmetries then define $\sigma\tau$ to be the symmetry 'do τ then apply σ to the result'. That this gives us a group is not difficult to see: we have just noted that closure holds; e is the identity element; the inverse of any symmetry is clearly a symmetry ('reverse the action of the symmetry'); and associativity follows because

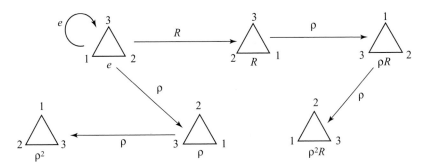

Fig. 4.8

symmetries are transformations (so functions). For the equilateral triangle there
are six different symmetries, namely e, ρ, ρ^2, R, ρR and $\rho^2 R$ (there can be at
most six symmetries since there are only six permutations of three vertices).
See Fig. 4.8. We note some 'relations': $\rho^3 = e$; $R^2 = e$; $\rho^2 R = R\rho$. You may
observe that there are others, but it turns out that they can all be derived from
the three we have written down.

This group is often denoted as $D(3)$ and we may write down its group table,
either by making use of the relations above or by calculating the effect of
each product of symmetries on the triangle with labelled vertices. For instance,
Fig. 4.9 gives us the relation $\rho^2 R = R\rho$. To compute, for example, the product
$\rho R \cdot \rho^2 R$ we may either compute the effect of this symmetry on the triangle,
using a sequence of pictures such as that in the figures, or we may use the
relations above as follows:

$$\rho R \cdot \rho^2 R = \rho \cdot R\rho \cdot \rho R = \rho \cdot \rho^2 R \cdot \rho R = \rho^3 \cdot R\rho \cdot R = \rho^3 \cdot \rho^2 R \cdot R$$
$$= e\rho^2 R^2 = \rho^2 e = \rho^2.$$

Note in particular that this group is not Abelian

	e	ρ	ρ^2	R	ρR	$\rho^2 R$
e	e	ρ	ρ^2	R	ρR	$\rho^2 R$
ρ	ρ	ρ^2	e	ρR	$\rho^2 R$	R
ρ^2	ρ^2	e	ρ	$\rho^2 R$	R	ρR
R	R	$\rho^2 R$	ρR	e	ρ^2	ρ
ρR	ρR	R	$\rho^2 R$	ρ	e	ρ^2
$\rho^2 R$	$\rho^2 R$	ρR	R	ρ^2	ρ	e

We may obtain a permutation representation of this group by replacing each
symmetry by the permutation of the three vertices which it induces. Thus ρ is
replaced by (1 2 3), ρ^2 by (1 3 2), R by (1 2), ρR by (1 3) and $\rho^2 R$ by (2 3).

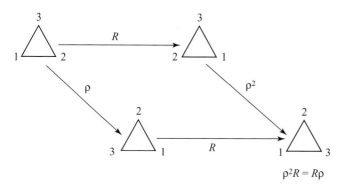

$$\rho^2 R = R\rho$$

Fig. 4.9

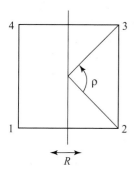

Fig. 4.10

With this relabelling, the above table becomes the table of the symmetric group $S(3)$ given in Section 4.1, p. 157.

Example 2 If we replace the triangle in Example 1 with a square, then we have a similar situation (Fig. 4.10). We take ρ to be rotation about the centre by $2\pi/4$ and R to be reflection in the perpendicular bisector of (say) the side joining vertices 1 and 2. This time we get a group with 8 elements (see Exercise 4.3.6), in which all the relations are consequences of the relations $\rho^4 = e$, $R^2 = e$, $\rho^3 R = R\rho$. This group is denoted by $D(4)$. (We can use the numbering of the vertices to obtain a faithful permutation representation of this group in $S(4)$.)

Examples 1 and 2 suggest a whole class of groups: the **dihedral group** $D(n)$ is the group of symmetries of a regular n-sided polygon. It has $2n$ elements and is generated by the rotation, ρ, anti-clockwise about the centre, by $2\pi/n$, together with the reflection, R, in the perpendicular bisector of any one of the sides, and these are subject to the relations $\rho^n = e$, $R^2 = e$, $\rho^{n-1} R = R\rho$.

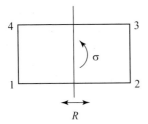

Fig. 4.11

Example 3 Let our geometric figure be a rectangle that is not a square (Fig. 4.11).

Then rotation by $2\pi/4$ is no longer a symmetry, although the rotation σ about the centre by $2\pi/2$ is. If, as before, we let R be reflection in the perpendicular bisector of the line joining vertices 1 and 2, then we see that the group has 4 elements e, σ, R, τ, where τ is σR and the table is as shown below.

	e	σ	R	τ
e	e	σ	R	τ
σ	σ	e	τ	R
R	R	τ	e	σ
τ	τ	R	σ	e

To conclude this section, we make some historical remarks.

The emergence of group theory is one of the most thoroughly investigated developments in the history of mathematics. In [Wussing] three rather different sources for this development are distinguished: solution of polynomial equations and symmetric functions; number theory; geometry.

The best known source is the study of exact solutions for polynomial equations. The solution of a general quadratic equation

$$ax^2 + bx + c = 0$$

was possibly known to the Babylonians and certainly was known to the early Greeks, although the lack of algebraic symbolism and their over-reliance on geometric interpretations meant that their solution seems over-complicated today. The solution was also known to the Chinese. The Greeks considered only positive real solutions. Brahmagupta (c. 628) was quite happy to deal with negative numbers as solutions of equations (and in general) but it would be another thousand years before such solutions were accepted in Europe.

The Arab mathematician al-Khwarizmi (c. 800) presented in his *Al-jabr wa'l muqābalah* (hence 'algebra') the systematic solution of quadratic equations, though he disallowed negative roots (and of course, complex roots). He also pointed out that a difficulty arises when b^2 is less than $4ac$: 'And know that, when in a question belonging to this case you have halved the number of roots and multiplied the half by itself, if the product be less than the number connected with the square, then the instance is impossible' (adapted from [Fauvel and Gray]).

The solution of the general cubic equation

$$ax^3 + bx^2 + cx + d = 0$$

was given by Cardano in his *Ars Magna* of 1545. He stated that the hint for the solution was given to him by Tartaglia. Scipione del Ferro had previously found the solution in some special cases.

Cardano's solution was split into a large number of cases because negative numbers were not then used as coefficients: for instance, we would think of $x^3 + 3x^2 + 5x + 2 = 0$ and $x^3 - 4x^2 - x + 1 = 0$ as both being instances of the general equation $ax^3 + bx^2 + cx + d = 0$ whereas, in those times, the second would necessarily have been presented as $x^3 + 1 = 4x^2 + x$. Still, Cardano had difficultly with the fact that the solutions need not always be real numbers. He did, however, have some inkling of the idea of 'imaginary' numbers: one of the problems he studied in *Ars Magna* is to divide 10 into two parts, the product of which is 40. The solutions of this problem are $5 + \sqrt{-15}$ and $5 - \sqrt{-15}$. Cardano says '...you will have to imagine $\sqrt{-15}, \ldots$ putting aside the mental tortures involved, a solution is obtained which is truly sophisticated'. The solution of the general quartic equation

$$ax^4 + bx^3 + cx^2 + dx + e = 0$$

(as we would write!) was found by Ferrari at Cardano's request and was also included in his *Ars Magna*.

We turn now to the general quintic equation

$$ax^5 + bx^4 + cx^3 + dx^2 + ex + f = 0.$$

Several attempts were made to find an algebraic formula which would give the roots in terms of the coefficients a, b, c, d, e, f. In fact, there can be no such general formula. Ruffini gave a proof for this, but his argument contained gaps which he was unable to fill to the satisfaction of other mathematicians, and the first generally accepted proof was given in 1824 by Abel (1802–29). Of course there still remained the problem of deciding whether any particular polynomial equation has a solution in 'radicals' (one expressed in terms of the coefficients, using addition, subtraction, multiplication, division and the extraction of nth roots). This problem was solved in 1832 by Galois (1811–32), and it is here that groups occur (some of the ideas already appeared in the work of Lagrange and Ruffini). The key idea of Galois was to associate a group to a given polynomial. The group is that of all permutations ('substitutions') of the roots of the polynomial which leave the polynomial invariant: the operation is just composition of permutations. Having translated the problem about equations into one about groups, Galois then solved the group theory problem and deduced the solution to the problem for polynomials. Galois was killed in a duel when he was 20, and one wonders what else he might have achieved had he lived.

It would be wrong to imply that Galois' work immediately revolutionised mathematics and produced a vast interest in 'group theory'. Both during his life, and initially after his death, the work of Galois went almost completely unappreciated. Partly this was due to a series of misfortunes which befell papers which he submitted for publication but also his work was very innovatory and difficult to understand at the time. Some of his main results were included in a letter written the night before the duel.

It should be emphasised that Galois did not understand the term 'group' in the (abstract, axiomatic) way in which we defined it at the beginning of this section. Galois used the term 'group' in a much more informal way and in a much more specific context – for Galois, groups were groups of substitutions. Indeed, although Cayley took the first steps towards defining an abstract group around 1850, his work was premature and groups were always groups of something, related to a definite context, until late in the century. It was Kronecker who, in 1870, first defined an abstract Abelian group. Then in 1882 von Dyck and Weber independently gave the definition of an abstract group.

In discussing the origins of group theory, one should also mention the work of Cauchy (1789–1857). In his work, groups occurred in a somewhat different way than they had in Galois'. Cauchy was interested in functions, $f(x_1, x_2, \ldots, x_n)$, of several variables, and in the permutations π in $S(n)$ which fix f (so that $f(x_1, x_2, \ldots, x_n) = f(x_{\pi(1)}, x_{\pi(2)}, \ldots, x_{\pi(n)})$). The set of permutations fixing a given function is a group under composition, and the group and the function f are closely related. Cauchy published an important paper in 1815 on groups and, between 1844 and 1846, developed and systematised the area significantly.

In 1846 Liouville published some of Galois' work. Serret lectured on it and gave a good exposition in his *Cours d'Algèbre supérieure* (1866). But it was only with the publication in 1870 of the book by Jordan – *Traité des substitutions et des équations algébriques* – that the subject came to the attention of a much wider audience.

There were other sources of groups. On the geometric side, Jordan had considered groups of transformations of a geometry. Number theory, of course, was another source, providing examples of the form $(\mathbb{Z}_n, +)$ and $(G_n, \cdot)$. Also groups (represented as linear transformations of a vector space) appeared in the work of Bravais on the possible structures of crystals.

In the late nineteenth century, the ideas of group theory began to pervade mathematics. A particularly notable development was the Erlanger Programme of the geometer Klein (1849–1925): the development of geometry (and geometries) in terms of the group of transformations which leave the particular geometry invariant.

Exercises 4.3

1. Decide which of the following sets are groups under the given operations:
   (i)   the set $\mathbb{Q}$ of rational numbers, under multiplication;
   (ii)   the set of non-zero complex numbers, under multiplication;
   (iii)   the non-zero integers, under multiplication;
   (iv)   the set of all functions from $\{1, 2, 3\}$ to itself, under composition of functions;
   (v)   the set of all real numbers of the form $a + b\sqrt{2}$ where a and b are integers, under addition;
   (vi)   the set of all 3×3 matrices of the form

$$\begin{pmatrix} 1 & a & b \\ 0 & 1 & c \\ 0 & 0 & 1 \end{pmatrix}$$

   where a, b, c are real numbers, under matrix multiplication;
   (vii)   the set of integers under subtraction;
   (viii) the set of real numbers under the operation $*$ defined by
   $$a * b = a + b + 2.$$

2. Let G be a group and let a, b be elements of G. Show that

$$(ab)^{-1} = b^{-1}a^{-1},$$

   and give an example of a group G with elements a, b for which

$$(ab)^{-1} \neq a^{-1}b^{-1}.$$

3. Let G be a group in which $a^2 = e$ for all elements a of G. Show that G is Abelian.

4. Let G be a group and let c be a fixed element of G. Define a new operation '$*$' on G by

$$a * b = ac^{-1}b.$$

   Prove that the set G is a group under $*$.

5. Let G be the set of all 3×3 matrices of the form

$$\begin{pmatrix} a & a & a \\ a & a & a \\ a & a & a \end{pmatrix}$$

where a is a non-zero real number. Find a matrix A in G such that, for all X in G, $AX = X = XA$. Prove that G is a group.

6. It was seen in Example 1 of Section 4.3.4 above that each possible permutation of the labels of the three vertices of an equilateral triangle is induced by a symmetry of the figure. On the other hand there are $4! = 24$ permutations of the labels of the vertices of a square, but there turn out to be only 8 symmetries of the square. Can you find a (short) argument which shows why only one-third of the permutations in $S(4)$ are obtained?

7. Write down the multiplication table for the group $D(4)$ of symmetries of a square.

8. Fill in the remainder of the following group table (the identity element does not necessarily head the first column). When you have done this, find all solutions of the equations:

$$\text{(i)}\, ax = b; \ \text{(ii)}\, xa = b; \ \text{(iii)}\, x^2 = c; \ \text{(iv)}\, x^3 = d.$$

	a	b	c	d	f	g
a	c			f		
b		f			c	
c	a					
d			c			
f						
g		d				c

[Hint: when you are constructing the table, remember that each group element appears exactly once in each row and column.]

9. Consult the literature to find out how to solve cubic equations 'by radicals'.

4.4 Algebraic structures

In the previous section, we saw that a group is defined to be a set together with an operation satisfying certain conditions, or **axioms**. These axioms were not chosen arbitrarily so as to generate some kind of intellectual game. The axioms were chosen to reflect properties common to a number of mathematical structures and we were able to present many examples of groups. In this section we consider briefly some of the other commonly arising algebraic structures.

Definition A **semigroup** is a set S, together with an operation $*$, which satisfies the following two properties (closure and associativity):

(S1) for all elements x and y of S, $x * y$ is in S; ➤ *associative*
(S2) for all elements x, y and z of S, we have $x * (y * z) = (x * y) * z$.

Since these are two of the group axioms, it follows that every group is a semigroup. But a semigroup need not have an identity element, nor need it have an inverse for each of its elements. *NO id or inverse needed.*

Example 1 The set of integers under multiplication, $(\mathbb{Z}, \cdot)$, is a semigroup. This semigroup has an identity element 1 but not every element has an inverse, so it is not a group.

Example 2 For any set X, the set, $F(X)$, of all functions from X to itself is a semigroup under composition of functions, since function composition is associative. For a specific example, take X to be the set $\{1,2\}$ (we considered this example in Section 2.2, but gave the functions different names there). There are four functions from X to X; the identity function e (so, $e(1) = 1$ and $e(2) = 2$), the function a with $a(1) = 2$ and $a(2) = 1$, the function b with $b(1) = b(2) = 1$ and the function c with $c(1) = c(2) = 2$. Here is the multiplication table for these functions under composition.

idempotent $x^2 = x$.

	e	a	b	c
e	e	a	b	c
a	a	e	c	b
b	b	b	b	b
c	c	c	c	c

You can see that this is not a group table: for example, b has no inverse. In fact, since $b^2 = b$, b is an example of an **idempotent** element: an element which satisfies the equation $x^2 = x$ (such elements figure prominently in Boole's *Laws of Thought* – part of the definition of a Boolean algebra, see below, is that every element is idempotent under '$\wedge$' and '$\vee$' – and the term was introduced by B. Peirce in 1870 in his *Linear Associative Algebras*). In a group G, if $g^2 = g$ then we can multiply each side by g^{-1} to obtain $g = e$; hence e is the only element in a group which is idempotent.

Example 3 As in Example 2, let X be a set and let $F(X)$ be the semigroup of all functions from X to itself, under composition. Let $f, g \in F(X)$. If f is a bijection hence, by Theorem 2.2.3, has an inverse, then, given an equation

$fg = fh$ we may compose with f^{-1} to get

$$f^{-1}(fg) = f^{-1}(fh);$$

hence

$$(f^{-1}f)g = (f^{-1}f)h;$$

that is

$$\mathrm{id}_X g = \mathrm{id}_X h \text{ and so } g = h.$$

But it is not necessary that f be invertible: in fact f is an injection if and only if whenever $fg = fh$ we must have $g = h$. For let $x \in X$. Then $fg = fh$ implies $(fg)(x) = (fh)(x)$. That is

$$f(g(x)) = f(h(x)).$$

So, since f is injective, it follows that $g(x) = h(x)$. This is true for every $x \in X$, so $g = h$.

Suppose, conversely, that $f \in F(X)$ is such that for all $g, h \in F(X)$ we have that $fg = fh$ implies $g = h$: we show that f is injective. For, if not, then there would be distinct $x_1, x_2 \in X$ such that $f(x_1) = f(x_2) = z$ (say). Take g to be the function on X that interchanges x_1 and x_2 and fixes all other elements: g is the permutation $(x_1 \ x_2)$. Take h to be the identity function on X. Then $fg = fh$, yet $g \neq h$ – contradiction.

So we have shown that f is injective if and only if $fg = fh$ implies $g = h$ for all $g, h \in F(X)$. Similarly f is surjective if and only if $gf = hf$ implies $g = h$ for all $g, h \in F(X)$ (you are asked to prove this as an exercise at the end of the section).

Example 4 A further class of examples of semigroups is provided by the finite state machines we discussed in Section 2.4. We illustrate this by determining the semigroup associated with the machine M which has states $S = \{0, 1, 2\}$ and input alphabet $\{a, b\}$ and whose state diagram is as shown in Fig. 4.12.

For any word w in $\{a,b\}$, we define a function $f_w \colon S \to S$ by

$f_w(0)$ is the state M would end up in, if it started in state 0 and read w,

$f_w(1)$ is the state M would end up in, if it started in state 1 and read w, and

$f_w(2)$ is the state M would end up in, if it started in state 2 and read w.

Since there are only a finite number (27) of possible functions from S to S, there are only a finite number of different functions f_w. These distinct functions are the elements of the semigroup of M. In our example, f_a is the identity

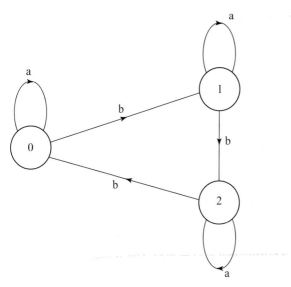

Fig. 4.12

function, taking each element of S to itself. The function f_b takes 0 to 1, 1 to 2 and 2 to 0. The function f_{bb} takes 0 to 2, 1 to 0 and 2 to 1. These are, in fact, the only distinct functions for M, so its semigroup has just three elements. The operation in the semigroup of M is that the 'product' of f_u and f_v is f_{uv}. We draw up the multiplication table for the semigroup, which is as shown.

	f_a	f_b	f_{bb}
f_a	f_a	f_b	f_{bb}
f_b	f_b	f_{bb}	f_a
f_{bb}	f_{bb}	f_a	f_b

(Note that, in our example, the final table is that of a group with three elements. That it is a group and not just a semigroup is just by chance.)

We now consider sets with two operations. These operations will be referred to as addition and multiplication although they need not be familiar 'additions' and 'multiplications': they need only satisfy the conditions listed below.

Definition A **ring** is a set R with two operations, called **addition** and **multiplication** and denoted in the usual way, satisfying the following properties:

add
comm
assoc
inv
identity
add is —
mult
product assoc
dist.

(R1)	for all x and y in R, $x + y$ is in R	closure under addition;
(R2)	for all x, y and z in R,	
	$x + (y + z) = (x + y) + z$	associativity of addition;
(R3)	there is an element, 0, in R such that for all x in R	
	$x + 0 = x = 0 + x$	existence of zero element;
(R4)	for every element x of R there is an element $-x$ in R such that	
	$x + (-x) = 0 = (-x) + x$	existence of negatives;
(R5)	for all x and y in R,	
	$x + y = y + x$	commutativity of addition;
(R6)	for all x, y in R, xy is in R	closure under multiplication;
(R7)	for all x, y and z in R,	
	$x(yz) = (xy)z$	associativity of multiplication;
(R8)	for all x, y and z in R,	
	$x(y + z) = xy + xz$, and	
	$(x + y)z = xz + yz$	distributivity.

The first five axioms say that R is an Abelian group under addition: axioms (R6) and (R7) say that R is a semigroup under multiplication. The eighth axiom is the one that says how the two operations are linked. The above list of axioms can therefore be summarised by saying that a ring is a set, equipped with operations called addition and multiplication, which is an additive Abelian group, is also a multiplicative semigroup, and in which multiplication distributes over addition.

Example 1   The set $\mathbb{Z}$ of integers with the usual addition and multiplication is a ring. Notice that this ring has an identity element, 1, with respect to multiplication, and also has commutative multiplication.

The set $2\mathbb{Z}$ of all even integers also is a ring, but it has no multiplicative identity: clearly 0 is not a multiplicative identity and if $n = 2m$ ($m \in \mathbb{Z}$) were an identity in this ring it would, in particular, be idempotent and so we would have $2m = (2m)^2 = 4m^2$ and hence $2m = 1$, contrary to m being an integer.

Example 2   The set $M_2(\mathbb{R})$ of 2×2 matrices with real coefficients is a ring under matrix addition and multiplication. This ring also has a multiplicative identity (the 2×2 identity matrix), but the multiplication is not commutative.

Many of the common examples of rings exhibit various significant special properties. Recall from Section 1.4 that an element x is a **zero-divisor** if x is not zero and if there is a non-zero element y with $xy = 0$. There are zero-divisors in Example 2 above: e.g.

$$\begin{pmatrix} 1 & 0 \\ 0 & 0 \end{pmatrix} \cdot \begin{pmatrix} 0 & 0 \\ 0 & 1 \end{pmatrix} = \begin{pmatrix} 0 & 0 \\ 0 & 0 \end{pmatrix}$$

Example 3   The set $\mathbb{Z}_n$ of congruence classes modulo n is a ring under the usual addition and multiplication. As we saw in Section 1.4 (Theorem 1.4.3 and Corollary 1.4.5), this set has zero-divisors unless n is a prime, in which case every non-zero element has a multiplicative inverse.

Example 4   Define $\mathbb{Z}[\sqrt{2}]$ to be the set of all real numbers of the form $a + b\sqrt{2}$ where a and b are integers. Then, equipped with the operations of addition and multiplication which are inherited from $\mathbb{R}$, this is a ring. Specifically, the operations are

$$(a + b\sqrt{2}) + (c + d\sqrt{2}) = (a + c) + (b + d)\sqrt{2},$$
$$(a + b\sqrt{2}) \times (c + d\sqrt{2}) = (ac + 2bd) + (ad + bc)\sqrt{2}.$$

We have just noted that the set is closed under the operations; clearly the set is closed under taking additive inverses (i.e. negatives); the other properties – associativity, distributivity, etc. – are inherited from $\mathbb{R}$ (they are true in $\mathbb{R}$ so certainly hold in the smaller subset $\mathbb{Z}[\sqrt{2}]$).

A similar example is $\mathbb{Z}[i]$ (where $i^2 = -1$): we obtain a subset of the ring $\mathbb{C}$ of complex numbers which is a ring in its own right.

We will consider rings of polynomials in Chapter 6. Also see Example 1 on p. 191.

Definition   A **field** is a set F equipped with two operations ('addition' and 'multiplication'), under which it is a commutative ring with identity element $1 \neq 0$ in which every non-zero element has a multiplicative inverse.

Example 1   For any prime p, the set $\mathbb{Z}_p$ is a field by Corollary 1.4.6.

Example 2   The sets $\mathbb{Q}$, $\mathbb{R}$ and $\mathbb{C}$ all are fields. The study of more general fields arose from the work of Galois (see below and the historical notes at the end of this section) and from that of Dedekind and Kronecker on number theory. The abstract study of fields was initiated by Weber (c. 1893) and Hensel and Steinitz made fundamental contributions at the beginning of the twentieth century.

Fields arose in Galois' work as follows. Given a polynomial $p(x)$ with rational coefficients, in the indeterminate x, the 'Fundamental Theorem of Algebra' says that it can be factorised completely into linear factors with complex coefficients. If we take the roots of this polynomial, we can adjoin them to the field $\mathbb{Q}$ of rational numbers and form the smallest extension field of $\mathbb{Q}$ containing them. The groups that Galois introduced are intimately connected with such extension fields of the rationals. (The connection is studied under the name 'Galois Theory'.)

For an example of this adjunction of roots, consider Example 3 below.

Example 3   The ring $\mathbb{Z}[\sqrt{2}]$ defined above is not a field but, if we define the somewhat larger set $\mathbb{Q}[\sqrt{2}]$ to be the set of all real numbers of the form $a + b\sqrt{2}$ where a and b are rational numbers, then we do obtain a field. The main point to be checked is that this set does contain a multiplicative inverse for each of its non-zero elements (checking the other axioms for a field is left as an exercise). So let $a + b\sqrt{2}$ be non-zero (thus at least one of a, b is non-zero and hence $a^2 - 2b^2 \neq 0$ since $\sqrt{2}$ is not rational). Then $(a + b\sqrt{2}) \times (c + d\sqrt{2}) = 1$ where $c = a/(a^2 - 2b^2)$ and $d = -b/(a^2 - 2b^2)$.

Observe, in connection with the comments in Example 2 above, that the polynomial $x^2 - 2$ factorises if we allow coefficients from the field $\mathbb{Q}[\sqrt{2}]$: $x^2 - 2 = (x - \sqrt{2})(x + \sqrt{2})$. But it does not factorise over $\mathbb{Q}$ since $\sqrt{2}$ is not a rational number. We may think of $\mathbb{Q}[\sqrt{2}]$ as having been obtained from $\mathbb{Q}$ by adjoining the roots $\sqrt{2}$ and $-\sqrt{2}$ then closing under addition, multiplication (and forming inverses of non-zero elements) so as to obtain the smallest field containing $\mathbb{Q}$ together with the roots of $x^2 - 2$.

As with groups, the axioms for our various algebraic systems have significant consequences. For example, the zero element in any ring is unique. It follows also (cf. proof of Corollary 1.4.5) that a field has no zero-divisors. We give an example of the way in which some of these consequences may be deduced.

Theorem 4.4.1   *Let R be any ring and let x be an element of R. Then*:

$$x0 = 0 = 0x.$$

Proof   Write y for $x0$. Then

$$y + y = x0 + x0 = x(0 + 0) \text{ using (R8)}$$
$$= x0 \qquad\qquad\qquad \text{using (R3)}$$
$$= y.$$

Thus $y + y = y$. Add $-y$ (which exists by condition (R4)) to each side of this equation, to obtain (using R2)) $y = 0$, as required. The proof for $0x$ is similar. □

One of the most commonly arising algebraic structures is composed of a field together with an Abelian group on which the field acts in a certain way: the Abelian group is then called a vector space over the field.

Definition   Given a field F, a **vector space** V over F is an additive Abelian group which also has a **scalar multiplication**. The scalar multiplication is an operation which takes any $\lambda \in F$ and $v \in V$ and gives an element, written λv, of V. The following axioms are to be satisfied:

(V1) for all v in V and λ in F, λv is an element of V;
(V2) for all v in V and λ, μ in F, $(\lambda \mu)v = \lambda(\mu v)$;
(V3) for all v in V, $1v = v$;
(V4) for all v in V and λ, μ in F, $(\lambda + \mu)v = \lambda v + \mu v$;
(V5) for all u, v in V and λ in F, $\lambda(u + v) = \lambda u + \lambda v$.

The elements of V are called **vectors** and the elements of the field F are called **scalars**. Vector spaces are usually studied in courses or books with titles such as 'Linear Algebra'.

Example   The most familiar examples of vector spaces occur when F is the field $\mathbb{R}$ of real numbers. Taking V to be $\mathbb{R}^2$, the set of ordered pairs (x, y) with x and y real numbers, addition and scalar multiplication are defined by

$$(x_1, y_1) + (x_2, y_2) = (x_1 + x_2, y_1 + y_2), \quad \text{and}$$
$$\lambda(x, y) = (\lambda x, \lambda y).$$

This makes the real plane $\mathbb{R}^2$ into a vector space.

We consider some more well known mathematical objects in the light of the structures we have introduced.

Example 1   Consider the set, $\mathbb{R}[x]$, of polynomials with real coefficients in the variable x. Clearly, we can add two polynomials by adding the coefficients of each power of x, and the result will be in $\mathbb{R}[x]$. For example

$$(x^2 + 3x + \pi) + (-5x^2 + 3) = -4x^2 + 3x + (\pi + 3).$$

The set is an additive Abelian group under this operation. We can also multiply two polynomials by collecting together powers of x. For example

$$(\sqrt{3} \cdot x - 1) \times (175x^2 + x + 1) = 175\sqrt{3} \cdot x^3 + (\sqrt{3} - 175)x^2 + (\sqrt{3} - 1)x - 1.$$

It is straightforward to check that $\mathbb{R}[x]$ is a ring under these operations.

The ring $\mathbb{R}[x]$ is not a field, since the polynomial $x + 1$ (for instance) does not have a multiplicative inverse. However $\mathbb{R}[x]$ is a commutative ring with identity that has no zero-divisors (a ring with these properties is called an **integral domain**).

The set $\mathbb{R}[x]$ has yet more structure: we can define a scalar multiplication by real numbers on the elements of $\mathbb{R}[x]$, by multiplying each coefficient of a given polynomial by a given scalar. For example

$$3\pi \cdot (x^2 + 3x) = 3\pi x^2 + 9\pi x.$$

Then $\mathbb{R}[x]$ becomes a vector space over $\mathbb{R}$. A ring that is at the same time a vector space (over a field K) under the same operation of addition, is known as a $(K\text{-})$**algebra**. Thus $\mathbb{R}[x]$ is an $\mathbb{R}$-algebra. (Rings of) polynomials will be discussed at length in Chapter 6.

Example 2   Given a prime number p, we consider the set $M_2(\mathbb{Z}_p)$ of 2×2 matrices whose entries are in $\mathbb{Z}_p$. Under the usual addition and multiplication of matrices, this is a ring with identity. However, this ring is not commutative and it does have zero-divisors. Again, we can define a scalar multiplication on $M_2(\mathbb{Z}_p)$ by setting, for $\lambda \in \mathbb{Z}_p$,

$$\lambda \begin{pmatrix} a & b \\ c & d \end{pmatrix} = \begin{pmatrix} \lambda a & \lambda b \\ \lambda c & \lambda d \end{pmatrix} \quad \text{(where we are computing modulo } p\text{)}.$$

This gives $M_2(\mathbb{Z}_p)$ the structure of a vector space over the field $\mathbb{Z}_p$ and so $M_2(\mathbb{Z}_p)$ is also an algebra (a $\mathbb{Z}_p$ - algebra).

Example 3   Consider the set C of 2×2 matrices of the form $a\mathbf{I} + b Y$ where a and b are real numbers and $\mathbf{I}$ and Y are respectively the matrices

$$\begin{pmatrix} 1 & 0 \\ 0 & 1 \end{pmatrix}, \begin{pmatrix} 0 & -1 \\ 1 & 0 \end{pmatrix}.$$

Equip this set with the usual matrix addition and multiplication. It is easily checked that this set is an $\mathbb{R}$-algebra.

In fact, we have just given one way of constructing the complex numbers. Regard $a\mathbf{I} + b Y$ as being '$a \cdot 1 + bi$'. You may check that $Y^2 = -\mathbf{I}$ and that these matrices add and multiply in the same way as expressions of the form $a + bi$ where $i^2 = -1$. The details of checking all this are left as an exercise for the reader.

Finally in this section, we consider one more type of structure. A **Boolean algebra** is a set B, together with operations '$\wedge$', '$\vee$' and '$\neg$' which will be called 'meet', 'join' and 'complement', such that for all $a, b \in B$ we have $a \wedge b$,

$a \vee b$ and $\neg a$ all in B, together with two distinguished elements, denoted 0 and 1, which are different ($0 \neq 1$). The axioms which must be satisfied are as follows.

Let a, b and c denote elements of B, then

$a \wedge a = a$ and	
$a \vee a = a$	idempotence,
$a \vee \neg a = 1$ and	
$a \wedge \neg a = 0$	complementation,
$a \wedge b = b \wedge a$ and	
$a \vee b = b \vee a$	commutativity,
$a \wedge (b \wedge c) = (a \wedge b) \wedge c$ and	
$a \vee (b \vee c) = (a \vee b) \vee c$	associativity,
$\neg(a \wedge b) = \neg a \vee \neg b$ and	
$\neg(a \vee b) = \neg a \wedge \neg b$	De Morgan laws,
$a \wedge (b \vee c) = (a \wedge b) \vee (a \wedge c)$ and	
$a \vee (b \wedge c) = (a \vee b) \wedge (a \vee c)$	distributivity,
$\neg\neg a = a$	double complement,
$a \wedge 1 = a$	property of 1,
$a \vee 0 = a$	property of 0.

We have already seen some examples of Boolean algebras.

Example 1 Let U be any set. Let B be the set of all subsets of U. Equip B with the operations of intersection, union and complement for its Boolean meet, join and complement. Let U play the role of 1 and the empty set that of 0. Now consult Theorem 2.1.1. It may be seen that the requirements that B must satisfy in order to be a Boolean algebra are precisely the first 13 properties mentioned in the theorem, together with one property of the universal set and one of the empty set. The theorem goes on to mention another property of the universal set, one of the empty set and two absorption laws. These are not required in our 'abstract' definition since they follow from the other laws. Indeed, our list of requirements above is itself redundant: a shorter list of axioms is implicit in Exercise 4.4.16.

It also follows from Theorem 3.1.1 that logical equivalence classes of propositional terms form Boolean algebras.

The period from the mid-nineteenth century to the early twentieth century saw the spectacular rise of abstract algebra. In that period of about a hundred years, the meaning of 'algebra' to mathematicians was completely transformed.

Still, in the early nineteenth century 'algebra' meant the algebra of the integers, the rationals, the real numbers and the complex numbers (the last still having dubious status to many mathematicians of the time). Moreover, the rules of algebra, such as $(a + b) + c = a + (b + c)$ and $ab = ba$, were regarded as fixed and given. The suggestion that there might be 'algebraic systems' obeying different laws would have been almost unintelligible at the time.

Nevertheless, the by then well established use of symbolic notation (as opposed to the earlier ways of presenting arguments, which used far more words) could not help but impress on mathematicians that many elementary arguments involved nothing more than manipulating symbols according to certain rules, and that the rules could be extracted and stated as axioms. For instance, it was noted that such arguments performed with the real numbers in mind also yielded results which were true of the complex numbers. At the time, this was regarded as somewhat puzzling, whereas we would now say that it is because both are fields (of characteristic zero, see Exercise 4.4.11 at the end of the section).

Peacock separated out, to some extent, the abstract algebraic content of such manipulations. But the actual laws were those applicable to the real and the complex numbers: in particular, the commutativity of multiplication was seen as necessary. Gregory and De Morgan continued this work and De Morgan further separated manipulations with symbols from their possible interpretations in particular algebraic systems. Still, the axioms were essentially those for a commutative integral domain. Indeed, the position of many mathematicians was almost that these laws were in some sense universal and necessary axioms, their form deriving simply from the manipulation of symbols.

But this point of view became untenable after Hamilton's development of the quaternions. Hamilton developed the point, between 1829 and 1837, that the meaning of the '+' appearing in the expression of a complex number such as $2 + 3i$ is quite distinct from the meaning of the symbol '+' in an expression such as $2 + 5$: one could as well write (a, b) for the complex number $a + bi$ and give the rules for addition and multiplication in terms of these ordered pairs. Thus complex numbers were pairs of real numbers, or 'two-dimensional' numbers, with an appropriately defined addition and multiplication. Because complex numbers could be used to represent forces (for instance) in the plane, there was interest in finding 'three-dimensional' numbers which could be used to represent forces and the like in 3-space. In fact, Gauss had already considered this problem and had, around 1819, come up with an algebra (in which the multiplication was not commutative). But the algebra turned out to be unsuitable for the representation of forces and, as with much of his work, he did not publish it, so that it remained unknown until the publication of his diaries in the later part of the century.

Hamilton searched for many years for such 'three-dimensional' numbers. On the 16th of October 1843, Hamilton and his wife were walking into Dublin by the Royal Canal. Hamilton had been thinking over the problem of 'three-dimensional numbers' and was already quite close to the solution. Then, in a flash of inspiration, he saw precisely how the numbers had to multiply. (Hamilton later claimed that he scratched the formulae for the multiplication into the stone of Brougham Bridge.) Hamilton had abandoned two preconceptions: that the answer was a three-dimensional algebra, in fact four dimensions were needed, and that the multiplication would be commutative – you can see from the group table (Example 5 of Section 4.3.1) that the quaternions have a non-commutative multiplication. Actually, Grassmann in 1844 published somewhat related ideas but his work was couched in rather obscure terms and this lessened its immediate influence.

Despite Hamilton's hopes, the use of quaternions to represent forces and other physical quantities in 3-space was not generally adopted by physicists: a formalism (essentially vector analysis), due to Gibbs and based on Grassmann's ideas, was eventually preferred. But the effect on mathematics was profound. For this was an algebra with properties very different from the real and complex numbers.

Somewhat later, Boole's development of the algebras we now term Boolean algebras (see above) provided other examples of new types of algebraic systems. Also, the development of matrix algebra and, more particularly, its recognition as a type of algebraic system (by Cayley and B. and C.S. Peirce) provided the, probably more familiar, example of the algebra of $n \times n$ matrices (with, say, real coefficients).

The effect of all this was to free algebra from the presuppositions which had limited its domain of applicability. In the later part of the century many algebras and kinds of algebras were found, and the shift towards abstract algebra – defining algebras in terms of the conditions which they must satisfy rather than in terms of some particular structure – was well under way.

Exercises 4.4

1. Which of the following sets are semigroups under the given operations:
   (i)   the set $\mathbb{Z}$ under the operation $a * b = s$, where s is the smaller of a and b;
   (ii) the set of positive integers under the operation $a * b = d$ where d is the gcd of a and b;
   (iii) the set $P(X)$ of all subsets of a set X under the operation of intersection;
   (iv) the set $\mathbb{R}$ under the operation $x * y = x^2 + y^2$;
   (v)   the set $\mathbb{R}$ under the operation $x * y = (x + y)/2$.

2. Prove the characterisation of surjections stated at the end of Example 3 (on p. 186).

3. Which of the following are rings? For those which are, decide whether they have identity elements, whether they are commutative and whether they have zero-divisors.

   (i)   The set of all 3×3 upper-triangular matrices with real entries, under the usual matrix addition and multiplication.

   (ii)   The set of all upper-triangular real matrices with 1s along the diagonal (the form is shown), with the usual operations.

$$\begin{pmatrix} 1 & a & b \\ 0 & 1 & c \\ 0 & 0 & 1 \end{pmatrix}$$

   (iii) The set, $P(X)$, of all subsets of a set X, with intersection as the multiplication and with union as the addition.

   (iv)   The set, $P(X)$, of all subsets of a set X, with intersection as the multiplication and with symmetric difference as the addition: the **symmetric difference**, $X \triangle Y$, of two sets X and Y is defined to be $X \triangle Y = (X \backslash Y) \triangle (Y \backslash X)$ (see Exercise 2.1.4).

   (v)   The set of all integer multiples of 5, with the usual addition and multiplication.

   (vi)   The set, $[4]_{24}\mathbb{Z}_{24}$, of all multiples, $[4]_{24}[a]_{24}$, of elements of $\mathbb{Z}_{24}$ by $[4]_{24}$, equipped with the usual addition and multiplication of congruence classes.

   (vii) As (vi), but with $[8]_{24}\mathbb{Z}_{24}$ in place of $[4]_{24}\mathbb{Z}_{24}$.

   (viii) The set of all rational numbers of the form m/n with n odd, under the usual addition and multiplication.

4. Use the axioms for a ring to prove the following facts.
   For all x, y, z in a ring,

   (i) $x(-y) = -xy = -x(y)$,

   (ii) $-1 \cdot x = -x$,

   (iii) $(x - y)z = xz - yz$ (where $a - b$ means, as usual, $a + (-b)$).

5. Show that if the ring R has no zero-divisors, and if a, b and $c \neq 0$ are elements of R such that $ac = bc$, then $a = b$.

6. Show that if x is an idempotent element of the ring R (that is, $x^2 = x$) then $1 - x$ also is idempotent.

7. Find an example of a ring R and elements x, y in R such that $(x + y)^2 \neq x^2 + 2xy + y^2$.

8. Find an example of a ring R and non-zero elements x, y in R such that $(x + y)^2 = x^2 + y^2$.

9. Which of the following are vector spaces over the given field?

(i) The set of all 2×2 matrices over $\mathbb{R}$, with the usual addition, and multiplication given by

$$\lambda \begin{pmatrix} a & b \\ c & d \end{pmatrix} = \begin{pmatrix} \lambda a & \lambda b \\ \lambda c & \lambda d \end{pmatrix}.$$

(ii) The set of all 2×2 matrices over $\mathbb{R}$, with the usual addition, and multiplication given by

$$\lambda \begin{pmatrix} a & b \\ c & d \end{pmatrix} = \begin{pmatrix} \lambda a & b \\ c & \lambda d \end{pmatrix}.$$

(iii) The set of all 2×2 matrices over $\mathbb{R}$, with the usual addition, and multiplication given by

$$\lambda \begin{pmatrix} a & b \\ c & d \end{pmatrix} = \begin{pmatrix} \lambda^{-1} a & b \\ c & \lambda^{-1} d \end{pmatrix}$$

for $\lambda \neq 0$ and

$$0 \begin{pmatrix} a & b \\ c & d \end{pmatrix} = \begin{pmatrix} 0 & 0 \\ 0 & 0 \end{pmatrix}.$$

10. Deduce the following consequences of the axioms for a vector space. For any vectors x, y and scalar λ,

(i) $0 \cdot x = 0$ ('0' here is the scalar zero).

(ii) $\lambda \cdot 0 = 0$ (here, '0' means the zero vector),

(iii) $\lambda(x - y) = \lambda x - \lambda y$,

(iv) $-1 \cdot x = -x$.

11. Let F be any field. The **characteristic** of F is defined to be the least positive integer n such that $1 + 1 + \cdots + 1 = 0$ (n 1s). If there is no such n then the characteristic of F is said to be 0. Show that if the characteristic of F is not zero then it must be a prime number.

12. Suppose that R is an integral domain. Let X be the set of all pairs (r, s) of elements of R with $s \neq 0$. Define a relation $\sim$ on X by $(r, s) \sim (t, u)$ iff $ru = st$. Show that $\sim$ is an equivalence relation. Let Q be the set of all equivalence classes $[(r, s)]$ of $\sim$. Define an addition and multiplication on Q by $[(r, s)] + [(t, u)] = [(ru + st, su)]$ and $[(r, s)] \times [(t, u)] = [(rt, su)]$. Show that these operations are well defined (i.e. do not depend on the chosen representatives of the equivalence classes). Show that Q is a commutative ring under these operations. Show that every non-zero element of Q has an inverse and hence that Q is a field.

Define a function $f: R \to Q$ by sending $r \in R$ to the equivalence class of $(r, 1)$. Show that this is an injection, that $f(r + t) = f(r) + f(t)$, that $f(rt) = f(r) \times f(t)$.

Thus there is a copy of the ring R sitting inside the field Q. Finally, show that every element of Q has the form $f(r) \cdot f(s)^{-1}$ for some $r, s \in R$ with $s \neq 0$. That is, Q is essentially the field consisting of all fractions formed from elements of R: Q is called the **field of fractions** (or **quotient field**) of R. You can check that if the initial ring R is the ring $\mathbb{Z}$ of integers then Q can be thought of as the field $\mathbb{Q}$ of rational numbers; if we start with the integral domain $R = \mathbb{Z}[\sqrt{2}]$ then we end up with a copy of the field $\mathbb{Q}[\sqrt{2}]$. [Hint: think of the pair (r, s) as being a 'fraction' r/s.]

13. Let F be the following set of matrices with entries in $\mathbb{Z}_2$, under matrix addition and multiplication (we write '0' for $[0]_2$, '1' for $[1]_2$):

$$\begin{pmatrix} 0 & 0 \\ 0 & 0 \end{pmatrix}, \begin{pmatrix} 1 & 0 \\ 0 & 1 \end{pmatrix}, \begin{pmatrix} 1 & 1 \\ 1 & 0 \end{pmatrix}, \begin{pmatrix} 0 & 1 \\ 1 & 1 \end{pmatrix}.$$

Show that F is a field and is also a $\mathbb{Z}_2$-algebra, where the elements λ of $\mathbb{Z}_2$ act by

$$\lambda \begin{pmatrix} a & b \\ c & d \end{pmatrix} = \begin{pmatrix} \lambda a & \lambda b \\ \lambda c & \lambda d \end{pmatrix}.$$

You should check that this is a field of characteristic 2 in the sense of Exercise 4.4.11.

14. Combine Example 4 of Section 4.3.3 with Example 3 on p. 192 to realise the ring $\mathbb{H}$ of quaternions as an $\mathbb{R}$-algebra of 4×4 matrices with real entries.

15. Let B be a Boolean algebra. Show that
    (1) $a \wedge 0 = 0$,
    (2) $a \vee (a \wedge b) = a$,
    (3) $a \vee (\neg a \wedge b) = a \vee b$.

16. (a) Let $B = (B; \wedge, \vee, \neg, 0, 1)$ be a Boolean algebra. Define new operations, '+' and '$\cdot$', on the set B by

$$a \cdot b = a \wedge b,$$
$$a + b = (a \vee b) \wedge \neg(a \wedge b).$$

(Observe that, in the context of algebras of sets, $a + b$ is the symmetric difference of a and b, see Exercise 4.4.3 (iv) above.)
    (i) Show that, with these operations, B is a commutative ring.
    (ii) Identify the zero element of this ring and the identity element.

:mpotent and that for every $a \in B$

ımutative ring in which every
$a = 0$ (such a ring is termed a
~~~~~~~ ~~~~~. ~~~~~~ ~~~~~~~~~~~ $\wedge$' and '$\vee$' on R by

$$a \wedge b = a \cdot b$$
$$a \vee b = a + b + a \cdot b$$

Show that $(R, \wedge, \vee, 0, 1)$ is a Boolean algebra.

Show also that if we start with a Boolean algebra, produce the ring as in (a), and then go back to a Boolean algebra as in (b), then we recover the Boolean algebra with which we began. Similarly the process (b) applied to a Boolean ring, followed by (a), recovers the original ring.

Thus one may say that Boolean algebras and Boolean rings are equivalent concepts.

17. Suppose that $(B, \wedge, \vee, \neg)$ is a Boolean algebra. Define a relation $\leq$ on B by $a \leq b$ iff $a = a \wedge b$. Show that $a = a \wedge b$ iff $b = a \vee b$. Prove that $\leq$ is a partial order on the set B and that $0 \leq a \leq 1$ for all $a \in B$. What is the relation $\leq$ in the case that B is a Boolean algebra $(B, \cap, \cup, {}^c)$ of sets?

In fact (you may try to show this as a further exercise) the Boolean operations $\wedge$, $\vee$ and $\neg$ may be defined in terms of this partial order and hence Boolean algebras may be regarded as certain kinds of partially ordered sets.

Summary of Chapter 4

In Section 4.1 we discussed permutations, including their cycle decomposition. The order and sign of a permutation were considered in the next section. In Section 4.3 we introduced the definition of a group and gave many examples of this concept. In the fourth section we gave a brief discussion of various other algebraic structures which appear in this book.

5 Group theory and error-correcting codes

By now we have met many examples of groups. In this chapter, we begin by considering the elementary abstract theory of groups. In the first section we develop the most immediate consequences of the definition of a group and introduce a number of basic concepts, in particular, the notion of a subgroup. Our definitions and proofs are abstract, but are supported by many illustrative examples. The main result in this chapter is Lagrange's Theorem, which is established in Section 5.2. This theorem says that the number of elements in a subgroup of a finite group divides the number of elements in the whole group. The result has many consequences and provides another proof of the theorems of Fermat and Euler which we proved in Chapter 1. In the third section we define what it means for two groups to be isomorphic: to have the same abstract form. Then, after describing a way of building new groups from old, we move on to describe, up to isomorphism, all groups with up to eight elements. The final section of the chapter gives an application of some of the ideas we have developed, to error-detecting and error-correcting codes.

5.1 Preliminaries

We introduced the idea of an abstract group in Section 4.3 and then gave many examples of groups. In this section we will prove a number of results which hold true for all these examples. For instance, we do not, every time we wish to refer to the inverse of an element of a group, prove that the inverse of that element is unique. Rather, we proved once and for all that if a is an element of a group then there is a unique inverse for a (Theorem 4.3.1). Then the result applies to any particular element in any particular group. This is one of the main advantages of working in the abstract: we may deduce a result once and for all without having to prove it again and again in particular cases. In this section, we will consider some elementary deductions from the definition of a group. Then the central concept of a subgroup is introduced.

We start by making some observations which may, at first sight, seem of little significance.

We are going to use the four group axioms to make further deductions. In doing this, we will only employ the usual rules of reasoning together with our four axioms. However, we need to be clear as to what the rules of reasoning say in this context. Suppose we have an equation such as $a = b$ for certain expressions, a and b in a group G. A correct deduction from this would be achieved by performing the same operation to both sides of our equation $a = b$. Care is needed when we carry this out. If, for example, we multiply both sides by a group element x, we must either multiply both sides by x on the right (to obtain $ax = bx$), or by x on the left (to obtain $xa = xb$). It would be incorrect to deduce that $ax = xb$. The underlying reason for this is that we cannot assume that our group is Abelian.

Example Suppose that in a given group, G, we know that the given elements g and h commute, so $hg = gh$. Then we can multiply both sides of this equation on the right by g^{-1} to obtain $g^{-1}hg = g^{-1}gh = (g^{-1}g)h = eh = h$ (we have made use of group axioms (G2), (G3) and (G4) from p. 170 in these equations). In fact this argument could be reversed: if $g^{-1}hg = h$ then $hg = gh$ (multiply by g on the right). A further deduction from $h = g^{-1}hg$ could be made by multiplying on the right again by h^{-1} to obtain $e = h^{-1}h = h^{-1}g^{-1}hg$ with this last argument again being reversible. Now work from the left by multiplying by g^{-1}, to obtain (after simplification) $g^{-1} = h^{-1}g^{-1}h$. Once again this last argument could be reversed. As a final step, multiply through by h^{-1} to obtain $g^{-1}h^{-1} = h^{-1}g^{-1}$ and we have shown that $hg = gh$ if and only if $h^{-1}g^{-1} = g^{-1}h^{-1}$. That is, two elements commute if and only if their inverses commute.

Our first result, Theorem 5.1.1 below, concerns solvability of certain equations in groups. Since it is our first really abstract result, it may be appropriate (in the spirit of Chapter 1) to give a more detailed commentary on both its statement and its proof. For the statement, note that this is a result about an arbitrary group. Since we are now operating at a higher level of abstraction than in Chapter 1, there are now two ways in which we can illustrate the statement of Theorem 5.1.1. In the first place, we could apply the statement in the context where we have made a particular choice for the group G mentioned in the statement. Alternatively, we could continue to keep our group G as a completely general one, but choose specific values for the elements a or b (such as $a = e$ or $b = a$). In the second sentence, we are concerned with solutions of equations like $a = bx$. Two points are worth making here. One is a point easily missed in a first reading: we say x is an element of G! Although it may seem clear that if $ax = b$, then (multiplying on the left by a^{-1} and simplifying), we can find a value for x, we must show that x is an element of our group G. The second point is that we claim this solution is unique: we do not just need to find some solution, we must show that it is the only one. Note also that we discuss solutions to two equations $a = bx$ and $a = yb$. We use different letters x and y to emphasise that in a general group the solutions of these two equations could well be different elements of G. After the proof, we will state a corollary before making some comments about the method of proof.

Theorem 5.1.1 *Let G be a group and let a and b be elements of G. Then there are unique elements x and y in G such that $a = bx$ and $a = yb$.*

Proof   We first consider the equation $a = bx$ and show that this equation does indeed have a solution. Then we show that there is only one solution. To see that a solution exists, we take x to be $b^{-1}a$. Since $b^{-1} \in G$ and G is closed under products this is an element of G. Then

$$bx = b(b^{-1}a)$$
$$= (bb^{-1})a \text{ by associativity (G2) in the definition of a group,}$$
$$= ea \text{ by existence of inverse (G4),}$$
$$= a \text{ by existence of identity (G3),}$$

so a solution exists. If c and d both are solutions of $a = bx$, so $a = bc = bd$, then multiply both sides of the equation $bc = bd$ on the left by the inverse of b to obtain

$$b^{-1}(bc) = b^{-1}(bd) \text{ hence}$$
$$(b^{-1}b)c = (b^{-1}b)d \text{ by associativity, and so}$$
$$ec = ed \text{ by (G4), giving}$$
$$c = d$$

as required. The proof for the equation $a = yb$ is similar and is left as an exercise for the reader. $\square$

Remark   The theorem allows us to 'cancel' in a group, provided we do this 'on the same side'. If g, h and b are elements of a group G and $bg = bh$, then g and h must be equal. Similarly, if $gb = hb$ we can deduce $g = h$. This is why no element of a group occurs twice in any row (or column) of a group multiplication table (as we saw in Section 4.3).

[handwritten: $X\,abad = ca.$ $xabb^{-1}= cab^{-1}$ $xa=cab^{-1}$ $X=cab^{-1}a^{-1}$]

Example 1 Let a, b, c be elements of a group G. Find a group element x such that $xaba^{-1} = c$.

To do this, remember to multiply consistently (on the right in this case) and also take the argument a step at a time. First multiply (on the right) by a to obtain

$$ca = (xaba^{-1})a = ((xab)(a^{-1}))a = (xab)(a^{-1}a) = xabe = xab.$$

Now multiply by b^{-1} on the right to obtain (after simplification) $xa = cab^{-1}$. Finally, multiply by a^{-1} on the right to obtain our solution $x = cab^{-1}a^{-1}$. Notice that there is no way to simplify the expression $cab^{-1}a^{-1}$ in general.

We can also solve equations using 'mixed' (right and left) terms but again care is required.

[handwritten: $baxb \neq b^{-1}c$ $baxb =c$]

Example 2 Let a, b, c be elements of a group G. Find a group element x such that $axb = b^{-1}c$.

Multiply by b^{-1} on the right to obtain $ax = b^{-1}cb$. Now multiply by a^{-1} on the left to get that $x = a^{-1}b^{-1}cb$.

Corollary 5.1.2 *Let G be a group. Then the identity element of G is unique, inverses are unique, $(a^{-1})^{-1}$ is a and $(ab)^{-1}$ is $b^{-1}a^{-1}$.*

[handwritten: $\neg$ solving what $=\ell$.]

Proof The first two parts have already been established in Theorem 4.3.1 (they may also be viewed as consequences of Theorem 5.1.1: take $a = b$ to deduce that the identity is unique, and take $a = e$ to deduce that inverses are unique). The fact that $(a^{-1})^{-1} = a$ follows since $(a^{-1})^{-1}$ and a both are solutions of $a^{-1}x = e$. Also $(ab)^{-1} = b^{-1}a^{-1}$ since both solve the equation $(ab)x = e$. □

[handwritten: solve for X.]

Comment on the proof of Theorem 5.1.1 As we have already noted, there are two ways to specialise a proof in order to come to terms with its abstractness. We could see what the proof would say in a particular circumstance we already know. Thus, in Theorem, 5.1.1, we could choose $a = b$, and then see why (as indicated in the Corollary 5.1.2) the proof provides us with a proof that inverses are unique in a group. Alternatively, we could focus on a specific group we feel familiar with, and try out the detailed steps of the argument in the context of that group.

Once we start on the details of the proof, note that most lines of the proof are, not just an equation, but include some comment on how the equation was obtained. These steps are all either uses of elementary logic (such as performing the same operation to both sides of a known equation, or substituting a known equality into a given equation), or are justified by using one of the four group axioms. It is possible that the reader might feel quite confident about the steps making up the proof, but wonder how it is possible to prove something for oneself without a model to imitate. The first requirement is really to understand every detail of an argument like this. Then try to see the proof as a whole and try to understand how the informal idea(s) behind the proof can, step-by-step, be transformed into a rigorous, formal proof.

We have already met the idea of taking powers of elements in several contexts, so the following definition should come as no surprise.

Definition   Let g be an element of a group G. The positive **powers** of g are defined inductively by setting $g^1 = g$ and $g^{k+1} = gg^k$. We can also define zero and negative powers by putting $g^0 = e$ and $g^{-k} = (g^{-1})^k$ for $k > 0$.

Note that this definition implies, for example that g^2 means $g \cdot g$. This may again seem a trivial point, but if g is itself a product of two other group elements, say $g = xy$, then g^2 will mean $(xy)(xy)$. This is not the same as $x^2 y^2$ in general.

The next result gives the index laws for group elements.

Theorem 5.1.3   *Let G be a group and let g and h be elements of G. For any integers r and s we have*

(i) $g^r g^s = g^{r+s}$,
(ii) $(g^r)^s = g^{rs}$,
(iii) $g^{-r} = (g^r)^{-1} = (g^{-1})^r$, *and*
(iv) *if* $gh = hg$, *then* $(gh)^r = g^r h^r$.

Proof   (iv) For non-negative integers, the proof of part (iv) is just like the proof of Theorem 4.2.1 (iv). If r is negative, say $r = -k$ for some positive integer k, then we have

$$g^r h^r = (g^{-1})^k (h^{-1})^k \text{ by definition}$$
$$= (g^{-1}h^{-1})^k \text{ since } k \text{ is positive}$$
$$= ((hg)^{-1})^k \text{ by Corollary 5.1.2 and since, as we verified above,}$$
$$gh = hg \text{ implies } h^{-1}g^{-1} = g^{-1}h^{-1}$$
$$= ((gh)^{-1})^k \text{ by assumption}$$
$$= (gh)^{-k} \text{ by definition}$$
$$= (gh)^r \text{ as required.}$$

(iii) Apply part (iv) with $h = g^{-1}$ to get $e = e^r = (gg^{-1})^r = g^r(g^{-1})^r$ for every integer r. Therefore, $(g^r)^{-1} = (g^{-1})^r$ and the latter, by definition, is g^{-r}.

(i) The case where both r and s are non-negative is proved just as in 4.2.1 (i). So, in treating the other cases, we may suppose that at least one of r, s is negative. We split this into three further cases: (a) the case when $r + s > 0$, (b) the case when $r + s = 0$, and (c) the case when $r + s < 0$.

(a) If $r + s > 0$ then at least one of r, s must be strictly greater than 0: say $r > 0$ (the argument supposing that $s > 0$ is similar). So, by assumption, $s < 0$ and hence $-s > 0$. Then, by the case where both integers are positive we have

$$g^{r+s}g^{-s} = g^{r+s-s} = g^r$$

so, multiplying on the right by g^s, we obtain

$$g^{r+s} = g^r g^s,$$

as required.

$$g^{r+s-s} \quad g^r \quad g = g^{r+s-s} = g^r g^s$$

$$g^{r+s}.$$

(b) When $r + s = 0$, then $s = -r$ so $g^r g^s = g^r g^{-r} = e$ by part (iii). But, by definition, $e = g^0 = g^{r+s}$.

(c) When $r + s < 0$, then $-r + (-s) > 0$. So, by the case where both integers are positive, we have

$$g^{-(r+s)} = g^{-s+(-r)} = g^{-s}g^{-r}.$$

By part (iii), the inverse of $g^{-(r+s)}$ is g^{r+s}: by 5.1.2 and part (iii), the inverse of $g^{-s}g^{-r}$ is $g^r g^s$. So we conclude $g^{r+s} = g^r g^s$.

(ii) The proof of this part follows by induction from part (i). □

It is a consequence of the first part of the above result that if g is an element of a group G and r, s are integers then

$$g^r g^s = g^{r+s} = g^{s+r} = g^s g^r.$$

That is, the powers of an element g commute with each other.

Next we define the order of an element of a group, another idea which should be familiar from Chapters 1 and 4.

Definition An element g of a group G is said to have **infinite order** if there is no positive integer n for which $g^n = e$. Otherwise, the **order** of g is the smallest positive integer such that $g^n = e$.

$g^n = e$, $+$ n ord of group.

The following result is proved in precisely the same way as is Theorem 4.2.3.

$\pi' = \pi^s$ iff $r \equiv s \bmod n$

Theorem 5.1.4   *Let g be an element of a group G and suppose that g has finite order n. Then $g^r = g^s$ if and only if r is congruent to s modulo n.*

$g^r = g^s$ iff $r \equiv s \bmod n$

Example 1   The order of a permutation, as defined in Section 4.2, is, of course, a special case of the general definition above. As we saw in Section 4.2, the order of a permutation π in the group $S(n)$ may easily be calculated in terms of the expression of π as a product of disjoint cycles. In a general group G, however, there will be no easy way to predict the order of an element.

Example 2   If G is a finite group then every element must have finite order (the proof is just like that for Theorem 4.2.2). So, to find elements of infinite order we must go to infinite groups. Let $GL(2, \mathbb{R})$ be the group of invertible 2×2 matrices with real entries. The matrix

$$A = \begin{pmatrix} 1 & 1 \\ 0 & 1 \end{pmatrix}$$

has infinite order since (as may be proved by mathematical induction) A^n is the matrix

$$A^n = \begin{pmatrix} 1 & n \\ 0 & 1 \end{pmatrix}.$$

If, in this example, we were to replace the field $\mathbb{R}$ by the field $\mathbb{Z}_p$ for some prime p, so we would consider 2×2 invertible matrices with entries in $\mathbb{Z}_p$, then the matrix

$$A = \begin{pmatrix} [1]_p & [1]_p \\ [0]_p & [1]_p \end{pmatrix}$$

would have order p:

$$A^p = \begin{pmatrix} [1]_p & [p]_p \\ [0]_p & [1]_p \end{pmatrix} = \begin{pmatrix} [1]_p & [0]_p \\ [0]_p & [1]_p \end{pmatrix} = I.$$

We now come to one of the key ideas in elementary group theory.

Definition   A non-empty subset H of a group G (more precisely, of $(G, *)$) is a **subgroup** of G if H is itself a group under the same operation $(*)$ as that of G (or, more precisely, under the operation of G restricted to H).

In particular, it must be that a subgroup H of a group G contains the identity element e of G (for if $f \in H$ acts as an identity in H then, working in G, from $ff = f = ef$, we deduce $f = e$) and the inverse of any element of H lies in H and is just its inverse in G.

In order to check whether or not a given subset of G is a subgroup, it would appear that we need to check the four group axioms. However, the next result shows that it is sufficient to check rather less. After establishing this, we will consider some standard examples of subgroups.

Theorem 5.1.5 *The following conditions on a non-empty subset H of a group G are equivalent.*

(i) *H is a subgroup of G.*
(ii) *H satisfies the following two conditions:*
    (a) *if h is in H then h^{-1} is in H; and*
    (b) *if h and k are in H then hk is in H.*
(iii) *If h and k are in H then hk^{-1} is in H.*

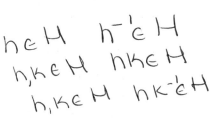

Proof It has to be shown that the three conditions are equivalent. What we will do is show that (i) implies (ii), (ii) implies (iii), (iii) implies (i). It then follows, for example, that (i) implies (iii) and indeed that the three conditions are equivalent.

That (i) implies (ii) follows directly, since the conditions in (ii) are two of the group axioms. It is also easy to see that (ii) implies (iii). For if h and k are in H then h and k^{-1} are in H (H is closed under taking inverses by (ii)(a)) and then hk^{-1} is in H (by (ii)(b)). So it only remains to show that (iii) implies (i).

We check the four group axioms for H. First note that the associativity axiom holds since if g, h and k are elements of H then certainly g, h and k are elements of G and so $(gh)k = g(hk)$. Next, we show that H contains the identity element of G. To see this take any h in H (this is possible since H is non-empty) and apply (iii) with $h = k$ to obtain that $e = hh^{-1}$ is in H. Now let g be any element of H and apply (iii) with h being e (which we now know to be in H) and k being g to see that (iii) implies that g^{-1} must be in H. Finally we check the closure axiom for H. Given x and y in H, we have just seen that y^{-1} must also be in H: applying (iii) with $h = x$ and $k = y^{-1}$ gives $xy \in H$ (since $(y^{-1})^{-1} = y$). □

Example 1 Let H be the set of even integers, considered as a subset of $G = (\mathbb{Z}, +)$.

In order to apply Theorem 5.1.5 we need first to check that the subset we are considering is not the empty set. In this case, there seems very little to say: of course there are even numbers! However, in general we might need to be more careful. A good habit to acquire is to check if the identity element of the whole group G is in the subset H (if so, this proves that H is non-empty, but we have already seen that every subgroup of G must contain the identity element of G, so we have not done unnecessary work). In this case, the identity element of G is the number 0 (since G is a group under addition). Also 0 is even because it is divisible by 2 (since $0 = 0 \cdot 2$).

make sure to check identity of G is in H.

We now check conditions (a) and (b) of (ii): they follow because the sum of two even integers is an even integer and the negative of an even integer is an even integer. We have shown that H is a subgroup of G

Example 2   The set $(\mathbb{Z}, +)$ is itself a subgroup of $(\mathbb{R}, +)$ which is in turn a subgroup of $(\mathbb{C}, +)$.

Example 3   Let H be the set, $A(n)$, of even permutation in the group $S(n)$ under composition of permutations.

The identity element of $S(n)$ is an even permutation, and so is in $A(n)$. By Theorems 4.2.8 and 4.2.9, the inverse of an even permutation is an even permutation and the sign of a product of two permutations is the product of the signs. It follows that $A(n)$ is a subgroup of $S(n)$.

Example 4   Next let H be the set of invertible diagonal 2×2 matrices, considered as a subset of the group, $GL(n, \mathbb{R})$, of all invertible 2×2 matrices under matrix multiplication.

Again it is easy to check that the identity element for G (the identity matrix) is in H (because the identity matrix is diagonal). Since the product of two diagonal matrices is a diagonal matrix and the inverse of a diagonal matrix is also a diagonal matrix, we deduce that H is a subgroup of G.

Example 5   Let H be the set of all $n \times n$ matrices with determinant 1, considered as a subset of $GL(n, \mathbb{R})$.

Again, we use part (ii) of the above theorem and check conditions (a) and (b). Note that the identity element of G has determinant 1, so is in H. If A, B are matrices with determinant 1, then AB also has determinant 1. The determinant of the inverse of an invertible matrix A is equal to 1 over the determinant of A, so if A has determinant 1, the determinant of A^{-1} is also equal to 1. Thus H is a subgroup, which is usually denoted $SL(n, \mathbb{R})$.

Remarks   (1) The advantage of checking for a subgroup in this systematic way is that we will immediately detect (if any of our checks fails) if a given subset is not a subgroup of G.

(2) Every group has obvious subgroups, namely the group G itself (any other subgroup is said to be **proper**) and the **trivial** or **identity** subgroup $\{e\}$ containing the identity element only. These will be distinct provided G has more than one element.

As a simple application of this result we show the following.

[handwritten: H ∩ K subgroup if H + K are subgroups]

Theorem 5.1.6 *Let G be a group and let H and K be subgroups of G. Then the intersection H ∩ K is a subgroup of G.*

Proof Note that $H \cap K$ is non-empty since both H and K contain e. We show that $H \cap K$ satisfies condition (iii) of Theorem 5.1.5. Take x and y in $H \cap K$: then x and y are both in H and so, since H is a subgroup, xy^{-1} is in H. Similarly, since x and y are both in K and K is a subgroup, xy^{-1} is in K. Hence xy^{-1} is in $H \cap K$. So, by Theorem 5.1.5, $H \cap K$ is a subgroup of G. □

A good source of subgroups is provided by the following.

Theorem 5.1.7 *Let G be a group and let g be an element of G. The set $\langle g \rangle = \{g^n : n \in \mathbb{Z}\}$ of all distinct powers of g is a subgroup, known as the subgroup **generated by** g. It has n elements if g has order n and it is infinite if g has infinite order.*

Proof To see that $\langle g \rangle$ is a subgroup, note first that it is non-empty since it contains g. If h and k are in $\langle g \rangle$ then h is g^i and k is g^j for some integers i and j, and so

$$hk^{-1} = g^i(g^j)^{-1}$$
$$= g^i g^{-j}$$
$$= g^{i-j} \text{ by Theorem 5.1.3.}$$

[handwritten: hk = g^i(g^j)^{-1}, g^i g^j = g^{i-j}]

Thus hk^{-1} is in $\langle g \rangle$ and so, by Theorem 5.1.5, $\langle g \rangle$ is a subgroup of G as required.

If g has infinite order, then the positive powers $g, g^2, \ldots, g^n, \ldots$ are all distinct (see the proof of Theorem 4.2.2) so $\langle g \rangle$ is an infinite group. If g has order n, then Theorem 5.1.4 shows that there are n distinct powers of g and hence that $\langle g \rangle$ has exactly n elements. □

Definition A group of the above type, that is, of the form $\langle g \rangle$ for some element g in it, is said to be **cyclic, generated by** g.

Remark It follows from Theorem 5.1.3 that a cyclic group is Abelian.

Example 1 To find all the cyclic subgroups of $S(3)$, we may take each element of the group $S(3)$ in turn and compute the cyclic subgroup which it generates. In this way, we obtain a complete list (with repetitions) as follows:

$$\langle \text{id} \rangle = \{\text{id}\}; \langle (12) \rangle = \{\text{id}, (12)\};$$
$$\langle (13) \rangle = \{\text{id}, (13)\}; \langle (23) \rangle = \{\text{id}, (23)\};$$
$$\langle (123) \rangle = \{\text{id}, (123), (132)\} = \langle (132) \rangle.$$

[handwritten: cyclic group, g is generated, can be written as power, so addition = mult, mult = power]

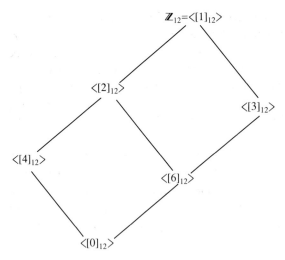

Fig. 5.1

Since we have listed all the cyclic subgroups in $S(3)$ and since $S(3)$ itself is not on the list, it follows that the group $S(3)$ is not cyclic.

Note that, in principle, we can find all the cyclic subgroups of a given finite group by taking each element of the group in turn and computing the cyclic subgroup it generates, then deleting any duplicates.

Example 2   As we saw above, the subgroup of $GL(2, \mathbb{R})$ generated by

$$\begin{pmatrix} 1 & 1 \\ 0 & 1 \end{pmatrix}$$

is an infinite cyclic group.

Example 3   In the additive group $(\mathbb{Z}, +)$, the cyclic subgroup generated by the integer 1 consists of all multiples (not powers, because the group operation is addition!) of 1 and so is the whole group. The subgroup $\langle 2 \rangle$ is the set of even integers. The subgroup $\langle 3 \rangle$ consists of all integer multiples of 3. The intersection is easily computed: $\langle 2 \rangle \cap \langle 3 \rangle = \langle 6 \rangle$.

Example 4   The group $(\mathbb{Z}_{12}, +)$ is itself cyclic, as are all its subgroups. The set of all its subgroups forms a partially ordered set under inclusion. The Hasse diagram of this set is shown in Fig. 5.1.

The reader may be familiar with vector spaces, where the change from structures generated by one element to those generated by two is not very great.

The situation for groups is very different. A group generated by two elements may well be immensely complicated. In Section 4.3.4 we saw some of the simpler examples: the dihedral groups (groups of symmetries of regular n-sided polygons) can be generated by two elements. These elements are the reflection R in the perpendicular bisector of any one of the sides and the rotation ρ, through $2\pi/n$ radians. Thus $R^2 = e = \rho^n$ and there is also the relation $\rho^{n-1}R = R\rho$. The resulting group has $2n$ elements which can all be written in the form $\rho^i R^j$ where j is 0 or 1 and $0 \le i < n$.

We also saw in Section 4.2 that the symmetric group $S(n)$ can be generated by its transpositions. In fact $S(n)$ can be generated by the $n - 1$ transpositions $(1\ 2), (2\ 3), \ldots, (n - 1\ n)$: for every element of $S(n)$ can be written as a product of these. (This was set as Exercise 4.2.10.) However, the relations between these generators are much more complicated than in the case of the dihedral groups.

Exercises 5.1

1. Prove that for any elements a and b of a group G, $ab = ba$ if and only if $(ab)^{-1} = a^{-1}b^{-1}$.

2. Let a, b be elements of a group G. Find (in terms of a and b) an expression for the solution x of the equation $axba^{-1} = b$.

3. Take G to be the cyclic group with 12 elements. Find an element g in G such that the equation $x^2 = g$ has no solution.

4. Use Theorem 5.1.1 to decide which of the following subsets of the given groups are subgroups:
   (i)   the subset of the symmetries of a square consisting of the rotations;
   (ii) the subset of $(\mathbb{R}, +)$ consisting of $\mathbb{R}\setminus\{0\}$ under multiplication;
   (iii) the subset $\{\mathrm{id}, (1\ 2), (1\ 3), (2\ 3)\}$ of $S(3)$;
   (iv) the subset of $(\mathrm{GL}(3, \mathbb{R}), \cdot)$ consisting of matrices of the form

$$\begin{pmatrix} 1 & a & b \\ 0 & 1 & c \\ 0 & 0 & 1 \end{pmatrix}.$$

5. Give an example of a group G and elements a, b, and c in G, such that a is different from b but $ac = cb$.

6. Let G be a cyclic group generated by x. Note that for any positive integer k, the set $\langle x^k \rangle$ is a subgroup of G. If x has finite order 12, consider the possible values for k (from 0 to 11) and, in each case, show that $\langle x^k \rangle$ is generated by x^d where d is the greatest common divisor of k and 12. Deduce that $\langle x^k \rangle$ has $12/d$ elements.

7. Let G be a cyclic group generated by x with x of order greater than 1. Let H be a subgroup of G with H neither G nor $\{e\}$. Let $m \geq 1$ be minimal such that x^m is in H. Use the division algorithm to show that x^m generates H and deduce that every subgroup of a cyclic group is cyclic.

8. Let g and x be elements of a group G. Show that for all positive integers k,

$$(g^{-1}xg)^k = g^{-1}x^k g.$$

Deduce that x has order 3 if and only if (for all $g \in G$) $g^{-1}xg$ has order 3. Show that the same is true for any integer $n \geq 1$ in place of 3.

9. Let G be any group and define the relation of conjugacy on G by aRb if and only if there exists $g \in G$ such that $b = g^{-1}ag$. Show that this is an equivalence relation on G.

10. Find $a \in G_{23}$, the group of invertible congruence classes modulo 23, such that every element of G_{23} is a power of a: that is, show that G_{23} is a cyclic group by finding a generator for it. Similarly show that G_{26} is cyclic by finding a generator for it. Is every group of the form G_n cyclic?

5.2 Cosets and Lagrange's Theorem

If H is a subgroup of a group G then G breaks up into 'translates', or *cosets*, of H. This notion of a coset is a key concept in group theory and we will make use of cosets in this section by proving Lagrange's Theorem. The remainder of the section is devoted to deriving consequences of this theorem, which may be regarded as the fundamental result of elementary group theory.

Definition   Let H be a subgroup of the group G, and let a be any element of G. Define aH to be the set of all elements of G which may be written as ah for some element h in H: $aH = \{ah : h \in H\}$. This is a (**left**) **coset** of H (in G): it is also termed the left coset of a with respect to H. Similarly define the **right coset** $Ha = \{ha : h \in H\}$. We make the convention that the unqualified term 'coset' means 'left coset'.

Notes   (1) The subgroup H is a coset of itself, being equal to eH.

(2) The element a is always a member of its coset aH since $a = ae \in aH$ (for e is in H, since H is a subgroup). Similarly, a is a member of the right coset Ha.

(3) If b is in aH then $bH = aH$. To see this suppose that $b = ah$ for some h in H. A typical member of bH has the form bk for some k in H. We have:

$$bk = (ah)k = a(hk).$$

Since H is a subgroup, hk is in H and so we have that bk is in aH. Thus $bH \subseteq aH$.

For the converse, note that from the equation $b = ah$ we may derive $a = bh^{-1}$, and h^{-1} is in H. So we may apply the argument just used to deduce $aH \subseteq bH$. Hence $aH = bH$ as claimed.

It follows that each coset of a given subgroup is determined by any one of the elements in it. Such an element is known as a **representative** for the coset.

(4) Unless a is in H the coset aH is not a subgroup (for if it were a subgroup it would have to contain the identity, so we would have $e = ah$ for some h in H, necessarily $h = a^{-1}$, but then since a^{-1} is in H we would have that $a = (a^{-1})^{-1}$ is a member of H).

(5) There are two 'trivial' cases: if $H = G$ then there is only one coset of H, namely $H = G$ itself; if $H = \{e\}$ then for every a in G the coset aH consists of just a itself.

Example 1 Take $G = (\mathbb{Z}, +)$ and let $n \geq 2$ be a positive integer. Let H be the set of all integer multiples of n: note that H is a subgroup of $\mathbb{Z}$ (proved as in Section 5.1). (We exclude the values $n = 1$ and $n = 0$ since they correspond to the two trivial cases mentioned in Note (5) above.) What are the cosets of H in G? We have already met them! For example H consists of precisely the multiples of n and so is just the congruence class of 0 modulo n. Similarly the coset $1 + H$ (we use additive notation since the operation in G is '+') is none other than the congruence class of 1 modulo n: and in general the coset $k + H$ of k with respect to H is just the congruence class of k modulo n. You may note that in this example right and left cosets coincide: $k + H = H + k$, since the group is Abelian.

Example 2 Take $G = (\mathbb{Z}_6, +)$ and let H be the subgroup with elements 0 and 3 (more precisely $[0]_6$ and $[3]_6$): this set of two elements does form a subgroup, being the subgroup generated by 3. The cosets are as follows:

$$0 + \{0, 3\} = \{0, 3\} = 3 + \{0, 3\};$$
$$1 + \{0, 3\} = \{1, 4\} = 4 + \{0, 3\};$$
$$2 + \{0, 3\} = \{2, 5\} = 5 + \{0, 3\}.$$

Observe that each of the three cosets of H in G has the same number (two) of elements as H (this is a general fact, see Theorem 5.2.2 below). We may also note that each coset has two representatives.

Example 3 Take G to be the symmetric group $S(3)$ with the usual composition of permutations, and let H be the subgroup consisting of the identity element together with the transposition $(1\ 2)$ (in the terminology of Section 4.1, this is the subgroup generated by (1 2)). To find the complete list of cosets of H in G, we may consider all sets of the form gH for $g \in G$. In this way we get a list of six cosets of H in G, but these are not all distinct:

$$\text{id} \cdot H = \text{id} \cdot \{\text{id}, (1\ 2)\} = \{\text{id}, (1\ 2)\};$$
$$(1\ 2)H = (1\ 2)\{\text{id}, (1\ 2)\} = \{(1\ 2), (1\ 2)(1\ 2)\} = H;$$
$$(1\ 3)H = \{(1\ 3), (1\ 3)(1\ 2)\} = \{(1\ 3), (1\ 2\ 3)\};$$
$$(2\ 3)H = \{(2\ 3), (2\ 3)(1\ 2)\} = \{(2\ 3), (1\ 3\ 2)\};$$
$$(1\ 2\ 3)H = \{(1\ 2\ 3), (1\ 2\ 3)(1\ 2)\} = \{(1\ 2\ 3), (1\ 3)\};$$
$$(1\ 3\ 2)H = \{(1\ 3\ 2), (1\ 3\ 2)(1\ 2)\} = \{(1\ 3\ 2), (2\ 3)\}.$$

We see that id $\cdot H = (1\ 2)H$, $(1\ 3)H = (1\ 2\ 3)H$ and $(2\ 3)H = (1\ 3\ 2)H$. Again we note that each coset contains two elements and so has two representatives. Notice that the right and left cosets of a given element with respect to H need not coincide:

$$H(1\ 3) = \{(1\ 3), (1\ 3\ 2)\} \neq (1\ 3)H.$$

In fact the right coset $H(1\ 3)$ is not the left coset of any element.

Example 4 Take G to be Euclidean 3-space ($\mathbb{R}^3$) with addition of vectors as the operation. Let H be the xy-plane (defined by the equation $z = 0$). Note that H is a subgroup of G. You should check that the cosets of H in G are the horizontal planes: indeed the coset of a vector $\mathbf{v}$ is just the plane which contains $\mathbf{v}$ and is parallel to H (the horizontal plane containing $\mathbf{v}$).

Next, we see why (as you may have noticed in our examples) if two cosets of a given subgroup H have an element in common, then they are equal. Indeed, in the proof we show that the relation on G defined by

$$xRy \text{ if and only if } y^{-1}x \in H$$

is an equivalence relation which partitions G into the distinct cosets of H.

Theorem 5.2.1 *Let H be a subgroup of the group G and let a, b be elements of G. Then either $aH = bH$ or $aH \cap bH = \emptyset$.*

Proof This follows from Note (3) at the beginning of this section. We must show that if aH and bH have at least one element in common then $aH = bH$. So suppose that there is an element c in $aH \cap bH$. By that note we have $aH = cH$ and $cH = bH$, as required.

There is an alternative way to present this proof, using the notion of equivalence relation. Define a relation R on G by xRy if and only if $y^{-1}x$ is in H. Then R is an equivalence relation:

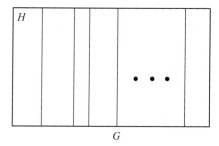

Fig. 5.2

clearly it is reflexive since e is in H;

it is symmetric since if $y^{-1}x$ is in H then so is $(y^{-1}x)^{-1} = x^{-1}y$;

it is transitive, since if $y^{-1}x$ and $z^{-1}y$ are in H then so is $z^{-1}x = (z^{-1}y)(y^{-1}x)$.

The equivalence class, $[x]$, of x is given by

$$\begin{aligned}
[x] &= \{y \in G : yRx\} \\
&= \{y \in G : x^{-1}y \in H\} \\
&= \{y \in G : x^{-1}y = h \text{ for some } h \text{ in } H\} \\
&= \{y \in G : y = xh \text{ for some } h \text{ in } H\} \\
&= xH,
\end{aligned}$$

so the result follows by Theorem 2.3.1.   □

It follows that if H is a subgroup of the group G then every element belongs to one and only one left coset of H in G. Thus the (left) cosets of H in G form a partition of G. For instance in Example 4 above the cosets of the xy-plane in 3-space partition 3-space into a 'stack' of (infinitely many) parallel planes (note that two such planes are equal or have no point in common). In Example 1 we partitioned $\mathbb{Z}$ into the congruence classes of integers modulo n.

So we have the picture of G split up into cosets of the subgroup H (Fig. 5.2).

The next important point is that these pieces all have the same size.

Theorem 5.2.2   *Let H be a subgroup of the group G. Then each coset of H in G has the same number of elements as H.*

Proof   (See Fig. 5.3.) Let aH be any coset of H in G. We show that there is a bijection between H and aH. Define the function $f : H \to aH$ by $f(h) = ah$.

(1) f is injective: for if $f(h) = f(k)$ we have $ah = ak$ and, multiplying on the left by a^{-1}, we obtain $h = k$ as desired.

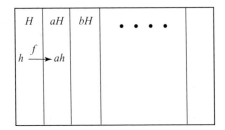

Fig. 5.3

(2) f is surjective: for if b is in aH then, by definition, b may be expressed in the form ah for some h in H; thus $b = f(h)$ is in the image of f, as required.

Therefore there is a bijection between H and aH, so they have the same number of elements (cf. Section 2.2), as claimed. (Observe that we did not assume that H is finite: the idea of infinite sets having the same numbers of elements was discussed in Section 2.2 where bijections were introduced and discussed.) □

Example   In the example on p. 167, we showed, by a special case of the above argument, that the two cosets of the alternating group $A(n)$ in the symmetric group $S(n)$ have the same size.

This leads us to a key result, which is usually named after Joseph Louis Lagrange (1736–1813) who essentially established it. At least, he proved it in a special case: the general idea of a group did not emerge until around the middle of the nineteenth century. A second special case was shown by Cauchy. The general result was given by Jordan (who attributed it to Lagrange and Cauchy: the proof is the same in each case).

Definition   Let G be a finite group. The **order** of G, $o(G)$, is the number of elements in G.

By Theorem 5.1.7, if G is cyclic, say $G = \langle g \rangle$, then the order of G is the order of the element g in the sense of the definition before 5.1.4 (the two uses of the term 'order' are distinct but related).

Theorem 5.2.3   (Lagrange's Theorem) *Let G be a finite group and let H be a subgroup of G. Suppose that H has m distinct left cosets in G. Then $o(G) = o(H) \cdot m$. In particular, the order of H divides the order of G.*

Proof   We have only to put together the pieces that we have assembled. By Theorem 5.2.1, and since G is finite, G may be written as a disjoint union of cosets of H: say

$$G = a_1 H \cup a_2 H \cup \ldots \cup a_m H$$

where $a_1 = e$ and $a_i H \cap a_j H = \emptyset$ whenever $i \neq j$.

By Theorem 5.2.2, the number of elements in each coset $a_i H$ is $o(H)$ (that is, the number of elements in H). So, since the union is disjoint (by Theorem 5.2.1), the number of elements in G is $m \cdot o(H)$, as claimed. In particular, the number of elements of G is a multiple of the number of elements of H: in other words $o(H)$ must divide $o(G)$, as required. $\square$

Comment Repeating the above proof using the distinct right cosets of H in G would also show that the number of these, n say, satisfies the same equation: $o(G) = n \cdot o(H)$. It therefore follows that $n = m$, so there are the same number of distinct right or left cosets of any given subgroup of a group. As we have seen, in general the list of distinct right cosets is not equal to the list of distinct left cosets, though the lists contain the same number of cosets.

We may quickly derive several important corollaries from Lagrange's Theorem.

Corollary 5.2.4 *Let g be an element of the group G. Then the order of g divides the order of the group G.*

$$|g| \mid |G|$$

Proof The order of g is equal to the number of elements in the cyclic subgroup which it generates (by Theorem 5.1.7). By Lagrange's Theorem, this number divides the number of elements in G. $\square$

Corollary 5.2.5 *Let G be a group of prime order p. Then G is cyclic.*

$$|G| = p \quad G \text{ is cyclic}$$

Proof Let x be any element of G other than the identity. By Theorem 5.1.7, $\langle x \rangle$ is a subgroup of G and it certainly contains more then one element ($x \neq e$). By Lagrange's Theorem the number of elements in $\langle x \rangle$ divides p so, since it is greater than 1, it must be p. Thus $\langle x \rangle$ must be G. $\square$

The next two corollaries of the result have been seen already in Section 1.6.

Corollary 5.2.6 (Fermat) *Let p be a prime number and a be any integer not divisible by p. Then $a^{p-1} \equiv 1 \bmod p$.*

Proof The group G_p of invertible congruence classes modulo p has $p - 1$ elements and consists of the congruence classes of integers not divisible by p.

Since the congruence class of a is in G_p it follows, by Corollary 5.2.4, that the order of a divides $p - 1$. The result now follows by Theorem 5.1.4 since $[1]_p$ is the identity element of G_p. □

Corollary 5.2.7 (Euler) *Let n be any integer greater than 1 and let a be relatively prime to n. Then $a^{\phi(n)} \equiv 1 \bmod n$.*

Proof The proof is similar to that of Corollary 5.2.6 since the number of elements in G_n is $\phi(n)$. □

elements $= \operatorname{ord} G_n = \phi(n)$

Remark Even from these few corollaries, one may appreciate that Lagrange's Theorem is very powerful: also illustrated is the strong connection between group theory and arithmetic.

Suppose that G is a group with n elements and let d be a divisor of n: it need not be the case that G has an *element* of order d (for instance, take G to be non-cyclic and take $d = n$). Indeed, the converse of Lagrange's Theorem is false in general. That is, if G is a group with n elements and if d is a divisor of n, G need not even have a *subgroup* with d elements. The simplest example here is the alternating group $A(4)$ of order 12 which has no subgroup with six elements. This is given as an exercise, with hints, in Section 5.3.

Corollary 5.2.4 shows an unexpected relationship between group theory and number theory. The order of a group influences its structure. This theme recurs throughout finite group theory which is, in fact, a very arithmetical subject. As an example of a way in which this relationship appears, we note that groups whose order is a power of a prime number p (p-groups) play a special role in the theory. The arithmetic also helps us understand the subgroup structure of a group. Although the converse of Lagrange's Theorem is false, each group of finite order divisible by a prime p does have certain important subgroups that are p-groups, and their existence (Sylow's Theorems) is one of the most significant results in the theory.

Exercises 5.2

1. Let G be the group G_{14} of invertible congruence classes modulo 14. Write down the distinct left cosets of the subgroup $\{[1]_{14}, [13]_{14}\}$.
2. Show that for $n \geq 3$, $\phi(n)$ is divisible by 2.
   [Hint: note that $(-1)^2 = 1$.]
3. Let G be the group $D(4)$ of symmetries of a square and τ be any reflection in G. Describe the left cosets of of the subgroup $\{1, \tau\}$ of G.

4. Let H be a subgroup of the group G and let a be an element of G. Fix an element b in aH (so b is of the form ah for some h in H). Show that

$$H = \{b^{-1}c : c \in aH\}.$$

5. Let G be the group G_{20}. What, according to Corollary 5.2.4, are the *possible* orders of elements of G and which of these integers are *actually* orders of elements of G?

5.3   Groups of small order

In this section we introduce the ideas of isomorphism and direct product. These will then be used to describe, up to isomorphism, all groups of order no more than 8.

Informally, we regard two groups G and H as being **isomorphic** ('of the same shape') if they can be given the 'same' multiplication table. More precisely, we require the existence of a bijection θ from G to H such that, if G is listed as $\{g_1, g_2, \ldots\}$ and if H is listed as $\{\theta(g_1), \theta(g_2), \ldots\}$ and if the multiplication tables are drawn up, then, if the (g_i, g_j) entry in the table for G is g_k, the $(\theta(g_i), \theta(g_j))$ entry in the table for H will be $\theta(g_k)$. This condition on the tables is simply that $\theta(g_i g_j) = \theta(g_i)\theta(g_j)$ for all i, j. (Notice that we used another Greek letter θ known as 'theta' to denote our bijection.)

Another way to understand the idea of two groups being isomorphic is to imagine the tables for the two groups G and H each being written on transparent slides. Do this using a different coloured pen for each element of the group G, replacing each occurrence of a given element g in the table (including the header row and column) by an appropriately coloured square. Use the colour associated with g for the element $\theta(g)$ in H and draw up another multi-coloured slide, replacing $\theta(g)$ wherever it occurs with a square coloured in the colour of g. Also, if necessary, rearrange the order of the entries in the header row and column so that they occur in the same order as on the slide with the table for G. The groups G and H are then isomorphic if the multi-coloured slide for G can be superimposed on the multi-coloured slide for H with no detectable differences. Thus any features of a group which may be determined from its multiplication table (such as being Abelian) must be shared with any isomorphic group.

There is actually no need to refer explicitly to the multiplication tables, and we make the following precise definition.

Definition   Let G and H be groups. A function $\theta : G \longrightarrow H$ is an **isomorphism** (from G to H) if it is a bijection and if

isomorphic — bijection
$$\theta(xy) = \theta(x)\theta(y)$$
220   Group theory and error-correcting codes

for all $x, y \in G$ we have $\theta(xy) = \theta(x)\theta(y)$ (∗)
('θ preserves the group structure').

Groups G and H are **isomorphic** if there exists an isomorphism from G to H.

Notes   (1) If G is any group then the identity function from G to itself is an isomorphism from G to G. That is, G is isomorphic to itself.

(2) If $\theta : G \longrightarrow H$ is an isomorphism then, as you are invited to verify, the inverse map $\theta^{-1} : H \longrightarrow G$ is an isomorphism. Thus, if G is isomorphic to H then so is H isomorphic to G.

(3) If $\theta : G \longrightarrow H$ and $\psi : H \longrightarrow K$ are isomorphisms, then the composition $\psi\theta : G \longrightarrow K$ is an isomorphism. Thus the relation of being isomorphic is transitive (ψ is the Greek letter 'psi').

(4) Taken together, (1), (2) and (3) show that the relation of being isomorphic is an equivalence relation.

Theorem 5.3.1   *Let θ be an isomorphism from G to H. Then $\theta(e_G)$ is the identity element of H and the inverse of $\theta(g)$ is $\theta(g^{-1})$. That is, $\theta(e_G) = e_H$ and $\theta(g^{-1}) = (\theta(g))^{-1}$.*

Proof   By e_G we mean the identity element of G; similarly by e_H is meant the identity element of H. For every g in G, we have that

$$e_H \cdot \theta(g) = \theta(g) = \theta(e_G \cdot g) = \theta(e_G)\theta(g).$$

Using Theorem 5.1.1, it follows that $\theta(e_G)$ is the identity element e_H of H. Similarly, Theorem 5.1.1 applied to the equation

$$\theta(g) \cdot \theta(g)^{-1} = e_H = \theta(e_G) = \theta(gg^{-1}) = \theta(g)\theta(g^{-1})$$

shows that $\theta(g^{-1})$ is $\theta(g)^{-1}$. □

Example 1   We have seen several examples of groups with four elements, namely $(\mathbb{Z}_4, +)$; the multiplicative group, G_5, of invertible congruence classes modulo 5; the subgroup of $S(4)$ consisting of the four permutations id, (1 2)(3 4), (1 3)(2 4), (1 4)(2 3); the group of symmetries of a rectangle.

Of these, two are cyclic: G_5 is generated by $[2]_5$ and the additive group $\mathbb{Z}_4$ is generated by $[1]_4$. We may define a function θ from G_5 to $\mathbb{Z}_4$ by sending $[2]_5$ to $[1]_4$. If this function is to be structure preserving then it must be that $[4]_5 = [2]_5^2$ is sent to $[2]_4 = [1]_4 + [1]_4$ and that, more generally, that nth power of $[2]_5$ is sent to the nth power (or rather, sum, since $(\mathbb{Z}_4, +)$ is an additive group) of $[1]_4$. Thus, specifying where the generator is sent also determines where all its powers are to be sent, assuming that the function is to satisfy condition (∗). Since the generators $[2]_5$ and $[1]_4$ have the same order, this function is well defined and, one may check, is an isomorphism (to see this directly, draw up the two group tables, then rearrange one of them using θ).

This shows that the groups G_5 and $\mathbb{Z}_4$ are isomorphic. The argument applies more generally to show that any two cyclic groups with the same number of elements are isomorphic (send a generator of one group to a generator of the other and argue as above).

For this reason, a cyclic group with n elements is often denoted simply by C_n (the fact that it is cyclic and has n elements determines it up to isomorphism). This notation is used when the group operation is written multiplicatively, whereas $\mathbb{Z}_n$ tends to be used in conjuction with additive notation.

The other two groups with four elements are isomorphic to each other, as may be seen be inspecting their multiplication tables (in Section 4.3). Specifically, an isomorphism from the given subgroup of $S(4)$ to the group of symmetries of a rectangle may be given by taking (1 2)(3 4) to σ, (1 3)(2 4) to R and (1 4)(2 3) to τ (and of course, id to e). However, neither of these is isomorphic to (either of) the cyclic groups. One way to see this is to observe that, in the second two examples, each element is its own inverse, but in the cyclic groups there are elements which are not their own inverse. (Any property defined solely in terms of the group operation must be preserved by any isomorphism.)

Example 2   We have seen two examples of non-Abelian groups with six elements: the symmetric group $S(3)$ and the dihedral group $D(3)$ of symmetries of an equilateral triangle. These are also isomorphic (see Example 1 in . Section 4.3.4). It should be remarked that it may be very difficult to determine whether or not two (large) groups are isomorphic: even it they are isomorphic, there may be no 'obvious' isomorphism.

Example 3   You may check that the function $f : (\mathbb{R}, +) \longrightarrow (\mathbb{C}, \cdot)$ which takes $r \in \mathbb{R}$ to $e^{2\pi i r}$ satisfies the condition (∗) of the definition of isomorphism, but is not 1-1, since for any integer n one has $e^{2\pi i n} = 1$. Hence it is not an isomorphism.

Let us define the relation R on $\mathbb{R}$ by rRs if and only if $e^{2\pi i r} = e^{2\pi i s}$. Then R is an equivalence relation. It is straightforward to check that the equivalence class of 0 is a subgroup of $\mathbb{R}$: indeed it is just the set of all integers. The equivalence classes are just the cosets of this subgroup in $\mathbb{R}$. A group operation may be defined on the set G of equivalence classes (cosets), by setting $[r] + [s] = [r + s]$, where $[r]$ denotes the coset of r: you should check that this is well defined. Then define the function $g : G \longrightarrow \mathbb{C}$ by setting $g([r]) = f(r)$: again, you should check that the definition of g does not depend on the representative chosen. Finally, let S be the image of the function g: it is the set of all complex numbers of the form $e^{2\pi i r}$, the circle in the complex plane with centre the origin and radius 1. Then g is an isomorphism from G to S.

Now we describe a way of obtaining new groups from old.

Definition   Given groups G and H, the **direct product** $G \times H$ is the set of all ordered pairs (g, h) with g in G and h in H, equipped with the following multiplication

$$(g_1, h_1)(g_2, h_2) = (g_1 h_1, g_2 h_2).$$

Comment   We have used the notation $X \times Y$ in Chapter 2 to denote the Cartesian product of X and Y. Here, our sets are actually groups so each has a operation which, although often written multiplicatively, should not be confused with the operation, $\times$, on sets. The direct product means 'the set of ordered pairs' and does not involve combining, in any way, the elements of G with those of H. Indeed, our ordered pair notation keeps elements of G apart from elements of H.

Theorem 5.3.2   *For any groups G and H, the direct product $G \times H$ is a group. In the case that G and H are finite, the order of this group is the product of the orders of G and H.*

Proof   One checks the group axioms for $G \times H$: closure is clear; associativity follows from that for G and H; the identity is (e_G, e_H) and the inverse of (g, h) is (g^{-1}, h^{-1}). For the last part, see Exercise 2.1.8.   □

Notes   (1) Given groups G and H, the product groups $G \times H$ and $H \times G$ are isomorphic (define the isomorphism to take (g, h) to (h, g)).

(2) Forming direct products of groups is an associative operation in the sense that, given groups G, H, and K, there is an isomorphism from $(G \times H) \times K$ to $G \times (H \times K)$ (given by sending $((g, h), k)$ to $(g, (h, k))$) so we may write, without real ambiguity, $G \times H \times K$. The notations G^2, G^3, etc. are often used for $G \times G$, $G \times G \times G$, and so on.

Example 1   Let G and H both be the group G_3 of invertible congruence classes modulo 3. The group $G \times H$ has four elements:

$$([1]_3, [1]_3), ([1]_3, [2]_3), ([2]_3, [1]_3) \text{ and } ([2]_3, [2]_3).$$

It should be clear that $([1]_3, [1]_3)$ is the identity and that, for all a and b,

$$([a]_3, [b]_3)^2 = ([a^2]_3, [b^2]_3) = ([1]_3, [1]_3).$$

It is easy to check that $G \times H$ is Abelian.

Example 2 The direct product $S(3) \times S(3)$ has $36 (= 6 \times 6)$ elements. This group is not Abelian (since $S(3)$ is not Abelian).

Example 3 We may now explain a point which arose in Section 1.4. Let a be a generator for C_4 and let b be a generator for C_2. Then $C_4 \times C_2$ has eight elements:

$$(e, e), (e, b), (a, e), (a, b), (a^2, e), (a^2, b), (a^3, e) \text{ and } (a^3, b).$$

(Note that the 'e' appearing in the first coordinate is the identity element of C_4 whereas that appearing in the second coordinate is the identity element of C_2.)

It may be checked that this group is isomorphic to the group G_{20} by the function given by

$$(e, e) \rightarrow [1], (e, b) \rightarrow [11], (a, e) \rightarrow [3], (a, b) \rightarrow [13],$$
$$(a^2, e) \rightarrow [9], (a^2, b) \rightarrow [19], (a^3, e) \rightarrow [7] \text{ and } (a^3, b) \rightarrow [17].$$

This explains why the table for G_{20}, given in Section 1.4 splits into four blocks. The blocks are obtained by ignoring the first coordinate of the element of $C_4 \times C_2$ corresponding to a given element of G_{20}: so if we look only at the second coordinate then we obtain the structure of the multiplication table for C_2.

	(e, e)	(a, e)	(a^3, e)	(a^2, e)	(e, b)	(a, b)	(a^3, b)	(a^2, b)
(e, e)	(e, e)	(a, e)	(a^3, e)	(a^2, e)	(e, b)	(a, b)	(a^3, b)	(a^2, b)
(a, e)	(a, e)	(a^2, e)	(e, e)	(a^3, e)	(a, b)	(a^2, b)	(e, b)	(a^3, b)
(a^3, e)	(a^3, e)	(e, e)	(a^2, e)	(a, e)	(a^3, b)	(e, b)	(a^2, b)	(a, b)
(a^2, e)	(a^2, e)	(a^3, e)	(a, e)	(e, e)	(a^2, b)	(a^3, b)	(a, b)	(e, b)
(e, b)	(e, b)	(a, b)	(a^3, b)	(a^2, b)	(e, e)	(a, e)	(a^3, e)	(a^2, e)
(a, b)	(a, b)	(a^2, b)	(e, b)	(a^3, b)	(a, e)	(a^2, e)	(e, e)	(a^3, e)
(a^3, b)	(a^3, b)	(e, b)	(a^2, b)	(a, b)	(a^3, e)	(e, e)	(a^2, e)	(a, e)
(a^2, b)	(a^2, b)	(a^3, b)	(a, b)	(e, b)	(a^2, e)	(a^3, e)	(a, e)	(e, e)

Indeed, if we look at the blocks, then we have the multiplication table for the two cosets of the subgroup $C_4 \times \{e\}$ in $C_4 \times C_2$ (cf. Example 3 on p. 221).

Example 4   When both G and H are Abelian, we often write the group operation in $G \times H$ as addition. For example, the group $\mathbb{Z}_2 \times \mathbb{Z}_2$ has four elements

$$([0]_2, [0]_2), ([0]_2, [1]_2), ([1]_2, [0]_2) \text{ and } ([1]_2, [1]_2).$$

Since

$$([a]_2, [b]_2) + ([c]_2, [d]_2) = ([a + c]_2, [b + d]_2)$$
$$= ([c + a]_2, [d + b]_2) = ([c]_2, [d]_2) + ([a]_2, [b]_2),$$

we see that $\mathbb{Z}_2 \times \mathbb{Z}_2$ is an Abelian group. In fact, $\mathbb{Z}_2 \times \mathbb{Z}_2$ is isomorphic to the group in Example 1 as well as to the group of symmetries of a rectangle. This group with four elements is often referred to as the **Klein four group** and is denoted $\mathbb{Z}_2 \times \mathbb{Z}_2$ or $C_2 \times C_2$ depending on whether we wish to use additive or multiplicative notation.

Theorem 5.3.3   *Let m and n be relatively prime integers. Then the direct product $C_m \times C_n$ is cyclic.*

Proof   Let a and b be generators for C_m and C_n respectively. So, by 5.1.7, the order of a is m and that of b is n. Then, for any integer k,

$$(a, b)^k = (a^k, b^k)$$

(this is proved by induction, using the definition of the group operation in the direct product). Thus, if $(a, b)^k = (e, e) = (a, b)^0$ then we have, by Theorem 5.1.4, that both m and n divide k. Since m and n are relatively prime, mn divides k by Theorem 1.1.6.

We also have $(a, b)^{mn} = (a^{mn}, b^{mn}) = ((a^m)^n, (b^n)^m) = (e, e)$. It follows that the order of (a, b) is mn. Thus the distinct powers of (a, b) exhaust the group $C_m \times C_n$ (which has mn elements) and so the group is indeed cyclic.   □

We now start our classification of groups of small orders.

Groups of order 1   Any group contains an identity element e, so if G has only one element then G consists of only the identity element. Clearly any two such groups are isomorphic!

Groups of order 2   If G has two elements, we must have $G = \{e, g\}$ for some g different from e. Since G is a group and so is closed under the operation, g^2 is in G. Now, g^2 cannot be g, for $g^2 = g$ implies (multiply each side by g^{-1}) $g = e$. It must therefore be that $g^2 = e$. This lets us construct the group

$$g^{-1}g^2 = g^{-1}g$$
$$g \neq e$$

table and also shows that there is only one possibility for the shape of this table. Hence there is (up to isomorphism) just the one group of order 2.

	e	g
e	e	g
g	g	e

Groups of order 3 Suppose that G has three different elements e (identity), g and h. We must have $gh = e$ (otherwise $gh = g$ or $gh = h$ and cancelling gives a contradiction). Similarly $hg = e$. This is enough to allow us to construct the group table. For example, g^2 is not e (otherwise $g^2 = e = gh$ so cancelling gives $g = h$, contrary to assumption) nor is it g (otherwise $g = e$) and so g^2 must be h. Thus G has the table shown.

	e	g	g^2
e	e	g	g^2
g	g	g^2	e
g^2	g^2	e	g

Remark Since 2 and 3 are prime numbers, the above two cases are, in fact, covered by the Corollary 5.2.5 to Lagrange's Theorem and the remarks on pp. 220 and 221.

Groups of order 4 First suppose that there is an element g in G of order 4. Then G must consist of $\{e, g, g^2, g^3\}$, and the multiplication table is constructed easily (the first table below): we note that G is cyclic.

If there is no element of order 4 then, by Corollary 5.2.4, each non-identity element of $G = \{e, g, h, k\}$ must have order 2. Also, by the kind of cancelling argument that we have already used, it must be that $gh = k$. The following result is of general use.

Theorem 5.3.4 *Let G be a group in which the square of every element is 1. Then G is Abelian.*

Proof For all g in G, we have that $g^2 = e = gg$. Thus every element is its own inverse. Since the inverse of xy is also $(xy)^{-1} = y^{-1}x^{-1}$ by 5.1.2, we have

$$xy = (xy)^{-1} = y^{-1}x^{-1} = yx$$

and so G is Abelian. $\square$

This applies to our non-cyclic group $G = \{e, g, h, k\}$ and we construct the group table (the second below) easily using the facts that rows and columns of group tables can have no repeated elements. We have shown that any group with four elements is isomorphic to one of the two groups given by the tables below.

	e	g	g^2	g^3
e	e	g	g^2	g^3
g	g	g^2	g^3	e
g^2	g^2	g^3	e	g
g^3	g^3	e	g	g^2

	e	g	h	k
e	e	g	h	k
g	g	e	k	h
h	h	k	e	g
k	k	h	g	e

In other terms, we have shown that a group of order 4 is either cyclic or isomorphic to the Klein four group $C_2 \times C_2$.

Groups of order 5   Since 5 is a prime, we deduce (by Corollary 5.2.5) that G is cyclic, isomorphic to C_5, and consists of the powers of an element of order 5. Hence we can draw up the group table.

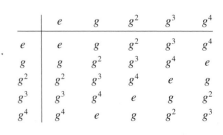

	e	g	g^2	g^3	g^4
e	e	g	g^2	g^3	g^4
g	g	g^2	g^3	g^4	e
g^2	g^2	g^3	g^4	e	g
g^3	g^3	g^4	e	g	g^2
g^4	g^4	e	g	g^2	g^3

Groups of order 6   If G contains an element of order 6 then there is no room in G for anything other than the powers of this element and so G is cyclic, isomorphic to C_6.

By Lagrange's Theorem, the only possible orders of elements of G are 1, 2, 3 and 6. Suppose then that G does not contain an element of order 6. If G contained no element of order 3 then all the non-identity elements of G would have to have order 2 and then, by Theorem 5.3.4, G would be Abelian. If that were the case, let a and b be non-identity elements of G. Then ab cannot be e, a or b (by 'cancelling' arguments). It follows that $\{e, a, b, ab\}$ is a subgroup of G (for this set is closed under products and inverses). But 4 does not divide 6, so Lagrange's Theorem (5.2.3) says that this is impossible.

Therefore G does have an element a (say) of order 3. Thus we have three of the elements of G: e, a, a^2. Let b be any other element of G. Then the six elements e, a, a^2, b, ba, ba^2 must be distinct: just note that any equation between them, on cancelling, leads to something contrary to what we have assumed. For example, if we have already argued that e, a, a^2, b and ba are distinct, then ba^2 is different from each:

if $ba^2 = e$, then b would be the inverse of a^2, so $b = a$;
if $ba^2 = a$, then $ba = e$ and b would be $a^{-1} = a^2$;
if $ba^2 = a^2$, then $b = e$;
if $ba^2 = b$, then $a^2 = e$; and
if $ba^2 = ba$, then $a = e$.

In a similar way, we can show that b^2 must be e, since if b^2 were equal to b, ba or ba^2, we could deduce that the elements would not be distinct. If b^2 were a or a^2, it would follow that the powers of b $(b, b^2, \ldots, b^5)$ would be distinct and hence G would be cyclic. Similar arguments show that ab must be ba^2 and that in fact the multiplication table of G is that shown on the right below.

	e	g	g^2	g^3	g^4	g^5
e	e	g	g^2	g^3	g^4	g^5
g	g	g^2	g^3	g^4	g^5	e
g^2	g^2	g^3	g^4	g^5	e	g
g^3	g^3	g^4	g^5	e	g	g^2
g^4	g^4	g^5	e	g	g^2	g^3
g^5	g^5	e	g	g^2	g^3	g^4

	e	a	a^2	b	ba	ba^2
e	e	a	a^2	b	ba	ba^2
a	a	a^2	e	ba^2	b	ba
a^2	a^2	e	a	ba	ba^2	b
b	b	ba	ba^2	e	a	a^2
ba	ba	ba^2	b	a^2	e	a
ba^2	ba^2	b	ba	a	a^2	e

The group on the left is cyclic of order 6, while that on the right is (necessarily) isomorphic to the symmetric group $S(3)$. An isomorphism f from the group on the right to $S(3)$ is given by

$$f(e) = \text{id}; \qquad f(a) = (1\ 2\ 3); \qquad f(a^2) = (1\ 3\ 2);$$
$$f(b) = (1\ 2); \qquad f(ba) = (2\ 3); \qquad f(ba^2) = (1\ 3).$$

Therefore a group of order 6 is isomorphic either to the cyclic group of order 6 or to the group $S(3)$.

Groups of order 7   As in the case of orders 3 and 5 we see that the only possibility is a cyclic group, C_7, the cyclic group with 7 elements, since 7 is prime. Drawing up the group table is left as an exercise.

Groups of order 8   At this point, we will just present the answer and leave the details of the calculations to Exercise 5.3.10 at the end of the section. By the kind of analysis we used for groups of order 6, it can be shown that there are five different types of group with eight elements. Three of these are Abelian and are C_8, $C_4 \times C_2$ and $C_2 \times C_2 \times C_2$. There are two types of non-Abelian group, the dihedral group $D(4)$ and the quaternion group $\mathbb{H}_0$. The tables for the last two can be found in the solution for Exercise 4.3.7, and in Example 5 of Section 4.3.1, respectively.

In the case of groups of order 8, it can be seen that the techniques we have developed become rather stretched and other methods are required to make progress on the problem of finding all groups with a given number of elements. The interested reader should consult one of the abundance of more advanced books devoted to group theory to see what techniques are available to study groups in general. However, it may be of interest to say a little more about the classification problem for finite groups.

In some sense, every finite group is built up from 'simple groups'. In order to define this term, we need to introduce the notion of a normal subgroup of a group. A subgroup N of a group G is **normal** if, for each g in G the left coset gN is equal to the right coset Ng. A group G is **simple** if the only normal subgroups of G are G itself and the trivial subgroup $\{e\}$. The importance of normal subgroups is on account of the following (cf. Example 3 before Theorem 5.3.2 and Example 3 before Theorem 5.3.3).

If N is a normal subgroup of the group G then the set of cosets of N in G may be turned into a group by defining $(gN) \cdot (hN) = ghN$ (normality of N is needed for this multiplication to be well defined). This group of cosets is denoted by G/N: it is obtained from G by 'collapsing' the normal subgroup to a single element (the identity) of the group G/N. One may say that the group G is built up from the normal subgroup N and the group G/N. So, if a group is not simple, then it may in some sense be decomposed into two smaller (so simpler) pieces. But a simple group cannot be so decomposed. Therefore the simple groups are regarded as the 'building blocks' of finite groups, in a way analogous to that in which the prime numbers are the 'building blocks' of positive integers.

The complete list of finite simple groups is known and its determination, which was essentially completed in the early 1980s, was one of the great achievements in mathematics. There are two aspects to this result. One is the production of a list of finite simple groups and the other is the verification that every finite simple group is on the list.

Most of the finite simple groups fall into certain natural infinite families of closely related groups such as groups of permutations or matrices: for instance, the alternating $A(n)$ for $n \geq 5$ are simple. But five anomalous, or sporadic, simple groups, groups which do not fit into any infinite family, were discovered by Mathieu between 1860 and 1873. No more sporadic simple groups were found for almost a hundred years, until Janko discovered one more in 1966. Between then and 1983 a further 20 were found, bringing the total of sporadic simple groups to 26. The last found and largest of these is the so-called Monster (or Friendly Giant). The possible existence of this simple group was predicted in the mid-1970s and a construction for it was given by Robert Greiss in 1983. This is a group in which the number of elements is

$$2^{46} \times 3^{20} \times 5^9 \times 7^6 \times 11^2 \times 13^3 \times 17 \times 19 \times 23 \times 29 \times 31 \times 41 \times 47 \times 59 \times 71.$$

Clearly, one cannot draw up the multiplication table for such a group, and it is an indication of the power of techniques in current group theory that a great deal is understood about this largest sporadic simple group.

With Greiss' construction in 1983, the last finite simple group has been found, and thus the classification is complete, since all other possibilities for new finite simple groups have been excluded. It is estimated that over 200 mathematicians have contributed to this classification of the finite simple groups, and that the detailed reasoning to support the classification occupies over 15 000 printed pages. Since the first proof a number of mathematicians have been working to simplify the details.

Exercises 5.3

1. For each of the following groups with four elements, determine whether it is isomorphic to $\mathbb{Z}_2 \times \mathbb{Z}_2$ or $\mathbb{Z}_4$:
   (i)   the multiplicative group G_8 of invertible congruence classes modulo 8;
   (ii) the cyclic subgroup $\langle \rho \rangle$ of $D(4)$ generated by the rotation ρ of the square through $2\pi/4$;
   (iii) the groups with multiplication tables as shown (where the identity element does not necessarily head the first row and first column).

	a	b	c	d
a	d	c	a	b
b	c	d	b	a
c	a	b	c	d
d	b	a	d	c

	a	b	c	d
a	b	a	d	c
b	a	b	c	d
c	d	c	b	a
a	c	d	a	b

2. Consider the subgroup of $S(4)$ and the group of symmetries of the rectangle discussed after Theorem 5.3.1. There we defined one isomorphism between these groups but this is not the only one. Find all the rest.
   [Hint: choose any two elements of order 2; what are the restrictions on where an isomorphism can take them? Having determined where these two

elements are sent by the isomorphism, is there any choice for the destinations of the other two elements?]

3. Let G be any group and let g be an element of G. Define the function $f : G \longrightarrow G$ by $f(a) = g^{-1}ag\,(a \in G)$ (thus f takes every element to its conjugate by g). Show that f is an isomorphism from G to itself. Show, by example, that f need not be the identity function.

4. Give an example of cyclic groups G and H such that $G \times H$ is not cyclic.

5. Show that $G \times H$ is Abelian if and only if both G and H are Abelian.

6. Show that the set $\{(g, e) : g \in G\}$ forms a subgroup of $G \times H$.

7. Write down the multiplication tables for the direct products
   (i) $\mathbb{Z}_4 \times \mathbb{Z}_2$;
   (ii) $G_5 \times G_3$;
   (iii) $\mathbb{Z}_2 \times \mathbb{Z}_2 \times \mathbb{Z}_2$;
   (iv) $G_{12} \times G_4$.
   Which of the above groups are isomorphic to each other?

8. Let G be a group with six elements and let H be a group with fourteen elements. What are the possible orders of elements in the direct product $G \times H$?

9. Use the classification of groups with six elements to show that $A(4)$ has no subgroup with 6 elements.
   [Hint: check that the product of any two elements of $A(4)$ of order 2 has order 2.]

10. Let G be a non-Abelian group with eight elements. Show that G has an element a say of order 4. Let b be an element of G which is not e, a, a^2 or a^3. By considering the possible values of b^2 and of ba and ab, show that G is isomorphic either to the dihedral group or to the quaternion group.

5.4   Error-detecting and error-correcting codes

Messages sent over electronic and other channels are subject to distortions of various sorts. For instance, messages sent over telephone lines may be distorted by other electromagnetic fluctuations; information stored on a disc may become corrupted by strong magnetic fields. The result is that the message received or read may be different from that originally sent or stored. Extreme examples are the pictures sent back from space probes, where a very high error rate occurs.

It is therefore important to know if an error has occurred in transmission: for then one may ask that the message be repeated. In some circumstances it may be impossible or undesirable for the message to be repeated. In that case, the message should carry a certain degree of redundancy, so that the original message may be reconstructed with a high degree of certainty. The way to do this is to add a number of check symbols to the message so that errors may be detected or even corrected. In general, the greater the number of check symbols, the more unlikely it is that an error will go undetected.

Notice that the codes discussed here are not designed to prevent confidential information from being read (in contrast to the public key codes of Chapter 1). The object here is to ensure the accuracy of the message after transmission.

Another point which should perhaps be made explicit here is that there can be no method which reconstructs a distorted message with 100 per cent accuracy. When we say that a received message m is 'corrected' to m_1, we mean that it is *most likely* that m_1 was the message originally sent. If there is a low frequency of errors, then the probability that an error is wrongly 'corrected' may be made extremely small. This point is discussed further in Example 2 below.

We introduce the general idea of a coding function and discuss the concepts of error detection and correction. Then we specialise to linear codes. One way to produce codes of this sort is to use a generator matrix. This matrix tells us how to build in some redundancy by adding check digits to the original message. The subsequent correction of the message can be carried out using a coset decoding table. In producing this table, we again encounter the idea of a coset which was fundamental to the proof of Lagrange's Theorem.

Example A well known example of error correction is provided by the ISBN (International Standard Book Number) of published books. This is a sequence of nine digits $a_1 a_2 \ldots a_9$ where each a_i is one of the numbers $0, 1, \ldots, 9$, together with a check digit which is one of the symbols $0, 1, 2, 3, 4, 5, 6, 7, 8, 9$ or X (with X representing 10). This last digit is included so as to give a check that the previous 9 digits have been correctly transcribed, and is calculated as follows. Form the integer $n = 10a_1 + 9a_2 + 8a_3 + \cdots + 2a_9$ and reduce this modulo 11 to obtain an integer b between 1 and 11 inclusive. The check digit is obtained by subtracting b from 11. Thus, for the number 0 521 35938, b is obtained by reducing

$$n = 0 + 45 + 16 + 7 + 18 + 25 + 36 + 9 + 16$$

modulo 11: the result is 7 and the check digit is $11 - 7 = 4$. The ISBN corresponding to 0 521 35938 is therefore 0 521 35938 4. If a librarian or bookseller made a single error in copying this number (say the IBSN was written as 0 521 35958 4) then the check digit obtained from the first nine digits would not be 4. (This is because an error of size k in position i would produce an error of k times $11 - i$ in n. Since 11 is prime, the error thereby introduced into the sum would not be divisible by 11.) This is an example of a code which *detects* a single error. Since there is no way of telling where the error is, the code is not error *correcting*.

The bar code used on many products also contains a check digit.

In the ISBN code, numbers are represented in the decimal system but we will consider from now on information which is stored or transmitted in binary form (of course the same general principles apply to other cases). This includes any information handled by a computer. Most other forms may be converted to binary: for example, English text may be converted by replacing each letter, numeral, space or punctuation mark by a suitable binary-based code (such as ASCII) for it. So from now on, we will consider only codes which apply to strings of 0s and 1s. We will think of the set $\{0,1\}$ as coming equipped with the operation of addition (and multiplication) mod 2: it is customary in this context to write $\mathbf{B}$ instead of $\mathbb{Z}_2$ (since $\mathbb{Z}_2$ with the operations of addition and multiplication is a Boolean ring, see Exercise 4.4.16).

Definition　A **word** of **length** n is a string of n binary digits. Thus 0001, 1110 and 0000 are words of length 4. We shall think of words of length n as members of $\mathbf{B}^n$, the Cartesian product of n copies of the binary set $\mathbf{B}$ regarded as an Abelian group under addition. In this notation, if w is a word in $\mathbf{B}^n$ and x is in $\mathbf{B}$ we use wx to denote the word of length $n + 1$ with its first n 'letters' being those of w and its last letter being x. This should not be confused with the product of w by x.

We formalise the idea of check symbols as follows. Suppose our original messages are composed of words of length m; we choose a 'coding function' $f : \mathbf{B}^m \longrightarrow \mathbf{B}^n$ and, instead of sending a word w, we send the word $f(w)$. Thus the messages we send are composed of words of length n (rather than m). Any word of length n in the image of f is called a **codeword**.

There is an obvious constraint on a suitable coding function f: f should be injective, otherwise there would be two different words of length m that would be sent as the same word of length n. Note that this means that n should be greater than or equal to m and, in practice, strictly greater than m since we wish to add some check digits ($n - m$ of them).

Here are two examples of coding functions.

Example 1　Define $f : \mathbf{B}^m \longrightarrow \mathbf{B}^{m+1}$ by $f(w) = wx$ where x is 0 if the number of non-zero digits in w is even, and x is 1 if the number of non-zero digits in w is odd. To give a more specific example, take $m = 3$: there follow the eight words in $\mathbf{B}^3$ and beneath each is its image under f.

$$
\begin{array}{cccccccc}
000 & 001 & 010 & 011 & 100 & 101 & 110 & 111 \\
0000 & 0011 & 0101 & 0110 & 1001 & 1010 & 1100 & 1111
\end{array}
$$

The last digit is a parity-check digit: any correctly transmitted word has an even number of 1s in it. Therefore this code enables one to detect any single error in the transmission of a codeword since, if a single digit is changed, the word received will then have an odd number of 1s in it and so not be a *codeword*. In fact any odd number of errors will be detected, but an even number of errors will fail to be detected. Another point about this code is that it does not allow one to correct an error without re-transmission of the word.

Example 2   Define $f : \mathbf{B}^m \longrightarrow \mathbf{B}^{3m}$ by $f(w) = www$; thus the word is simply repeated three times. So, for example, if $m = 6$ and if $w = 101111$ then $f(w) =$ 101111101111101111. You should convince yourself that this code will detect any single error or any two (non-cancelling) errors. For instance if $f(w)$ as above is received as 100111101011101111 (with two errors) then we can see at once that an error has occurred in transmission since the received message is not a six-letter word three times repeated. However this code does not necessarily detect three errors. If $f(w)$ were received as 001111001111001111 (three errors) then it looks as if the original word w was 001111, whereas the original word was 101111.

It should also be noted that although two errors can be *detected*, this code can *correct* only one error. For instance if $f(w)$ were received as 101011101011101111 then we could consider it most likely that the original word was 101011, not 101111. Let us be more explicit.

Suppose that www is the word sent and m is the word received: breaking it into three blocks of six letters each, we write $m = abc$ where a, b and c are words of length 6 and 'abc' means the word whose first six letters are those of a, whose next six letters are those of b and whose last six letters are those of c. If no errors have been made in transmission then $a = b = c = w$. If one error has been made, then two of a, b, and c are equal to each other (so, necessarily, to w), so we correct the message and conclude (correctly) that the original word is w. If two errors have been made then it could happen that $abc = w'ww'$ (say): we would then conclude (incorrectly) that the original word was w'. We say, therefore, that this code can correct one error (but not two).

Thus we correct errors on the basis of 'most likely message to have been sent, on the assumption that errors occur randomly'. Suppose, for illustration, that the probability that a given digit is transmitted wrongly is 1 in 1000 (0.001). Then the probability that a single transmitted word (18 digits) contains a single error is 0.017 696 436 (that is, about 1 in 60); the probability that a given word contains two (respectively three or more) errors is 0.000 150 570 (0.000 000 806). It follows that the probability of incorrectly 'correcting' a word containing an error is less than one in ten thousand and the probability of failing to detect that a received word is erroneous is about one in one hundred million (note that, even given that three errors have occurred, it is a small probability that the result consists of a six-letter word three times repeated). A book of 1000 pages, 40 lines to a page, around 60 characters (including spaces) to a line, contains about 2 400 000 characters. Six binary digits are easily sufficient to represent the alphabet plus numerals and punctuation, so the book may be represented by 2 400 000 binary words of length 6. So, with the above likelihood of error, the probability that even just one character of the book is transmitted wrongly and the error not detected is about 1 in 40.

The code in Example 2 above is superior to that in Example 1 in that it can detect up to two errors (rather than only one) and can even correct any single error. On the other hand it requires the sending of a message three times as long as the original one, whereas the first code involves only a slight increase in the length of message sent. Most of our examples below will be more efficient than this second one.

Definitions   The **weight** of a binary word w, wt(w), is defined to be the number of 1s in its binary expression. Thus, for example

$$\text{wt}(001101) = 3; \qquad \text{wt}(000) = 0; \qquad \text{wt}(111) = 3.$$

The **distance** between binary words v, w of the same length is defined to be the weight of their difference:

$$d(v, w) = \text{wt}(v - w).$$

You should note that, since we are working in a product of copies of **B** (in which $+1 = -1$), every element is its own additive inverse and so, for example $v - w = v + w$. Hence $d(v, w) = \text{wt}(v + w)$. It follows also that the ith entry of $v + w$ is 0 if the ith entries of v and w are the same, and is 1 if v and w differ at their ith places. Hence the distance between v and w is simply the number of places at which they differ.

Example

$$d(010101, 101000) = \text{wt}(010101 + 101000) = \text{wt}(111101) = 5;$$

$$d(1111, 0110) = \text{wt}(1111 + 0110) = \text{wt}(1001) = 2;$$

$$d(w, w) = \text{wt}(w + w) = 0 \text{ for any word } w.$$

If w is a word that is transmitted and is received as v then the number of errors which have occurred in transmission is (ignoring errors which cancel) the distance between v and w (for the alteration of a single digit of a word results in a word at a distance of 1 from the original word). Therefore a good coding

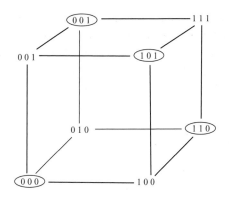

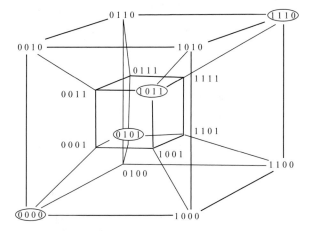

Fig. 5.4

function $f : \mathbf{B}^m \longrightarrow \mathbf{B}^n$ will be one which maximises the minimum distance between different codewords.

We can illustrate this by drawing a graph, rather like those in Section 2.3 but with undirected edges. For the vertices of the graph, we take all the binary words of length n, and we join two vertices by an edge if the distance between them is 1 (that is, if an error in a single digit can convert one to the other). Then the distance between two words is the number of edges in a shortest path from one to the other. A good coding function is one for which the codewords are well 'spread' through this graph.

We show a couple of examples (Fig. 5.4), in which the codewords have been ringed. We have limited ourselves to words of length 3 and 4, since words of length n would most naturally be represented as the vertices of an

n-dimensional cube, and representing a five-dimensional cube on a piece of paper would be messy. The first example shows the codewords for the parity-check
code $f : \mathbf{B}^2 \longrightarrow \mathbf{B}^3$. The second shows the codewords for a coding function
$f : \mathbf{B}^2 \longrightarrow \mathbf{B}^4$.

Theorem 5.4.1 *Let $f : \mathbf{B}^m \longrightarrow \mathbf{B}^n$ be a coding function. Then f allows the
detection of k or fewer errors if and only if the minimum distance between
distinct codewords is at least $k + 1$.*

Proof If a word w is obtained from a codeword by making k (or fewer) changes
then w cannot be another codeword if the minimum distance between distinct
codewords is $k + 1$. Thus the code will detect these errors. Conversely, if the
code detects k errors, then no two codewords can be at a distance k from each
other (for then k errors could convert one codeword to another and the change
would not be detected). □

Theorem 5.4.2 *Let $f : \mathbf{B}^m \longrightarrow \mathbf{B}^n$ be a coding function. Then f allows the
correction of k or fewer errors if and only if the minimum distance between
distinct codewords is at least $2k + 1$.*

Proof If the distance between the codewords v and w is $2k + 1$ then $k + 1$
errors in the transmission of v will indeed be detected. But the resulting word
may be closer to w than to v, and so any attempt at error correction could result
in the (incorrect) interpretation that w was more likely than v to have been the
word sent. □

Example 3 Suppose we define the coding function $f : \mathbf{B}^4 \longrightarrow \mathbf{B}^9$ by setting
$f(w) = wwx$ where x is 0 or 1 according as the weight of the word w is even
or odd (so our coding function repeats the word and also has a parity-check
digit). Opposite each word w in $\mathbf{B}^4$ we list $f(w)$:

0000	000000000	0001	000100011
0010	001000101	0011	001100110
0100	010001001	0101	010101010
0110	011001100	0111	011101111
1000	100010001	1001	100110010
1010	101010100	1011	101110111
1100	110011000	1101	110111011
1110	111011101	1111	111111110

You may check, by computing $d(u, v)$ for all $u \neq v$, that the minimum distance between codewords is 3 and so, by Theorems 5.4.1 and 5.4.2, the code detects up to two errors and can correct any single error.

Checking the minimum distance between codewords is a tedious task: but it can be circumvented for certain types of codes by using a little theory.

Definition Let $f : \mathbf{B}^m \longrightarrow \mathbf{B}^n$ be a coding function. We say that this gives a **linear code** if the image of f forms a subgroup of $\mathbf{B}^n$. (These codes are also referred to as group codes, but the word 'linear' helps to remind us that the group operation is addition.)

What do we have to check in order to show that we have a linear code? Since the group operation is addition, we must show that if words u and v are in the image of f then so is the word $u + v$. We do not have to check that if u is a codeword then $-u$ also is a codeword since in $\mathbf{B}^n$ every element is self-inverse! ($-u = u$): also '$\mathbf{0}$' may be obtained as $u + u$ for any u in the image of f (where $\mathbf{0}$ denotes the codeword with all entries '0').

One advantage of linear codes is that the minimum distance between codewords is relatively easily found.

Theorem 5.4.3 *Let $f : \mathbf{B}^m \longrightarrow \mathbf{B}^n$ be a linear code. Then the minimum distance between distinct codewords is the lowest weight of a non-zero codeword.*

Proof Let d be the minimum distance between distinct codewords: so $d(u, v) = d$ for some codewords u, v. Let x be the minimum weight of a non-zero codeword: so $\mathrm{wt}(w) = x$ for some codeword w. Then, since d is the minimum distance between codewords and w and $\mathbf{0}$ are codewords, we have:

$$d \leq d(w, \mathbf{0}) = \mathrm{wt}(w + \mathbf{0}) = \mathrm{wt}(w) = x.$$

On the other hand

$$d = d(u, v) = \mathrm{wt}(u + v).$$

Since we have a linear code, $u + v$ is a (non-zero) codeword so, by minimality of x, $\mathrm{wt}(u + v) \geq x$: thus $d \geq x$. Hence $d = x$ as claimed. $\square$

We now present a method of producing linear codes.

Definition Let m, n be integers with $m < n$. A **generator matrix** G is a matrix with entries in $\mathbf{B}$ and with m rows and n columns, the first m columns of which form the $m \times m$ identity matrix $\mathbf{I}_m$. We may write such a matrix as a partitioned matrix: $G = (\mathbf{I}_m A)$ where A is an $m \times (n - m)$ matrix.

For instance, the following are generator matrices (of various sizes):

$$\begin{pmatrix} 1 & 0 & 1 \\ 0 & 1 & 1 \end{pmatrix}; \qquad \begin{pmatrix} 1 & 0 & 0 & 1 \\ 0 & 1 & 0 & 1 \\ 0 & 0 & 1 & 1 \end{pmatrix};$$

$$\begin{pmatrix} 1 & 0 & 0 & 1 & 1 & 0 \\ 0 & 1 & 0 & 1 & 0 & 1 \\ 0 & 0 & 1 & 0 & 0 & 0 \end{pmatrix}; \qquad \begin{pmatrix} 1 & 0 & 0 & 0 & 1 & 0 \\ 0 & 1 & 0 & 0 & 1 & 0 \\ 0 & 0 & 1 & 0 & 0 & 1 \\ 0 & 0 & 0 & 1 & 0 & 1 \end{pmatrix}.$$

Given a generator matrix G with m rows and n columns, we may define the corresponding coding function $f = f_G : \mathbf{B}^m \longrightarrow \mathbf{B}^n$ by treating the elements of $\mathbf{B}^m$ as row vectors and setting $f_G(w) = wG$ for w in $\mathbf{B}^m$.

For example, suppose that G is the second matrix above: so $f_G : \mathbf{B}^3 \longrightarrow \mathbf{B}^4$. We have, for instance

$$f_G(011) = (011) \cdot G = (011) \begin{pmatrix} 1 & 0 & 0 & 1 \\ 0 & 1 & 0 & 1 \\ 0 & 0 & 1 & 1 \end{pmatrix} = (0110).$$

Observe that if $f = f_G : \mathbf{B}^m \longrightarrow \mathbf{B}^n$ arises from a generator matrix then the first m entries of the codeword $f(w)$ form the original word w: thus $f(w)$ has the form wv, where v contains the 'check digits'.

Theorem 5.4.4 *Let G be a generator matrix. Then f_G is a linear code and, in fact, $f_G(v + w) = f_G(v) + f_G(w)$ for all words v, w.*

Proof We have $f_G(\mathbf{0}_m) = \mathbf{0}_m G = \mathbf{0}_n$ where $\mathbf{0}_k$ denotes the k-tuple with all entries '0': so the set of codewords is non-empty. Suppose that u, u' are codewords: say

$$u = f_G(v) \text{ and } u' = f_G(w).$$

Then

$$f_G(v + w) = (v + w)G$$
$$= vG + wG \text{ (matrix multiplication distributes over addition)}$$
$$= f_G(v) + f_G(w)$$
$$= u + u'.$$

Thus the set of codewords is closed under addition. We have already noted that we do not need to check for inverses because every element is its own inverse. Thus the set of codewords is a group, as required.   □

Referring back to Example 3 above we see that we can use Theorem 5.4.3 to derive that the minimum distance between (distinct) codewords is 3, the minimum weight of a non-zero codeword. We can justify the use of Theorem 5.4.3 by checking that the code is given by the generator matrix below (then appealing to Theorem 5.4.4):

$$G = \begin{pmatrix} 1 & 0 & 0 & 0 & 1 & 0 & 0 & 0 & 1 \\ 0 & 1 & 0 & 0 & 0 & 1 & 0 & 0 & 1 \\ 0 & 0 & 1 & 0 & 0 & 0 & 1 & 0 & 1 \\ 0 & 0 & 0 & 1 & 0 & 0 & 0 & 1 & 1 \end{pmatrix}$$

Examples 1 and 2 on pp. 232 and 233 are also linear codes obtained from generator matrices. In Example 1, the matrix G is obtained from the $m \times m$ identity matrix by adding a column of 1s to the right of it. In Example 2, the generator matrix simply consists of three $m \times m$ identity matrices placed side by side. Let us consider some further examples. After giving the generator matrix G we list the words of $\mathbf{B}^m$ beside the corresponding codewords in $\mathbf{B}^n$.

Example 1 Consider the coding function $f : \mathbf{B}^2 \longrightarrow \mathbf{B}^4$ with generator matrix

$$G = \begin{pmatrix} 1 & 0 & 1 & 1 \\ 0 & 1 & 0 & 1 \end{pmatrix}$$

00	0000
01	0101
10	1011
11	1110

The minimum distance between codewords is 2 (being the minimum weight of a non-zero codeword), so the code can detect one error but cannot correct errors.

Example 2 Next take $f : \mathbf{B}^2 \longrightarrow \mathbf{B}^5$ with

$$G = \begin{pmatrix} 1 & 0 & 1 & 1 & 0 \\ 0 & 1 & 0 & 1 & 1 \end{pmatrix}$$

00	00000
01	01011
10	10110
11	11101

The minimum distance between codewords is 3, so the code can detect two errors and correct one error.

Example 3   Consider the coding function $f : \mathbf{B}^3 \longrightarrow \mathbf{B}^6$ with

$$G = \begin{pmatrix} 1 & 0 & 0 & 1 & 1 & 1 \\ 0 & 1 & 0 & 1 & 0 & 1 \\ 0 & 0 & 1 & 1 & 1 & 1 \end{pmatrix}$$

000	000000
001	001111
010	010101
011	011010
100	100111
101	101000
110	110010
111	111101

The minimum distance between codewords is 2 (being the minimum weight of a non-zero codeword), so the code can detect one error but cannot correct even all single errors.

Example 4   Finally consider the coding function $f : \mathbf{B}^3 \longrightarrow \mathbf{B}^{10}$ given by

$$G = \begin{pmatrix} 1 & 0 & 0 & 1 & 1 & 0 & 1 & 1 & 0 & 0 \\ 0 & 1 & 0 & 1 & 0 & 1 & 1 & 0 & 1 & 0 \\ 0 & 0 & 1 & 0 & 1 & 1 & 1 & 0 & 0 & 1 \end{pmatrix}$$

000	0000000000
001	0010111001
010	0101011010
011	0111100011
100	1001101100
101	1011010101
110	1100110110
111	1110001111

The minimum distance between codewords is 5, so the code can detect four errors and correct two errors. Note how much better this is than the code which simply repeats the word a total of four times: the latter code uses more digits (12 instead of 10) yet detects and corrects fewer errors (it detects three and can correct one error).

If you are familiar with 'linear algebra', you may have realised that in our use of matrices we are essentially doing linear algebra over the field $\mathbf{B} = \mathbb{Z}_2$ of 2 elements. More precisely, in the terminology of Section 4.4, we are regarding $\mathbf{B}^n$ as a vector space over the field $\mathbf{B} = \mathbb{Z}_2$.

Now we describe how to arrange the work of detecting and correcting errors in any received message. *Suppose throughout that we are using a linear code* $f : \mathbf{B}^m \longrightarrow \mathbf{B}^n$. Let W be the set of codewords in $\mathbf{B}^n$ (a subgroup of $\mathbf{B}^n$).

Suppose that the word w is sent but that an error occurs in, say, the last digit, resulting in the word v being received. So v agrees with w on each digit but the last, where it differs from w: that is $v = e_n + w$ where e_n is the word of length n which has all digits '0' except the last, which is '1'. Thus the set of words which may be received as the result of a single error in the last digit is precisely the set $e_n + W$. In the language of group theory, this is precisely the coset of e_n with respect to the subgroup W of $\mathbf{B}^n$. (Since the group operation is addition, the (left) coset of W containing e_n is written as $e_n + W$ rather than $e_n W$.)

In the same way we see that an error in the ith digit converts the subgroup W of codewords into the coset $e_i + W$ where e_i is the word of length n which has all entries '0' except the ith, which is '1'. Similarly two, or more, errors result in the set of codewords being replaced by a coset of itself. For instance if $n = 10$ then an error in the third digit combined with an error in the fifth digit transforms the codeword w into the word $0010100000 + w$, so replaces the subgroup W of codewords with the coset $0010100000 + W$.

Suppose then that we receive a word which is not a codeword. We know that at least one error has occurred: we wish to recover the word that was sent without having the message retransmitted. Of course there is a problem here: even if the word received is a codeword it is possible that it is not the original word sent (for example if the distance between two codewords is three, then three errors could conspire to convert the one to the other). So we must simply be content to recover the word most likely to have been sent. This is known as *maximum likelihood decoding*. What we do therefore is to 'correct' the message by replacing the received word v by the codeword to which it is closest. Thus we compute the distances between the received word v and the various codewords w, look for the minimum distance $d(v, w)$, and replace v by w. In the event of a tie we just choose any one of the closest codewords.

Rather than do the above computation every time we receive an erroneous word, it is as well to do the computations once and for all and to prepare a table showing the result of these computations. We will now describe how to draw up such a **coset decoding table**.

We are supposing that we have a coding function $f : \mathbf{B}^m \longrightarrow \mathbf{B}^n$ for which the associated code is a linear code and we define W to be the subgroup of $\mathbf{B}^n$ consisting of all codewords. List the elements of W in some fixed order with the zero n-tuple as the first element, then array these as the top row of the decoding table. Now look for a word of $\mathbf{B}^n$ of minimum weight among those *not* in W: there will probably be more than one of these, so just choose any one, v say. Beneath each codeword w on the top row place the word $v + w$. Thus the second row of our table lists the elements of the coset $v + W$. Next look for an element u of minimum weight which is *not already listed* in our table and list its coset, just as we listed the coset of v, to form the next line of the table (so beneath each word w of W we now have $v + w$ and, beneath that, $u + w$). Repeat this procedure until all elements of $\mathbf{B}^n$ have been listed.

This decoding table is used as follows: on receiving the word v we look for where it occurs in the table (looking for it in the top row first); having found it we replace it by the (code)word which lies directly above it on the top row.

Example   Consider the coding function $f : \mathbf{B}^2 \longrightarrow \mathbf{B}^5$ and generator matrix G as in Example 2 on p. 239 (so f is a linear code). We saw that the codewords are

$$00000 \quad 01011 \quad 10110 \quad 11101.$$

We will retain this order and array them as the first line of the table. Next, look for a word of minimum weight not in this list. There are five words $e_1, \ldots,$ e_5 of weight 1 and none of these is in this list. We just choose any one of them, say 00001. The second row of the table is formed by placing beneath each codeword w the word $00001 + w$ (which is as w but with its last digit changed):

$$
\begin{array}{cccc}
00000 & 01011 & 10110 & 11101 \\
00001 & 01010 & 10111 & 11100
\end{array}
$$

Now look for a word of minimum weight not yet in the table. There are four choices $\{e_1, \ldots, e_4\}$; let us be systematic and take 00010 to obtain

$$
\begin{array}{cccc}
00000 & 01011 & 10110 & 11101 \\
00001 & 01010 & 10111 & 11100 \\
00010 & 01001 & 10100 & 11111
\end{array}
$$

For our next three choices we may take e_3, e_2, and e_1:

$$
\begin{array}{cccc}
00000 & 01011 & 10110 & 11101 \\
00001 & 01010 & 10111 & 11100 \\
00010 & 01001 & 10100 & 11111 \\
00100 & 01111 & 10010 & 11001 \\
01000 & 00011 & 11110 & 10101 \\
10000 & 11011 & 00110 & 01101
\end{array}
$$

Since $\mathbf{B}^5$ has $2^5 = 32$ elements and the subgroup W has $2^2 = 4$ elements, there remain two rows to be added before every element of $\mathbf{B}^5$ is listed (making eight rows in all). Every element of $\mathbf{B}^5$ of weight 1 is now in the table (as well as some others of higher weight), so in our search for elements not in the table we must start looking among those of weight 2. A number of these are already in the table but, for example, 10001 is not:

```
00000   01011   10110   11101
00001   01010   10111   11100
00010   01001   10100   11111
00100   01111   10010   11001
01000   00011   11110   10101
10000   11011   00110   01101
10001   11010   00111   01100
```

Searching among words of length two we see that 00101 has not yet appeared, so its coset gives us the last row of the table:

```
00000   01011   10110   11101
00001   01010   10111   11100
00010   01001   10100   11111
00100   01111   10010   11001
01000   00011   11110   10101
10000   11011   00110   01101
10001   11010   00111   01100
00101   01110   10011   11000
```

So if a word is transmitted with one error it will appear in the second to sixth rows. If a word is transmitted with two errors then it may appear in any row but the top one (it need not appear in one of the last two rows since two errors may bring it within distance 1 of a codeword).

How do we use this table? Suppose that the message

$$00 \quad 01 \quad 01 \quad 00 \quad 10 \quad 11 \quad 11 \quad 01 \quad 00$$

is to be sent. This will actually be transmitted (after applying the generator matrix) as:

00000   01011   01011   00000   10110   11101   11101   01011   00000.

Suppose that it is received (with a very high number of errors!) as:

00000   00011   01011   00000   11100   11101   10101   11101   01000.

To apply the decoding table we replace each of these received words by the entry on the top row which lies above it in the table:

00000   01011   01011   00000   11101   11101   11101   11101   00000.

Then we recover what, we hope, is the original message by extracting the first two digits of each of these words: 00 01 01 00 11 11 11 11 00.

We see then that we have corrected all the single errors which have occurred (but not the double or triple errors). In practice the probability of even a single error occurring should be small, and the probability of two or more errors correspondingly much smaller.

The entries in the first column of our coset decoding table are called **coset leaders**. The maximum likelihood decoding assumption corresponds to the choice of coset leaders to be of minimum weight. The received word is decoded as the codeword to which it is closest. This is a reasonable way to proceed but, as in the last example, if the number of errors is too high we may be led to an incorrect decoding. In any given coset there may be more than one word of the minimum weight for that coset (as in the last two rows in the example). The choice of which of these is to be coset leader corresponds to the fact that words which contain a comparatively large number of errors may be of equal distance from more than one codeword. Thus in the example above, choosing 10001 as coset leader means that we decode 10001 as 00000 and 01100 as 11101. But if we had chosen (as we could have) 01100 as coset leader then we would decode 01100 as 00000 and 10001 as 11101. So there can be a certain arbitrariness when dealing with words which contain a large number of errors.

If one found in practice that one was having to use the last two rows of the above decoding table, one would conclude that the rate of errors was too high for the code to deal with effectively.

Let us construct the decoding table for Example 3 before considering how one may avoid having to construct (and store) the whole table in the case where the code is given by a generator matrix.

Example   Let $f : \mathbf{B}^3 \longrightarrow \mathbf{B}^6$ and let the generating matrix be as shown: we also list the codewords.

$$G = \begin{pmatrix} 1 & 0 & 0 & 1 & 1 & 1 \\ 0 & 1 & 0 & 1 & 0 & 1 \\ 0 & 0 & 1 & 1 & 1 & 1 \end{pmatrix}$$

000	000000	100	100111
001	001111	101	101000
010	010101	110	110010
011	011010	111	111101

We array the codewords along the top row and then look for words of minimum length not already included in the table, and array their cosets as described.

000000	001110	010101	011011	100111	101001	110010	111100
000001	001111	010100	011010	100110	101000	110011	111101
000010	001100	010111	011001	100101	101011	110000	111110
000100	001010	010001	011111	100011	101101	110110	111000
001000	000110	011101	010011	101111	100001	111010	110100
010000	011110	000101	001001	110111	111001	100010	101100
100000	101110	110101	111011	000111	001001	010010	011100
000011	001101	010110	011000	100100	101010	110001	111111

Then the message:

$$010111 \quad 111111 \quad 010000 \quad 101110 \quad 101110 \quad 011011$$

would be corrected as

$$010101 \quad 111100 \quad 000000 \quad 001110 \quad 001110 \quad 011011.$$

Definition   Given the $m \times n$ generator matrix $G = (I_m \, A)$ we define the corresponding **parity-check matrix** H to be the $n \times (n - m)$ matrix

$$\begin{pmatrix} A \\ I_{n-m} \end{pmatrix}.$$

For example, if G is the matrix in Example 2 above then the corresponding parity-check matrix H is

$$\begin{pmatrix} 1 & 1 & 0 \\ 0 & 1 & 1 \\ 1 & 0 & 0 \\ 1 & 0 & 0 \\ 0 & 1 & 0 \\ 0 & 0 & 1 \end{pmatrix}$$

Given a word w in $\mathbf{B}^n$ we define the **syndrome** of w to be the matrix product wH in $\mathbf{B}^{n-m}$.

Theorem 5.4.5   *Let H be the parity-check matrix associated with a given code. Then w is a codeword if and only if its syndrome wH is the zero element in $\mathbf{B}^{n-m}$.*

Proof To see this, note that w is a codeword if and only if w has the form uv where $v = uA$. This equation may be rewritten as

$$0 = uA - vI_{n-m} = uA + vI_{n-m} = (uv)H = wH$$

where '0' is the zero $(n - m)$-tuple: that is, $wH = 0$. Thus we see that w is a codeword if and only if wH is 0 □

Corollary 5.4.6 *Two words are in the same row of the coset decoding table if and only if they have the same syndrome.*

Proof Two words u and v are in the same row of the decoding table if and only if they differ by a codeword w, say $u = v + w$, that is, if and only if $u - v = w \in W$. Since $(u - v)H = uH - vH$ and since $wH = 0$ exactly if $w \in W$ (by Theorem 5.4.5), we have that u and v are in the same row exactly if $uH = vH$, as required.   □

We construct a decoding table with syndromes by adding an extra column at the left which records the syndrome of each row (and is obtained by computing the syndrome of any element on that row). This makes it easier to locate the position of any given word in the table (compute its syndrome to find its row). It is also quite useful in the later stages of constructing the table: when we want to check whether or not a given word already is in the table, we can compute its syndrome and see if it is a new syndrome or not. Also, it is not necessary to construct or record the whole table: it is enough to record just the column of syndromes and the column of coset leaders. Then, given a word w to decode, compute its syndrome, add to (subtract from, really) w the coset leader u which has the same syndrome – the word $w + u$ will then be the corrected version of w – finally read off the first m digits to reconstruct the original word.

This means that it is sufficient to construct a two-column decoding table, one which contains just the column of coset leaders and the column of syndromes. The advantage of using a table showing only coset leaders and syndromes is well illustrated by the next example.

Example Let $f : \mathbf{B}^8 \longrightarrow \mathbf{B}^{12}$ be defined using the generator matrix

$$G = \begin{pmatrix} 1 & 0 & 0 & 0 & 0 & 0 & 0 & 0 & 1 & 1 & 1 & 0 \\ 0 & 1 & 0 & 0 & 0 & 0 & 0 & 0 & 0 & 1 & 1 & 0 \\ 0 & 0 & 1 & 0 & 0 & 0 & 0 & 0 & 1 & 0 & 1 & 0 \\ 0 & 0 & 0 & 1 & 0 & 0 & 0 & 0 & 0 & 0 & 1 & 1 \\ 0 & 0 & 0 & 0 & 1 & 0 & 0 & 0 & 1 & 1 & 0 & 0 \\ 0 & 0 & 0 & 0 & 0 & 1 & 0 & 0 & 0 & 1 & 0 & 1 \\ 0 & 0 & 0 & 0 & 0 & 0 & 1 & 0 & 1 & 0 & 0 & 1 \\ 0 & 0 & 0 & 0 & 0 & 0 & 0 & 1 & 1 & 1 & 0 & 1 \end{pmatrix}$$

There are $2^8 = 256$ codewords and the minimum distance between codewords is 3. To see this, note that there is a codeword of weight 3, for example that given by the second row of G (it is $(01000000) \cdot G$). Any other codeword is obtained by adding together rows of G and so must have weight at least 2 in the first 8 entries. Since the entries of different rows in positions 9 to 12 are different, there can be no two codewords which are distance 2 from each other.

Thus the code detects two errors and corrects one error. This code is as effective as our examples from $\mathbf{B}^3$ to $\mathbf{B}^6$ but is considerably more efficient (we do not have to double the number of digits sent: rather just send half as many again). The parity check matrix H is the 12×4 matrix

$$\begin{pmatrix} 1 & 1 & 1 & 0 \\ 0 & 1 & 1 & 0 \\ 1 & 0 & 1 & 0 \\ 0 & 0 & 1 & 1 \\ 1 & 1 & 0 & 0 \\ 0 & 1 & 0 & 1 \\ 1 & 0 & 0 & 1 \\ 1 & 1 & 0 & 1 \\ 1 & 0 & 0 & 0 \\ 0 & 1 & 0 & 0 \\ 0 & 0 & 1 & 0 \\ 0 & 0 & 0 & 1 \end{pmatrix}$$

There will be 16 cosets in our table showing the syndrome then coset leader.

Syndrome	Coset leader
0000	000000000000
0001	000000000001
0010	000000000010
0100	000000000100
1000	000000001000
1101	000000010000
1001	000000100000
0101	000001000000
1100	000010000000
0011	000100000000
1010	001000000000
0110	010000000000
1110	100000000000
1011	000100001000
0111	000100000100
1111	100000000001

This table was produced by writing the 12 'unit error' vectors for the coset leaders and listing the appropriate syndromes (the rows of the parity check matrix). This gives 13 rows, counting the first. The final three rows are obtained by listing the three elements of $\mathbf{B}^4$ which do not occur as syndromes in the first 13 rows and seeing how to express these as combinations of the known syndromes. For example, we see by inspection that the syndrome 1011 does not occur in the first 13 rows. It may be expressed in several ways as combinations of the syndromes in the first 13 rows, for example

$$1011 = 1000 + 0011 = 1010 + 0001.$$

The first corresponds to the choice of 000100001000 as coset leader, the second to 001000000001. As we have seen, neither of these is 'correct', rather each is just a choice of how to correct words with more than one error.

Now to correct the message

$$000010000100 \quad 110110010000 \quad 001100101111$$

we compute the syndrome of each word. They are

$$1000 \quad 1010 \quad 1111$$

(thus none of the these is a codeword). The corresponding coset leaders are

$$000000100001 \quad 000000000001 \quad 100000000000.$$

Each of these is added to the corresponding received word so as to obtain a codeword, and thus we correct to get

$$000010100101 \quad 110110010001 \quad 101100101110.$$

Now we see how much more efficient it may be to compute and store only coset leaders with syndromes: the table above contains $16 \cdot (12 + 4) = 256$ digits; how many digits would the full decoding table contain? It would have 16 rows, each containing 256 twelve-digit words, that is $16 \cdot 256 \cdot 12 = 49152$ digits!

The key point which makes the above example so efficient is that the rows of the parity-check matrix are non-zero and distinct. Such a code clearly corrects one error (by the argument used at the beginning of the above example). The extreme example of such a code occurs when the rows of the parity-check matrix contain all the $2^n - 1$ non-zero vectors in $\mathbf{B}^n$. Whenever this is the case, every vector v is a single error away from being a codeword, say $v = w + e_i$, and its syndrome is that of e_i which is the ith row of H. The code associated with such a parity check matrix is known as a **Hamming code**. In such a code the spheres of radius 1 centred on the codewords partition the entire 'space' of words ('radius' here is measured with respect to the distance function $d(u, v)$ between words).

Hamming codes are examples of 'perfect codes': codes in which the codewords (of length n say) are evenly distributed throughout the words of length n.

Example   When $n = 3$ one possible parity-check matrix associated with a Hamming code is

$$\begin{pmatrix} 1 & 1 & 1 \\ 1 & 1 & 0 \\ 1 & 0 & 1 \\ 0 & 1 & 1 \\ 1 & 0 & 0 \\ 0 & 1 & 0 \\ 0 & 0 & 1 \end{pmatrix}$$

The corresponding generator matrix is

$$\begin{pmatrix} 1 & 0 & 0 & 0 & 1 & 1 & 1 \\ 0 & 1 & 0 & 0 & 1 & 1 & 0 \\ 0 & 0 & 1 & 0 & 1 & 0 & 1 \\ 0 & 0 & 0 & 1 & 0 & 1 & 1 \end{pmatrix}$$

and the syndrome plus coset leader decoding table is

Syndrome	Coset leader
000	0000000
111	1000000
110	0100000
101	0010000
011	0001000
100	0000100
010	0000010
001	0000001

Example   Let us give one more example of constructing the two-column decoding table. Consider the coding function $f : \mathbf{B}^3 \longrightarrow \mathbf{B}^6$ with generating matrix (and codewords) as shown.

$$G = \begin{pmatrix} 1 & 0 & 0 & 1 & 0 & 1 \\ 0 & 1 & 0 & 0 & 1 & 1 \\ 0 & 0 & 1 & 1 & 1 & 0 \end{pmatrix}$$

000	000000
001	001110
010	010011
011	011101
100	100101
101	101011
110	110110
111	111000

Since the minimum weight of a non-zero codeword is 3, the code detects two errors and corrects one error.

The parity-check matrix is

$$\begin{pmatrix} 1 & 0 & 1 \\ 0 & 1 & 1 \\ 1 & 1 & 0 \\ 1 & 0 & 0 \\ 0 & 1 & 0 \\ 0 & 0 & 1 \end{pmatrix}$$

There are $2^{6-3} = 8$ cosets of the group of 8 codewords in the group of 64 words of length 2^6, so there will be 8 rows in the decoding table, which is as shown below:

Syndrome	Coset leader
000	000000
001	000001
010	000010
100	000100
110	001000
011	010000
101	100000
111	100010

Now suppose that a message is encrypted using the number-to-letter equivalents

000	A	001	C	010	E	011	N
100	O	101	R	110	S	111	T

and then is sent after applying the coding function f. Suppose that the message received is

101110   100001   101011   111011   010011   011110   111000.

If we were not to attempt to correct the message and simply read off the first three digits of each word, we would obtain

101   100   101   111   010   011   111

which, converted to alphabetical characters, gives us the nonsensical

RORTENT.

But we note that errors have occurred in transmission, since some of the received words are not codewords, so we apply the correction process. The syndromes of the words received are obtained by forming the products wH where w ranges over the (seven) received words and H is the parity-check matrix above. They are

101   100   000   011   000   011   000.

The corresponding coset leaders are

100000   000100   000000   010000   000000   010000   000000.

The corrected message, obtained by adding the coset leaders to the corresponding received words, is therefore

001110   100101   101011   101011   010011   001110   111000.

Extracting the initial three digits of each word gives

001   100   101   101   010   001   111

and this, converted to alphabetical characters, yields the original message

CORRECT.

Error-correcting codes were introduced in the late 1940s in order to protect the transmission of messages. Although motivated by this engineering problem, the mathematics involved has become increasingly sophisticated. A context where they are of great importance is the sending of information to and from space probes, where retransmission is often impossible. Examples are the pictures sent back from planets and comets.

An application to group theory occurs in the idea of the group of a code. Given a code with codewords of length n, the group of the code consists of the permutations in $S(n)$ that send codewords to codewords. (A permutation π acts on a codeword by permuting its letters according to π.) There is a very important example of a code with words of length 24, known as the Golay code, whose group is the Mathieu group M_{24} which is one of the sporadic simple groups. This is just one example of the interaction between group theory and codes.

Exercises 5.4

1. Refer back to the example at the beginning of this section on ISBN numbers. One of the most commonly made errors in transcribing numbers is the interchange of two adjacent digits: thus 3 540 90346 could become 3 540 93046. Show that the check digit at the end of the ISBN will also detect this kind of error.

2. For each of the following generator matrices, say how many errors the corresponding code detects and how many errors it corrects:

$$\begin{pmatrix} 1 & 0 & 0 & 1 & 0 \\ 0 & 1 & 0 & 1 & 1 \\ 0 & 0 & 1 & 0 & 1 \end{pmatrix}; \quad \begin{pmatrix} 1 & 0 & 0 & 1 & 0 & 1 \\ 0 & 1 & 0 & 0 & 0 & 1 \\ 0 & 0 & 1 & 1 & 0 & 1 \end{pmatrix};$$

$$\begin{pmatrix} 1 & 0 & 0 & 0 & 0 & 1 & 0 & 1 \\ 0 & 1 & 0 & 0 & 1 & 0 & 1 & 0 \\ 0 & 0 & 1 & 0 & 1 & 1 & 0 & 1 \\ 0 & 0 & 0 & 1 & 0 & 1 & 1 & 1 \end{pmatrix}; \quad \begin{pmatrix} 1 & 0 & 0 & 0 & 1 & 0 & 1 & 1 \\ 0 & 1 & 0 & 0 & 0 & 0 & 0 & 0 \\ 0 & 0 & 1 & 0 & 1 & 1 & 1 & 1 \\ 0 & 0 & 0 & 1 & 0 & 0 & 1 & 1 \end{pmatrix}.$$

3. Let $f : \mathbf{B}^3 \longrightarrow \mathbf{B}^9$ be the coding function given by

$$f(abc) = abcabc\bar{a}\bar{b}\bar{c}$$

where $\bar{x}$ is 1 if x is 0 and $\bar{x}$ is 0 if x is 1. List the eight codewords of f. Show that f does not give a group code. How many errors does f detect and how many errors does it correct?

4. Give the complete coset decoding table for the code given by the generator matrix

$$\begin{pmatrix} 1 & 0 & 0 & 1 & 1 & 0 \\ 0 & 1 & 0 & 1 & 0 & 1 \\ 0 & 0 & 1 & 0 & 1 & 1 \end{pmatrix}.$$

5. For the code given by the 8×12 generator matrix on p. 247, correct the following message:

101010101010 111111111100 000001000000 001000100010 001010101000.

6. Write down the two-column decoding table for the code given by the generator matrix

$$\begin{pmatrix} 1 & 0 & 0 & 1 & 1 & 0 & 1 \\ 0 & 1 & 0 & 1 & 1 & 1 & 0 \\ 0 & 0 & 1 & 1 & 0 & 1 & 1 \end{pmatrix}.$$

Use this table to correct the message

1100011   1011000   0101110   0110001   1010110.

7. Let $f : \mathbf{B}^3 \longrightarrow \mathbf{B}^6$ be given by the generator matrix

$$\begin{pmatrix} 1 & 0 & 0 & 1 & 0 & 1 \\ 0 & 1 & 0 & 1 & 1 & 0 \\ 0 & 0 & 1 & 0 & 1 & 1 \end{pmatrix}.$$

Write down the two-column decoding table for f. A message is encoded using the letter equivalents

000	blank	100	A	010	E	001	T
110	N	101	R	011	D	111	H

011011   110000   010110   100000   110110   110111   011111.

Decode the received message.

Summary of Chapter 5

This chapter was an introduction to basic group theory. Although groups themselves were defined in Chapter 4, we did little beyond offering definitions and examples there. In the present chapter, we first investigated the power of the four group axioms and discussed equation-solving in groups as well as some simple facts about orders of elements in groups. An idea of great importance is that of a subgroup (a non-empty set which itself satisfies the four group axioms using the same law of composition as that of the group). Associated with each subgroup, there is a partition of the group into distinct (left or right) cosets. The fundamental result (Lagrange's Theorem) is that the number of distinct left cosets of a subgroup H in a group G is equal to $o(G)/o(H)$. This result has many elementary consequences and we saw some of the power of this result in the process of classifying (producing a list of) groups with a small number of elements (this was done in Section 5.3).

In the final section of the chapter, we considered the application of the decomposition of a group into cosets of a subgroup to the elementary theory of error-correcting codes. These codes provide a systematic way to send messages, with some extra information (check digits) in such a way that an error occurring in the original messages will not just be noticed (detected) by the receiver but, in many cases, may even be corrected.

6 Polynomials

6.1 Introduction

We have mentioned polynomials on a couple of occasions already, but now is the time to take a closer look at them.

A (**real**) **polynomial function** f is a map from the set $\mathbb{R}$ to itself, where the value, $f(x)$, of the function f at every (real) number x is given by a formula which is a (real) linear combination of non-negative-integral powers of x (the same formula for all values of x).

An example of a polynomial function is the function which cubes any number x and adds 1 to the result: we write $f(x) = 1 \cdot x^3 + 1$ or, more usually, $f(x) = x^3 + 1$ since a coefficient of 1 before a power of x is normally omitted.

An expression, such as $x^3 + 1$ or $x^6 - 3x^2 + \frac{1}{2}$, which is a (real) linear combination of non-negative-integral powers of x (and which, therefore, defines a polynomial function) is usually referred to as a **polynomial with coefficients in $\mathbb{R}$** (or **with real coefficients**). It is also, of course, possible to consider polynomials with other kinds of coefficients: for example we might wish to allow coefficients which are complex numbers; or we might wish only to consider polynomials with rational coefficients, etc. In such cases we refer to polynomials with coefficients from $\mathbb{C}$, or $\mathbb{Q}$, etc.

Notice that the following polynomial expressions all define the same function: $x^3 + 2x - 1, x^3 + 0x^2 + 2x - 1, -1 + 2x + x^3 + 0x^5$. That is, adding a term with 0 coefficient makes no difference (to the function defined) nor, because addition of real numbers is commutative, does rearranging terms. We wish to regard these three expressions (and all others we can get from them by adding terms with 0 coefficient and by rearranging terms) as being 'the same' polynomial. In other words, we regard two polynomial expressions as being equivalent if we can get from one to the other by rearranging terms and adding or deleting terms with 0 coefficient. It is normal to say and write that such expressions are 'equal' rather than 'equivalent': so we write, for example, $-1 + 2x + x^3 = x^3 + 2x - 1$.

A typical polynomial can, therefore, be written in the form

$$a_0 + a_1x + \cdots + a_ix^i + \cdots$$

If we want to make this look more uniform we may write

$$a_0x^0 + a_1x^1 + \cdots + a_ix^i + \cdots$$

We say that a_ix^i is a **term** of the polynomial and that a_i is the **coefficient** of x^i. We do require that a polynomial should only have finitely many non-zero terms, that is, $a_i = 0$ for all but a finite number of values of i. We say that the power x^i **appears** in the polynomial if $a_i \neq 0$. So x^3 appears in $x^3 + 0x^2 + 2x - 1$ but x^2 does not.

We use notation such as $f(x), g(x), r(x)$, etc. for polynomials but sometimes we drop the '(x)', writing f, g, r etc. We also use the same notation for the functions defined by polynomials.

Summation notation gives a compact way of writing polynomials: in this notation a typical polynomial $f(x)$ has the form $f(x) = \sum_{i=0}^{n} a_ix^i$; the other way of writing this is $a_0 + a_1x + \cdots + a_nx^n$ (where we have replaced a_0x^0 which equals $a_0 \cdot 1$ by a_0, and a_1x^1 is written more simply as a_1x). If $a_n \neq 0$, in other words if x^n is the highest power of x which appears in the polynomial, then we call a_nx^n and a_n the **leading term** and **leading coefficient** respectively and we say that that the **degree** of $f(x)$ is n and write $\deg(f(x)) = n$. For example the degree of $x^3 + 2x - 1$ is 3.

The zero polynomial is a special case since it has no non-zero coefficients. We shall use the convention that the degree of the zero polynomial is -1 (since doing so makes some things easier to state), although some authors prefer to say that it has degree $-\infty$ or simply say that its degree is undefined.

It is clear from our definition of degree that polynomials of degree one (also known as **linear** polynomials) are those of the form $f(x) = ax + b$ (where a, b are real numbers and $a \neq 0$). Polynomials of degree 2 (**quadratic** polynomials) are those of the form $f(x) = ax^2 + bx + c$ (where a, b, c are in $\mathbb{R}$ and $a \neq 0$). Polynomials of degrees 3, 4 and 5 are also referred to as **cubic, quartic** and **quintic** polynomials respectively. Notice that a polynomial of degree 0 is one of the form $f(x) = a$ (with $a \neq 0$); these, and the zero polynomial, are also referred to as **constant** polynomials. The function defined by a constant polynomial is, of course, a constant function (its value does not depend on x).

Example   Let $f(x) = -5x^7 - 6x^4 + 2$, $g(x) = 1 - 4x$ and $h(x) = 5$. Then $\deg(f) = 7$, $\deg(g) = 1$, $\deg(h) = 0$. The leading terms of f, g and h are, respectively, $-5x^7, -4x, 5$ and their leading coefficients are $-5, -4$ and 5.

We have indicated already that the coefficients of a polynomial may be drawn from a range of different mathematical structures, but for our purposes we shall initially regard the coefficients as coming from the set, $\mathbb{R}$, of real numbers. We denote the set of all polynomials whose coefficients are real numbers by $\mathbb{R}[x]$. The 'x' which occurs in the expression of a polynomial should be thought of as a variable which may be substituted by an arbitrary (real) number α. We describe this as **evaluating** the polynomial **at** $x = \alpha$ and we write $f(\alpha)$ for the (real) number which is obtained by replacing every occurrence of x in the expression of $f(x)$ by α. For example, if $f(x) = -5x^7 - 6x^4 + 2$ and $\alpha = -2$ then $f(\alpha) = -5(-2)^7 - 6(-2)^4 + 2 = 546$.

The reader is probably accustomed to drawing graphs to illustrate polynomial functions. Thus a linear function has, as its graph, a straight line. The slope, and any point of intersection of the line with the x-axis and with the y-axis, are easily determined from the coefficients a, b appearing in the formula, $f(x) = ax + b$, defining the values of $f(x)$. We will be concerned with the general question of where the graph of a polynomial function intersects the x-axis: we say that the real number α is a **zero** (or **root**) of the polynomial f if $f(\alpha) = 0$ (that is, if the graph of $y = f(x)$ crosses the x-axis where $x = \alpha$).

For a quadratic polynomial function, the corresponding graph may cross or touch the x-axis in two, one or no points depending on the values of the constants a, b and c in the formula $f(x) = ax^2 + bx + c$ defining the function. We have the well known formula for determining, in terms of the coefficients a, b and c, these points. Namely, the zeros of f are given by the formula

$$x = \frac{-b \pm \sqrt{b^2 - 4ac}}{2a}.$$

It follows from this that f will have a repeated zero when $b^2 = 4ac$, no real zero when $b^2 - 4ac$ is negative and two distinct real zeros when $b^2 - 4ac$ is positive.

There are similar, though more complicated, formulae which give the zeros of general polynomials of degree 3 and 4. It will not be necessary here to know, or use, these formulae. As we mentioned in the historical remarks at the end of Section 4.3, the search for a formula for the zeros of a general polynomial of degree 5 was one of the ideas which lay at the origins of group theory. Recall that Abel proved that there is no formula of this general sort (i.e. involving arithmetic operations and taking roots) which gives the zeros of a general quintic polynomial. Nevertheless, for *some* types of quintics (and higher-degree polynomials) there is a formula. Galois gave the exact conditions on a polynomial for there to be a formula for its zeros. He did this by associating a group to each polynomial and then he was able to interpret the existence of such a formula for the zeros of the polynomial in terms of the structure of this group.

Note that a constant polynomial cannot have a zero unless it is actually the zero polynomial (and then every number is a zero!) so, when we make general statements about zeros of polynomials we sometimes have to insert a clause excluding constant polynomials.

We noted above that not every polynomial has a (real) zero. For example, if x is a real number, x^2 is never negative (and is only zero if x is zero), so $x^2 + 1$ is never zero. Thus the polynomial $x^2 + 1$ has no real roots. (The reader may be aware that, in order to find a zero of this equation, we need to use complex numbers. It is a very general result, beyond the scope of this book, that every non-constant real polynomial has a zero, which may be a complex number. Another fact that we will need to use later is that if the real polynomial f has a complex zero α, then the complex conjugate of α is also a zero. A general introduction to and summary of some basic facts about complex numbers is given in the Appendix.)

In this chapter we will see that there are some remarkable similarities between the set, $\mathbb{Z}$, of integers with its operations of addition and multiplication and the set, $\mathbb{R}[x]$, of polynomials with its algebraic operations – these we now define.

Just as for the set of integers, we have two algebraic operations on the set $\mathbb{R}[x]$. You have almost certainly met these before, at least in the context of specific examples.

First we have addition of polynomials. Given f and g in $\mathbb{R}[x]$, say

$$f(x) = a_0 + a_1 x + \cdots + a_n x^n + \cdots$$

and

$$g(x) = b_0 + b_1 x + \cdots + b_n x^n + \cdots,$$

we define their **sum** by

$$f(x) + g(x) = (a_0 + b_0) + (a_1 + b_1)x + \cdots + (a_n + b_n)x^n + \cdots$$

Of course all but a finite number of the coefficients $(a_i + b_i)$ will be zero (because this is so for $f(x)$ and $g(x)$). In fact, if f has degree n, so

$$f(x) = a_0 + a_1 x + a_2 x^2 + \cdots + a_n x^n,$$

with $a_n \neq 0$ and g has degree m, so

$$g(x) = b_0 + b_1 x + \cdots + b_m x^m,$$

with $b_m \neq 0$ then we have

$$(f + g)(x) = (a_0 + b_0) + (a_1 + b_1)x + (a_2 + b_2)x^2 + \cdots + (a_k + b_k)x^k$$

where k is the larger of n and m.

There are a couple of points to make about this last formula. First, if, say $n > m$, so $k = n$, then all the coefficients, $b_{m+1} \cdots b_n$ of g beyond b_m, are 0. But it is useful to have them there so that we can write down a formula for the sum $f + g$ in a uniform way. We will see something similar when we define multiplication of polynomials.

The other point to make is that it is clear from the last formula that the degree of $f + g$ is less than or equal to k, the larger of n and m. It is possible for the degree of $f + g$ to be strictly smaller than this (for example, if $f = x^2 + x$ and $g = -x^2 - 1$) because the leading terms might cancel; this can only happen, though, if $\deg(f) = \deg(g)$.

The reader should be able to see that this definition is just a formal way of stating what is probably obvious: we obtain $f + g$ by adding together corresponding powers of x in f and g.

(We also remark that we have found it convenient to write general polynomials with 'lowest powers first' whereas, in numerical examples we usually write the highest powers of x first.)

Example   The sum of the quadratic polynomial $f(x) = 2x^2 - 5x + 3$ and the linear polynomial $g(x) = 5x - 2$ is the polynomial $(f + g)(x) = 2x^2 + 1$ (that is, $(2 + 0)x^2 + (-5 + 5)x + (3 - 2)$).

As we stated in Section 4.4, the set, $\mathbb{R}[x]$, of polynomials forms an Abelian group under addition, with the polynomial 0 as its identity and the inverse of the polynomial $f(x) = a_0 + a_1 x + a_2 x^2 + \cdots + a_n x^n$ being the polynomial $-f$ given by $-f(x) = -a_0 - a_1 x - a_2 x^2 - \cdots - a_n x^n$.

The second basic operation on the set of polynomials is multiplication. If

$$f(x) = a_0 + a_1 x + \cdots + a_n x^n$$

and

$$g(x) = b_0 + b_1 x + \cdots + b_m x^m$$

then

$$(fg)(x) = c_0 + c_1 x + \cdots + c_{n+m} x^{n+m}$$

where,

$$c_0 = a_0 b_0$$
$$c_1 = a_0 b_1 + a_1 b_0$$
$$c_2 = a_0 b_2 + a_1 b_1 + a_2 b_0$$
$$\vdots$$
$$c_i = a_0 b_i + a_1 b_{i-1} + a_2 b_{i-2} + \cdots + a_i b_0$$
$$\vdots$$

(Bear in mind our convention that any undefined coefficients should be taken to be zero: refer back to the comments after the definition of addition of polynomials.)

The definition might look quite complicated but it is only saying the obvious thing: to obtain the formula for fg, take the formulae for f and g and multiply them together, gathering together all the coefficients of the same power of x. If this is not obvious to you now then try multiplying together two polynomials with general coefficients, say $f(x) = a_0 + a_1 x + a_2 x^2 + a_3 x^3$ and $g(x) = b_0 + b_1 x + b_2 x^2$, and then gathering together terms with the same power of x to see where the above expressions for c_0, c_1, etc. come from. The formulae above for the coefficients, c_i, of a product are useful when dealing with polynomials of large degree but, for 'small' polynomials, in practice we use the procedure of multiplying out and gathering terms together.

One consequence of this definition is that, provided f and g are non-zero, the degree of fg is the sum of the degree of f and the degree of g because if, with notation as above, $\deg(f) = n$, so $a_n \neq 0$ and $\deg(g) = m$, so $b_m \neq 0$, then (you should check), every coefficient c_l with $l > n + m$ is 0 and the coefficient, c_{n+m}, of x^{n+m} is $a_n b_m$, which is non-zero.

Example   The product of the quadratic polynomial $f(x) = 2x^2 - 5x + 3$ and the linear polynomial $g(x) = 5x - 2$ is the polynomial of degree three $(fg)(x) = (2x^2 - 5x + 3)(5x - 2)$. Multiplying out and rearranging, we obtain

$$(fg)(x) = 10x^3 - 4x^2 - 25x^2 + 10x + 15x - 6 = 10x^3 - 29x^2 + 25x - 6.$$

The set, $\mathbb{R}[x]$, of polynomials equipped with these two operations satisfies most of the familiar laws of algebra. For example one can check that the distributive law, namely

$$f(x)h(x) + g(x)h(x) = (f(x) + g(x))h(x)$$

holds for polynomials. One may also check that we have the commutative law for multiplication of polynomials: $f(x)g(x) = g(x)f(x)$. These, and all other such elementary properties, will be used without comment from now on. We do not include their (straightforward) proofs but we do comment that they depend ultimately on the fact that the corresponding properties (such as commutativity and distributivity) hold for real numbers. These properties are summarised in the statement that $\mathbb{R}[x]$ is a commutative ring.

We define subtraction in the obvious way: the **difference** $f - g$ of the polynomials f and g, where $f(x) = a_0 + a_1 x + a_2 x^2 + \cdots + a_n x^n$ and

$g(x) = b_0 + b_1 x + \cdots + b_m x^m$ with, say, $m \leq n$ is given by
$$(f - g)(x) = (a_0 - b_0) + (a_1 - b_1)x + (a_2 - b_2)x^2 + \cdots + (a_n - b_n)x^n.$$
Note that $f - g = f + (-g)$.

The situation for division is much more complicated (and interesting), and will be discussed in the next section.

We emphasise that the facts we have needed to use in our discussion are the basic properties (such as commutativity and distributivity) of the algebraic operations on $\mathbb{R}$: it is these which underpin the corresponding properties of $\mathbb{R}[x]$. For this reason, we could equally have considered $\mathbb{C}[x]$, the set of polynomials with complex coefficients of a complex variable x, in place of $\mathbb{R}[x]$. The definition of addition and multiplication of polynomials would be as above and we would have exactly the same basic algebraic properties as for $\mathbb{R}[x]$. However, one major difference is that every non-constant polynomial, f, in $\mathbb{C}[x]$ has a zero: that is, there is a complex number α such that $f(\alpha) = 0$.

This process, changing coefficients, does not end here.

Given any prime number p, we can consider (as in Chapter 1) the set, $\mathbb{Z}_p$, of congruence classes modulo p. Again, this set is a ring and so we can consider $\mathbb{Z}_p[x]$, the set of polynomials **over** $\mathbb{Z}_p$ (that is, with coefficients in $\mathbb{Z}_p$). Everything (definitions and basic algebraic properties) is as before. We do, of course, when adding and multiplying such polynomials, have to calculate the coefficients of the sum and the product using arithmetic modulo our prime p. Thus if $p = 2$, $(x^2 + x + 1) + (x^2 + 1) = x$ and when $p = 3$, $(x + 2)^2 = x^2 + x + 1$. Similarly, when evaluating such polynomials, we have to calculate modulo p. You might wonder why we do this only for prime numbers p: surely in almost everything we have said so far we could replace the prime p by any integer $n \geq 2$. That is correct and it is really only when we come to division (in the next section) that we do need p to be prime. For remember that if n is not prime then $\mathbb{Z}_n$ is not a field and this means that we cannot always divide by non-zero elements. For the next section we certainly need to be in a context where we can always divide by non-zero coefficients.

There is an important feature of polynomials with coefficients in $\mathbb{Z}_p$. We can, in principle, find all the zeros of any polynomial. Because there are only a finite number of elements in $\mathbb{Z}_p$, we can simply substitute each element of $\mathbb{Z}_p$ in turn. For small p this process will be simple to apply and will give us the zeros of any given polynomial. For example, as a polynomial in $\mathbb{Z}_2$, the quadratic $x^2 + 1$ takes the value 1 when $x = 0$ and when $x = 1$ takes the value $1 + 1$, which is zero modulo 2, so 1 is a zero. In fact it is clear that, as a polynomial in $\mathbb{Z}_2$, $x^2 + 1$ is equal to $(x + 1)^2$, so this polynomial actually has a repeated zero. However the polynomial $f(x) = x^2 + x + 1$ has no zeros in $\mathbb{Z}_2$ because $f(0) = 1 = f(1)$.

Exercises 6.1

1. Add the following pairs of polynomials:

   (i)   the real polynomials $x^2 + 7x + 3$ and $x^2 - 5x - 3$;
   (ii) the real polynomials $x^3 - 2x^2 + x - 1$ and $-x^3 - x^2 + x + 1$;
   (iii) the complex polynomials $x^2 + 7x + 3$ and $x^2 - i5x - 3i$;
   (iv) the complex polynomials $x^3 - 2ix^2 + ix - i$ and $-x^3 - ix^2 + ix + i$;
   (v)   the polynomials over $\mathbb{Z}_3$, $x^2 + 2x + 1$ and $x^2 + 2x + 2$;
   (vi) the polynomials over $\mathbb{Z}_3$, $x^3 + 2x^2 + x - 1$ and $2x^3 - x^2 + x + 1$.

2. Multiply the following pairs of polynomials:

   (i)   the real polynomials $x^2 + 7x + 3$ and $x + 1$;
   (ii) the real polynomials $x^3 - 2x^2 + x - 1$ and $x^2 + x + 1$;
   (iii) the complex polynomials $x^2 + 7x + 3$ and $ix + 3$;
   (iv) the complex polynomials $ix^3 - 2x^2 + x - i$ and $ix^2 + ix - i$;
   (v)   the polynomials over $\mathbb{Z}_2$, $x^2 + x + 1$ and $x^2 + x + 1$;
   (vi) the polynomials over $\mathbb{Z}_2$, $x^3 + x^2 + x + 1$ and $x^2 + x + 1$.

3. Find a zero of each of the given polynomials:

   (i)   the real polynomial $x^3 - 3x^2 + 4x - 2$;
   (ii) the complex polynomial $x^3 - 7x^2 + x - 7$;
   (iii) the polynomial $x^3 + 4x^2 + 2x + 4$ over $\mathbb{Z}_5$.

6.2   The division algorithm for polynomials

Definition   We say that the polynomial g **divides** the polynomial f if there is a polynomial q such that $f = qg$, that is, $f(x) = q(x)g(x)$. We start with a result which gives us a partial answer to the question of when one polynomial divides another.

Proposition 6.2.1   *Let s and t be polynomials of degree n and m respectively, say*

$$s(x) = a_0 + a_1x + \cdots + a_nx^n \qquad and \qquad t(x) = b_0 + b_1x + \cdots + b_mx^m$$

with $a_n \neq 0$ and $b_m \neq 0$. Assume that $n \geq m$. Then the degree of the polynomial

$$u(x) = s(x) - (a_n/b_m)x^{n-m}t(x)$$

is strictly less than the degree of s.

Proof Since $s(x)$ and $x^{n-m}t(x)$ are both of degree n (note that the leading term of $x^{n-m}t(x)$ is $b_m x^m \cdot x^{n-m} = b_m x^n$), the degree of $u(x)$ can be at most n. However the coefficient of x^n in $s(x)$ is a_n and the coefficient of this power of x in the polynomial $(a_n/b_m)x^{n-m}t(x)$ is $(a_n/b_m)b_m = a_n$. Thus the coefficient of x^n in $u(x)$ is zero, and the degree of u is indeed less than n. □

We shall refer to the term $(a_n/b_m)x^{n-m}$ as a **partial term** in the division of s by t. Using the above result repeatedly is the key to dividing one polynomial, f say, by another, g. (Here we mean 'dividing' to obtain a quotient and a remainder: only if the remainder is zero do we say that 'g divides f'.)

Example Divide the polynomial $f(x) = x^4 - 3x^2 + 2x - 4$ by the polynomial $g(x) = x^2 - 3x + 2$.

We first apply Proposition 6.2.1 with $s = f$ and $t = g$, so $a_0 = -4$, $a_1 = 2$, $a_2 = -3$, $a_3 = 0$ and $a_4 = 1$. Also $b_0 = 2$, $b_1 = -3$ and $b_2 = 1$. According to Proposition 6.2.1, the polynomial $u_1(x) = f(x) - (1/1)x^2 g(x)$ will have degree less than 4. Indeed, we can calculate to see that $u_1(x) = 3x^3 - 5x^2 + 2x - 4$. Next apply Proposition 6.2.1 again, with u_1 in place of s, and g in place of t, to obtain a polynomial

$$u_2(x) = u_1(x) - (3/1)xg(x)$$
$$= \cdots$$
$$= 4x^2 - 4x - 4.$$

We can repeat this process once more. We obtain

$$u_3(x) = u_2(x) - 4g(x) = 4x^2 - 4x - 4 - 4(x^2 - 3x + 2) = 8x - 12.$$

We summarise this calculation by rearranging the equations above to obtain

$$f(x) = x^2 g(x) + u_1(x)$$
$$u_1(x) = 3xg(x) + u_2(x)$$
$$u_2(x) = 4g(x) + u_3(x)$$

and so

$$f(x) = x^4 - 3x^2 + 2x - 4$$
$$= x^2 g(x) + u_1(x)$$
$$= x^2 g(x) + (3xg(x) + u_2(x))$$
$$= (x^2 + 3x)g(x) + (4g(x) + u_3(x))$$
$$= (x^2 + 3x)g(x) + 4g(x) + (8x - 12)$$
$$= (x^2 + 3x + 4)g(x) + (8x - 12)$$
$$= (x^2 + 3x + 4)(x^2 - 3x + 2) + (8x - 12).$$

This type of calculation is often presented in the following 'long division' format.

$$
\begin{array}{r}
x^2 + 3x + 4 \\
\hline
x^2 - 3x + 2 \mid x^4 \qquad\quad - 3x^2 + 2x - 4 \\
x^4 - 3x^3 + 2x^2 \\
\hline
3x^3 - 5x^2 + 2x - 4 \\
3x^3 - 9x^2 + 6x \\
\hline
4x^2 - 4x - 4 \\
4x^2 - 12x + 8 \\
\hline
8x - 12
\end{array}
$$

We give some words of explanation about the way this calculation has been set out. The polynomial g is written on the second line to the left of the '|' sign with the polynomial f to its right. The top line records our partial terms. The line below f is obtained by multiplying g by the first partial term. The line below that is then obtained by subtracting polynomials. The next line arises from multiplying g by the next partial term and the following line is again obtained by subtraction. Continue in this way, alternating the products of g by the partial terms with the results of subtracting two polynomials, until we arrive at a polynomial of degree less than that of g (in this case the linear polynomial $8x - 12$). This is the remainder when f is divided by g.

The reader may need practice to be able to carry out this process with confidence, and several examples are provided in the end-of-section exercises. It is clear that we can repeat this process for any given polynomials f and g and hence we can write f in the form $qg + r$, where q is the polynomial obtained by adding together the partial terms and where r will either be the zero polynomial or have degree strictly less than the degree of g. Thus, in the above example $q(x) = x^2 + 3x + 4$ and $r(x) = 8x - 12$. The process we have just described for computing these polynomials q and r from f and g can be used as the basis of a proof of the next result. However, we prefer to give a different proof of this basic fact about division of polynomials in order to bring out more clearly the connection between polynomials and integers.

Theorem 6.2.2 (The Division Theorem for Polynomials) *Let f and g be polynomials (with real coefficients) with the degree of g being greater than zero (that is, g is not a constant polynomial). Then there are polynomials q and r, such that $f = qg + r$, where the degree of r is strictly less than that of g.*

Proof If the degree of g is greater than that of f, then we may take $q = 0$ and $r = f$. We may, therefore suppose that m, the degree of g is less than or equal to n, the degree of f. For definiteness, suppose that $f(x) = a_0 + a_1 x + a_2 x^2 + \cdots + a_n x^n$ and $g(x) = b_0 + b_1 x + \cdots + b_m x^m$. Consider the set S of those polynomials which are obtained by subtracting a polynomial multiple of g from f:

$$S = \{f - gt : t \text{ is in } \mathbb{R}[x]\}.$$

This set is non-empty since it contains $f (= f - 0 \cdot g)$. Now denote by D the set of degrees of those polynomials in S. Since f is in S, we see that D is also non-empty. Applying the well-ordering principle to D, a non-empty set of integers ≥ -1, we see that D has a least element k, say, so we can select a polynomial $r \in S$ of degree k. Since r is in S we have $r = f - qg$ for some q. Suppose that $r(x) = c_0 + c_1 x + \cdots + c_k x^k$.

Next we wish to show that the degree of r is less than m. To show this, we assume that this is not so and obtain a contradiction. To do this, apply Proposition 6.2.1 with the polynomial r here playing the role of the polynomial s in the lemma and the polynomial g here playing the role of t in the lemma. Then the polynomial

$$r - (c_k/b_m)x^{k-m} g = f - qg - (c_k/b_m)x^{k-m} g = f - (q + (c_k/b_m)x^{k-m})g$$

has degree less than k (the degree of r). Since this polynomial is clearly in S, its degree is in D, contrary to the definition of k. Thus, the degree of r must be strictly less than m. □

This proof is an almost word-for-word generalisation of the division theorem for integers (Theorem 1.1.1). We have a simple, but important consequence of this result.

Corollary 6.2.3 *Given a polynomial f, the value $x = \alpha$ is a zero of f if and only if f is divisible by $x - \alpha$.*

Proof Suppose first that α is a zero of f. We apply the division theorem with $g(x) = x - \alpha$, so $f = qg + r$ where r has degree less than 1 (the degree of g). Therefore r is a constant c, say, and so $f(x) = (x - \alpha)q(x) + c$. Evaluating at $x = \alpha$, we obtain $f(\alpha) = (\alpha - \alpha)q(\alpha) + c$. However, we know that $f(\alpha) = 0$ (α is a zero of f) so we obtain $c = 0$, and conclude that $x - \alpha$ divides f.

For the converse, if f is divisible by $x - \alpha$, we have $f(x) = (x - \alpha)q(x)$, for some polynomial $q(x)$. Put $x = \alpha$ to get $f(\alpha) = (\alpha - \alpha)q(\alpha) = 0 \cdot q(\alpha) = 0$, so α is a zero of f. □

Note that nothing in our proof of Theorem 6.2.2 or its corollary requires us to work over $\mathbb{R}$ and so both results hold over $\mathbb{C}$ and over $\mathbb{Z}_p$. The corollary is very useful (in conjunction with the quadratic formula) in factorising (writing as a product of polynomials of smaller degree) polynomials.

Example 1 Factorise the real polynomial $f(x) = x^3 + 6x^2 + 11x + 6$. A reasonable place to start is to compute the value of f at small values (positive, negative and zero) of x. In this case, $f(0) = 6$, $f(1) = 1 + 6 + 11 + 6 = 24$, $f(-1) = -1 + 6 - 11 + 6 = 0$. Therefore -1 is a zero of f and so by the corollary $x - (-1) = x + 1$ divides f. Carry out the division to see that $f(x) = (x + 1)(x^2 + 5x + 6)$. Next, we can factorise the quadratic $x^2 + 5x + 6$, either by using the formula to find the zeros, or by noticing that $x^2 + 5x + 6 = (x + 2)(x + 3)$. Thus $f(x) = (x + 1)(x + 2)(x + 3)$.

Of course, not all polynomials with integer coefficients will have small integer roots. It is difficult, in general, to find roots of real polynomials, which is why people looked for something like the quadratic formula (that is, a formula which gives an exact expression) which would work for polynomials of higher degree.

Example 2 Factorise the polynomial $f(x) = x^3 + 2x^2 + x + 2$ over $\mathbb{Z}_3$.

Since $\mathbb{Z}_3$ is conveniently small, it is easy to find all zeros of $f(x)$. We substitute the three possible values of x into f to find: $f(0) = 2$, $f(1) = 1 + 2 + 1 + 2 = 0$ and $f(2) = 2^3 + 2 \cdot 2^2 + 2 + 2 = 2$. Thus the only zero is $x = 1$ and so $(x - 1) = (x + 2)$ divides f. We can therefore write $f(x)$ in the form $(x + 2)(ax^2 + bx + c)$. This equals $ax^3 + (2a + b)x^2 + (2b + c)x + 2c$. Setting this equal to $x^3 + 2x^2 + x + 2$ and equating coefficients gives $a = 1$, $2a + b = 2$, so $b = 0$, and $2c = 2$, so $c = 1$. With these values we do also have $2b + c = 1$. Thus $f(x) = (x + 2)(x^2 + 1)$. The quadratic $x^2 + 1$ has value 1 when $x = 0$, value 2 when $x = 1$ and value 2 when $x = 2$, so this quadratic has no zeros over $\mathbb{Z}_p$. It follows that we cannot factorise this quadratic into a product of two linear factors, since such a factorisation would, by Corollary 6.2.3 lead to zeros of $x^2 + 1$. Since the degree of a product of two polynomials is the sum of their degrees, there is no further possible factorisation of $x^2 + 1$.

For our next result, we return to the order of development followed in Chapter 1 and, continuing along this road, introduce the greatest common divisor of two polynomials.

Proposition 6.2.4 *Let f and g be non-zero polynomials. Then there is a polynomial $d(x)$ such that*

(i) *$d(x)$ divides both $f(x)$ and $g(x)$, and*
(ii) *if $c(x)$ is any polynomial which divides both $f(x)$ and $g(x)$, then $c(x)$ divides $d(x)$.*

Proof Let S be the set of polynomials of degree ≥ 0 of the form $f(x)r(x) + g(x)s(x)$, as $r(x)$ and $s(x)$ vary over the set of all polynomials. Consider the set D of integers which are the degrees of the polynomials in S. Since $f(x) = 1 \cdot f(x) + 0$, the degree of $f(x)$ is in D, so D is a non-empty set of non-negative integers. Now apply the well-ordering principle to the set D to select a polynomial $d(x)$ in S such that the degree of $d(x)$ is the smallest integer in D. Since $d(x)$ is in S we have $d(x) = f(x)r(x) + g(x)s(x)$ for some polynomials r and s. We show that d has properties (i) and (ii).

If the polynomial $c(x)$ divides $f(x)$, say $f(x) = c(x)u(x)$ and $c(x)$ divides $g(x)$, say $g(x) = c(x)v(x)$ then

$$d(x) = f(x)r(x) + g(x)s(x)$$
$$= c(x)u(x)r(x) + c(x)v(x)s(x)$$
$$= c(x)(u(x)r(x) + v(x)s(x))$$

so $c(x)$ divides $d(x)$. (Therefore condition (ii) is satisfied.)

To show that $d(x)$ divides $f(x)$, we use Theorem 6.2.2 to write $f(x) = d(x)q(x) + r(x)$ where r has degree strictly less than the degree of $d(x)$. Then

$$r(x) = f(x) - q(x)d(x)$$
$$= f(x) - q(x)(f(x)r(x) + g(x)s(x))$$
$$= f(x)(1 - q(x)r(x)) - g(x)s(x)q(x)$$

is in S or is 0. If $r(x)$ were non-zero then its degree would be in D. But the degree of $d(x)$ was the smallest element of D. Therefore $r(x)$ must be zero and hence $d(x)$ divides $f(x)$. A similar proof shows that $d(x)$ divides $g(x)$. $\square$

Remark By now it should be clear, if you check back to Chapter 1, what our strategy in this chapter is. We are repeating much of that earlier chapter, with essentially the same definitions and very much the same proofs. The details of the proofs need some care and there is the odd change of emphasis in order to take account of the fact that we are dealing with polynomials rather than integers. The general principle to follow in making this change from integers to polynomials is to replace the size of a positive integer by the degree of a polynomial.

This situation, where strong similarities between different situations are seen, is a familiar one in mathematics and it leads to the search for the common content in what might initially seem like very different contexts. Typically this common content is abstracted into ideas and definitions which apply in various contexts, and usually not just in those which motivated the definitions. We have already seen this in basic group theory, where common features of permutations, number systems and other mathematical objects were extracted and developed in a general, more abstract, context.

In presenting mathematics one has a choice. It is possible to present the abstract mathematics first, develop theorems in that context, and then apply them to various special cases. That is an efficient approach (one does not prove the 'same' result over and over again in different contexts) but introducing the abstract ideas right at the beginning presents the student with a steep 'learning curve'. The alternative approach, which we have adopted in this book, is to develop various motivating examples and then extract their common content, so providing the reader with a somewhat longer, but less steep, path.

In the case of the material we are discussing now and the strongly similar material in Chapter 1 there is, indeed, a common generalisation: to what are called 'Euclidean rings'. To treat these would take us beyond the introductory character of this book but the interested reader should look at, for example, [Allenby] or [Fraleigh] listed in the references towards the end of the book.

We can now define a **greatest common divisor** of two polynomials f and g as a polynomial d which satisfies the two conditions of Proposition 6.2.4. Such a polynomial is not unique (which is why we say 'a' greatest common divisor rather than 'the' greatest common divisor). Nevertheless, the degree of any greatest common divisor of f and g is uniquely determined, since if c and d are both greatest common divisors, then c divides d (so the degree of c is less than or equal to that of d) and d divides c (so the degree of d is less than or equal to that of c). It follows that there is a non-zero polynomial of degree 0 (in other words a non-zero constant), λ, such that $c = \lambda d$. This shows that the only difference between any two greatest common divisors of f and g is that each is a multiple of the other by a non-zero constant.

As with integers, the process of finding a greatest common divisor of polynomials uses repeated application of the division theorem and is illustrated by the following example.

Example   Find a greatest common divisor $d(x)$ of the polynomials $f(x) = x^4 - 5x^3 + 7x^2 - 5x + 6$ and $g(x) = x^3 - 6x^2 + 11x - 6$.

First we use the division algorithm to write

$$x^4 - 5x^3 + 7x^2 - 5x + 6 = (x + 1)(x^3 - 6x^2 + 11x - 6) + 2x^2 - 10x + 12$$

(Although, we have chosen a reasonably easy first example, it is in general best not to try such calculations in your head: write them down on a sheet of paper. You can, and should, check by multiplying out the right-hand side.)

We now use the division algorithm again, this time applied to our original polynomial g in place of f and with the remainder, $2x^2 - 10x + 12$, in place of g, giving

$$x^3 - 6x^2 + 11x - 6 = (x/2 - 1/2)(2x^2 - 10x + 12) + 0.$$

Therefore $2x^2 - 10x + 12$ is one of the greatest common divisors of our given polynomials f and g ($x^2 - 5x + 6$ is another). The result which explains why this process yields a greatest common divisor for the initial polynomials is given next and its proof is, again, essentially the same as that of the corresponding result in Chapter 1.

Proposition 6.2.5   *Let f, g be polynomials with g non-zero. Suppose that $f = gq + r$. Then a greatest common divisor of f and g is equal to a constant multiple of any greatest common divisor of g and r.*

Proof   First suppose that d is a greatest common divisor of f and g. Since d divides both f and g, d divides $f - gq = r$. It follows that d is a common divisor of g and r. Therefore if e is the greatest common divisor of g and r, we deduce that d must divide e. As a consequence of this, the degree of d is less than or equal to the degree of e.

Next, e is a common divisor of g and r, so e divides $f = gq + r$ and hence is a common divisor of f and g. It then follows from the definition of d that e divides d, so the degree of e is less than or equal to the degree of d. We conclude that e and d have equal degrees and, since each divides the other, one is a constant multiple of the other, as required.   □

We now have the main result of this section.

Theorem 6.2.6   (The Euclidean algorithm for polynomials) *Let f, g be polynomials. If g divides f then g is a greatest common divisor for f and g. Otherwise apply Theorem 6.2.2 to obtain a sequence of non-zero polynomials $r_1, \ldots, r_n$ satisfying*

$$f = gq_1 + r_1$$
$$g = r_1 q_2 + r_2$$
$$r_1 = r_2 q_3 + r_3$$
$$\vdots$$
$$r_{n-2} = r_{n-1} q_n + r_n$$
$$r_{n-1} = r_n q_{n+1}$$

Then r_n is a greatest common divisor for f and g.

Proof   Apply Proposition 6.2.1 repeatedly, denoting by $r_1, \ldots, r_n$ the non-zero remainders. Since the degrees of the polynomials $g, r_1, \ldots, r_n$ form a strictly decreasing sequence of non-negative integers, this process must terminate. This implies that there exists an integer n such that r_n divides r_{n-1}. Then r_n is a greatest common divisor of r_n and r_{n-1}. Proposition 6.2.5 then implies that r_n is a greatest common divisor for r_{n-1} and r_{n-2}. Repeated application of Proposition 6.2.5 implies that r_n is a greatest common divisor for f and g. □

In Chapter 1, we explained a matrix method to record the calculations made in obtaining the greatest common divisor of two integers. Although this could be done in a similar way for polynomials, each step in the calculation would involve a long division of polynomials of the type already discussed. As it is much more difficult to do this calculation in one's head, it is just as easy actually to carry out the step-by-step process explained in Theorem 6.2.6 when dealing with polynomials.

   Also, just as with integers, it is possible, after having computed the greatest common divisor, to work back through the equations and to obtain an expression for any greatest common divisor d of f and g in the form $d(x) = f(x)s(x) + g(x)t(x)$ for some polynomials s, t. This reverse process is illustrated in the following examples.

Example 1   Find a greatest common divisor $d(x)$ for the polynomials

$$f(x) = x^5 + 3x^4 - x^2 + 3x - 6 \text{ and } g(x) = x^4 + x^3 - x^2 + x - 2$$

and express $d(x)$ as a polynomial combination of f and g.

   We start by writing f as a multiple of g plus a remainder. Only the result of this calculation is given, but it should be made clear that this was done by long division of polynomials, in a calculation which the reader should check. The result of this first step is to obtain

$$f(x) = (x + 2)g(x) - x^3 + 3x - 2.$$

Thus, in the notation of Theorem 6.2.6, $q_1 = x + 2$ and $r_1 = -x^3 + 3x - 2$. Then

$$g(x) = (-x - 1)(-x^3 + 3x - 2) + 2x^2 + 2x - 4$$

so $q_2 = -(x + 1)$ and $r_2 = 2x^2 + 2x - 4 = 2(x^2 + x - 2)$. Applying the process once more, we find that

$$-x^3 + 3x - 2 = 2(x^2 + x - 2)((-x + 1)/2).$$

Thus the calculation of Theorem 6.2.6 would appear in this case as

$$f(x) = (x + 2)g(x) - x^3 + 3x - 2$$
$$g(x) = (-x - 1)(-x^3 + 3x - 2) + 2x^2 + 2x - 4$$
$$-x^3 + 3x - 2 = (2x^2 + 2x - 4)((-x + 1)/2).$$

Thus a greatest common divisor for f and g is $d(x) = 2x^2 + 2x - 4$ (or we could, instead, take $\frac{1}{2}$ times this, that is, $x^2 + x - 2$). Make this the subject of the second equation to obtain $d(x) = g(x) + (x + 1)(-x^3 + 3x - 2)$. Now make the cubic, $-x^3 + 3x - 2$, the subject of the first equation to obtain $f(x) - (x + 2)g(x) = -x^3 + 3x - 2$. Finally substitute this expression for the cubic into the equation for $d(x)$ to obtain

$$d(x) = g(x) + (x + 1)(f(x) - (x + 2)g(x))$$
$$= (x + 1)f(x) + (1 - (x + 1)(x + 2))g(x)$$

which we can simplify to $d(x) = (x + 1)f(x) - (x^2 + 3x + 1)g(x)$.
   At this stage, one should multiply out, as a check.

Example 2   As a second example, we take $f(x)$ to be the quartic equation $x^4 + x^3 - x - 1$ and $g(x)$ to be the cubic $x^3 - 2x^2 + 2x - 1$. Then, going through the steps of Theorem 6.2.6 as in the previous example we obtain

$$f(x) = (x + 3)g(x) + 4x^2 - 6x + 2$$
$$g(x) = \frac{1}{8}(2x - 1)(4x^2 - 6x + 2) + \frac{3}{4}x - \frac{3}{4}$$
$$4x^2 - 6x + 2 = \frac{3}{4}(x - 1)\left(\frac{4}{3}(4x - 2)\right).$$

(Once again the reader needs to take an active part in checking these calculations.) Then $d(x) = \frac{3}{4}(x - 1)$ (or, if you prefer, $x - 1$). Working back through the equations we have

$$d(x) = \frac{3}{4}(x - 1)$$
$$= g(x) - \frac{1}{8}(2x - 1)(4x^2 - 6x + 2)$$
$$= g(x) - \frac{1}{8}(2x - 1)(f(x) - (x + 3)g(x))$$
$$= g(x)\left(1 + \frac{1}{8}(2x - 1)(x + 3)\right) - \frac{1}{8}(2x - 1)f(x)$$
$$= \frac{1}{8}(2x^2 + 5x + 5)g(x) - \frac{1}{8}(2x - 1)f(x)$$

(alternatively, $x - 1 = \frac{1}{6}(2x^2 + 5x + 5)g(x) - \frac{1}{6}(2x - 1)f(x)$).

We make a couple of remarks. First, it is quite possible for two polynomials to have greatest common divisor 1 (equivalently, any non-zero constant): for instance the greatest common divisor of the real polynomials $x - 1$ and $x + 2$ is 1. Second, when you express the greatest common divisor of f and g as a combination of f and g the only fractions which appear should be numerical fractions (like $\frac{3}{4}$) not polynomial fractions (nothing like $\frac{1}{x-1}$ for instance)!

Example 3   As a final example in this section, we consider a case where our polynomials are over $\mathbb{Z}_2$.

Find a greatest common divisor for the polynomials $g(x) = x^4 + x^3 + x + 1$ and $f(x) = x^3 + x + 1$.

The long division process proceeds exactly as over $\mathbb{R}$, giving

$$g(x) = (x + 1)f(x) + x^2 + x.$$

At the next step we obtain

$$f(x) = (x + 1)(x^2 + x) + 1.$$

Thus 1 is a greatest common divisor for $f(x)$ and $g(x)$. Also $1 = f(x) - (x + 1)(x^2 + x)$, so we obtain

$$\begin{aligned}
1 &= f(x) - (x + 1)(g(x) - (x + 1)f(x)) \\
&= f(x)(1 + (x + 1)^2) - (x + 1)g(x) \\
&= x^2 f(x) + (x + 1)g(x)
\end{aligned}$$

where, in the last line, we have used the fact that we are working over $\mathbb{Z}_2$ (so $-1 = 1, -x = x$ etc.).

Exercises 6.2

1. For each of the following pairs of polynomials, f, g, write f in the form $qg + r$ with either $r = 0$ or the degree of r less than that of g:
   (i)   the real polynomials $f(x) = x^4 + x^3 + x^2 + x + 1$ and $g(x) = x^2 - 2x + 1$;
   (ii) the real polynomials $f(x) = x^3 + x^2 + 1$ and $g(x) = x^2 - 5x + 6$;
   (iii) the polynomials $f(x) = x^3 + x^2 + 1$ and $g(x) = x^2 - 5x + 6$ over $\mathbb{Z}_5$.
2. Factorise, as far as possible, the given polynomials:
   (i) $x^3 - x^2 - 4x + 4$ over $\mathbb{R}$;
   (ii) $x^3 - 3x^2 + 3x - 2$ over $\mathbb{R}$;
   (iii) $x^3 - 3x^2 + 3x - 2$ over $\mathbb{C}$;
   (iv) $x^3 - 3x^2 + 3x - 2$ over $\mathbb{Z}_7$;
   (v) $x^3 + x^2 + x + 1$ over $\mathbb{Z}_2$.

3. Find a greatest common divisor, $d(x)$, for each of the following pairs of polynomials and express $d(x)$ as a polynomial combination of the given pair:
   (i)   the polynomials $x^3 + 1$ and $x^2 + x - 1$ over $\mathbb{R}$;
   (ii) the polynomials $x^4 + x + 1$ and $x^3 + x + 1$ over $\mathbb{Z}_2$;
   (iii) the polynomials $x^3 - ix^2 + 2x - 2i$ and $x^2 + 1$ over $\mathbb{C}$.
4. Prove that if f is any polynomial and α any number and if we write $f(x) = (x - \alpha)g(x) + r(x)$ then $r(x)$ is the constant $f(\alpha)$.

6.3 Factorisation

We start with an important definition.

Definition   A non-constant polynomial f is said to be **irreducible** if the only way to write f as a product of two polynomials, $f = gh$, is for one of g and h to be a (non-zero) constant polynomial.

Example   A polynomial of degree 1, that is, one of the form $x - \alpha$, must be irreducible.

Remark   The reason why a new word 'irreducible' is used here is that it will be convenient to use the word 'prime' to describe a rather different concept. Thus we say that a non-constant polynomial f is **prime** if, whenever f divides a product, rs, of polynomials, then either f divides r or f divides s. In fact we shall show that these two ideas, prime and irreducible, coincide for polynomials (as they do for integers, see Theorem 1.3.1). In some rings these concepts differ.

Proposition 6.3.1   *Let f be an irreducible polynomial and suppose that r and s are polynomials such that f divides rs. Then f divides either r or s. That is, every irreducible polynomial is prime.*

Proof   Since f is irreducible, a greatest common divisor for f and r is either a constant polynomial c, or is (a scalar multiple of) f. In the latter case f divides r, so the only case we need consider is when f does not divide r and hence when this greatest common divisor is a constant. Then, by the division algorithm, there are polynomials u, v such that $c = fu + rv$. Multiply this equation by s to obtain

$$c \cdot s = f \cdot u \cdot s + r \cdot s \cdot v$$

Since f divides rs by hypothesis, f divides the right-hand side, so f divides cs which, since c is a constant, implies that f divides s, as required. □

Again, it may be helpful to compare this proof with the corresponding proof 1.1.6(i) from Chapter 1.

As with integers, we may easily extend this result using induction. Its proof will be one of the end-of-section exercises.

Corollary 6.3.2 *Let f be an irreducible polynomial and suppose that f divides the product $f_1 \ldots f_r$. Then f divides at least one of $f_1, f_2, \ldots, f_r$.*

In Proposition 6.3.1 we see that an irreducible polynomial is prime. The converse is easy.

Proposition 6.3.3 *Let f be a prime polynomial, then f is irreducible.*

Proof Suppose that $f(x)$ is a prime polynomial and that $f(x)$ has a factorisation as $g(x)h(x)$. Then $f(x)$ divides $g(x)h(x)$, so f must divide one of g or h. If, say, f has degree n, g has degree m and h has degree k, then we have that $n = m + k$ (since $f = gh$), but also n is less than or equal to either m or k (since f divides either g or h). We conclude that one of m or k is zero, so either g or h is a non-zero constant polynomial. □

Remark Since primes and irreducibles coincide we can go on to consider the issue of unique factorisation. In doing this, we mimic the result, Theorem 1.3.3 in Chapter 1, and its proof. It may be an instructive exercise to re-read the proof of that earlier theorem before reading the proof below.

Theorem 6.3.4 *Every non-constant polynomial f can be written in the form*

$$f = f_1 \cdots f_r$$

where $f_1, \ldots, f_r$ are irreducible polynomials. Furthermore this decomposition is unique in the sense that if also $f = g_1 \cdots g_r$, then $r = s$, and we may renumber the polynomials g_i so that each g_i is a constant multiple of the corresponding f_i (for $1 \le i \le r$).

Proof The proof is in two parts, the first to show that such a decomposition into irreducibles exists and the second to show that this decomposition is unique in the sense explained in the statement of the theorem.

We first show that f has a decomposition into irreducibles, using strong induction on the degree of f. The base case is for polynomials of degree 1 (linear polynomials). Since these are clearly irreducible, this case holds. Now suppose, by strong induction, that if g is any polynomial of degree less than or equal to k, then g has a decomposition of the required form. Let f be a polynomial of degree $k + 1$. Then either f is irreducible (in which case the result holds with $r = 1$), or f has a factorisation as gh, with neither g nor h a constant polynomial. In that case, since $k + 1$ is the sum of the degrees of g and h, our inductive hypothesis implies that each of g and h has degree less than or equal to k and hence has a decomposition into irreducibles. Writing these two decompositions next to each other gives the required decomposition of f.

For the second part of the proof, we use standard mathematical induction, this time on r, to show that any non-constant polynomial which has a factorisation into r irreducibles has a unique factorisation.

To establish the base case, $r = 1$, we suppose that f is an irreducible polynomial which may also be expressed as a product of non-constant irreducible polynomials $f = g_1 \cdots g_s$. If $s \geq 2$, we would contradict the fact that f is irreducible, so $s = 1$. Thus f can only be 'factorised' as a constant multiple of an irreducible g, which would therefore itself be a constant multiple of f.

Now suppose that $r > 1$ and take as our inductive hypothesis the fact that any non-constant polynomial which has a decomposition into $r - 1$ irreducibles has a unique decomposition (in the above sense). Suppose that

$$f = f_1 \cdots f_r = g_1 \cdots g_s$$

are two decompositions of f into irreducible polynomials. Since f_1 divides $g_1 \cdots g_s$ and f_1 is irreducible, hence prime, f_1 divides g_i for some i. After renumbering, we may suppose that g_1 is divisible by f_1. Since f_1 and g_1 are both irreducible, this means that g_1 is a constant multiple $c_1 f_1$ say. Now divide throughout by f_1 to obtain

$$f_2 \cdots f_r = c_1 g_2 \cdots g_s.$$

Since the polynomial on the left-hand side is a product of $r - 1$ irreducibles (and, on the right, the non-zero constant term can be absorbed into g_2), the inductive hypothesis allows us to conclude that $r - 1 = s - 1$ (so $r = s$) and, after renumbering, each g_i is a constant multiple of f_i for $i = 2, \ldots, r$ and hence for $i = 1, \ldots r$. □

Example If $f(x) = x^3 + 2x^2 - x - 2$ then, noting that $f(1) = 0 = f(-1)$, we easily obtain the factorisation $f(x) = (x - 1)(x + 1)(x + 2)$ into irreducible polynomials.

Example   The question of whether a polynomial is irreducible depends on where the coefficients come from. Thus $f(x) = x^2 + 1$ is irreducible as a real polynomial, since the only way it could be reducible would be if it were a product of two linear polynomials. Then it would have two real zeros, which is impossible, as we have seen. However, over the complex numbers $x^2 + 1 = (x + i)(x - i)$. Again as we have seen, $f(x) = (x + 1)^2$ over $\mathbb{Z}_2$. However, in $\mathbb{Z}_3$, $f(0) = 1$, $f(1) = 2$ and $f(2) = 2$, so in this case, $f(x)$ has no linear factor (by Corollary 6.2.3) so f is irreducible over $\mathbb{Z}_3$.

A very deep result (the Fundamental Theorem of Algebra), well beyond the scope of this book, says that every non-constant polynomial f over $\mathbb{C}$ has a complex zero, α_1, say. Then by Proposition 6.2.4, we may write $f(x)$ as $(x - \alpha_1) f_1(x)$ with $f_1(x)$ a polynomial of degree one less than the degree of f. Clearly, by repeating this procedure, we may write each non-constant polynomial in $\mathbb{C}$ as a product of linear polynomials. Thus if f is an irreducible complex polynomial we deduce that f must be a linear polynomial.

Corollary 6.3.5   *Every irreducible polynomial in $\mathbb{C}[x]$ is linear. Thus, if f is a complex polynomial of degree n, with a_n being the coefficient of x^n, then there are complex numbers $\alpha_1, \alpha_2, \ldots, \alpha_n$ such that*

$$f(x) = a_n(x - \alpha_1)(x - \alpha_2)\ldots(x - \alpha_n).$$

The situation is different for real polynomials. As we have mentioned (and prove in our Appendix), it is easy to show that if $f(x)$ is a real polynomial and α is a complex zero for $f(x)$, then the complex conjugate $\bar{\alpha}$ will be a root of $f(x)$. By unique factorisation this means that

$$(x - \alpha)(x - \bar{\alpha}) = x^2 - (\alpha + \bar{\alpha})x + \alpha\bar{\alpha}$$

divides $f(x)$. Since $\alpha + \bar{\alpha}$ and $\alpha\bar{\alpha}$ are both real, this quadratic is over the reals. So, given a polynomial f in $\mathbb{R}$, we first regard it as a complex polynomial in disguise (so each real coefficient is now regarded as a complex number which happens to be real). Then there is a factorisation of f as a polynomial in $\mathbb{C}[x]$, expressing f as a product of linear (complex) factors. Since each complex root will occur together with its conjugate 'partner', we may group each such pair together, as above, to produce a real quadratic polynomial (with no real roots). In particular, this implies that the irreducible real polynomials are either linear, or certain quadratics (those with no real zeros).

Corollary 6.3.6   *Irreducible real polynomials are either linear or quadratic. Thus, if f is a real polynomial of degree n, there are integers r and m with*

$n = m + 2r$ such that, if a_n is the coefficient of x^n, then there are real numbers $\alpha_1, \alpha_2, \ldots, \alpha_m$ and irreducible quadratic polynomials $x^2 + b_i x + c_i$ $(1 \le i \le r)$ with b_i, c_i real numbers such that

$$f(x) = a_n(x - \alpha_1)(x - \alpha_2) \ldots (x - \alpha_n)(x^2 + b_1 x + c_1) \ldots (x^2 + b_r x + c_r).$$

The situation is nothing like as straightforward as this for polynomials over $\mathbb{Z}_p$. There are irreducible polynomials in this case of arbitrarily large degrees. Since there will only be a finite number of polynomials of a given degree over $\mathbb{Z}_p$, we can work (inductively) up to a given polynomial f of degree n by considering all the polynomials over $\mathbb{Z}_p$ of degree less than n and then checking which of these are irreducible (that is, not themselves divisible by any polynomials of smaller degree in our list) and checking which actually divide f. However this is by no means a short calculation, especially since it can be shown that for any integer n there is an irreducible polynomial of degree n over $\mathbb{Z}_p$. A few examples over $\mathbb{Z}_p$ will illustrate this.

Example 1   Consider $f(x) = x^4 + x^3 + x^2 + x + 1$ over $\mathbb{Z}_2$. Since $f(0) = 1 = f(1)$, f has no zeros over $\mathbb{Z}_2$. This means that $f(x)$ has no linear factor, so if $f(x)$ does factorise as $g(x)h(x)$, the only possibility is that g and h are both quadratic. Then these factors would have the forms $g(x) = ax^2 + bx + c$ and $h(x) = dx^2 + ex + f$, where each of a, b, c, d, e and f is 0 or 1. Then

$$x^4 + x^3 + x^2 + x + 1 = (ax^2 + bx + c)(dx^2 + ex + f).$$

Expanding gives

$$(ax^2 + bx + c)(dx^2 + ex + f) = adx^4 + (ae + bd)x^3$$
$$+ (be + af + cd)x^2 + (ce + bf)x + cf.$$

Equating coefficients of x^4 we see that $ad = 1$. Since each of a, d is 0 or 1 this implies that $a = d = 1$ Similarly, looking at the constant term shows that $cf = 1$, so $c = f = 1$. So now we have that

$$x^4 + x^3 + x^2 + x + 1 = (x^2 + bx + 1)(x^2 + ex + 1).$$

Now equate coefficients of x^3 to see that $(e + b) = 1$ (so one of e and b is 0, the other is 1), and of x^2 to obtain $be + 1 + 1 = 1$ (so $be = 1$). Since these equations, $e + b = 1$ and $be = 1$, have no common solution, there cannot be any such factorisation and we conclude that the polynomial f is irreducible.

Example 2   Consider $f(x) = x^4 + 1$ over $\mathbb{Z}_3$. Again we start by looking for zeros of f. Now, $f(0) = 1$, $f(1) = 2$ and $f(2) = 1 + 1 = 2$, so there are no linear factors. Suppose that $f(x)$ were a product of quadratic factors, so

$$x^4 + 1 = (ax^2 + bx + c)(dx^2 + ex + f).$$

As in our previous example, we see that $ad = 1$ (so neither a nor d can be zero, and in fact $a = d$). Also $cf = 1$ (so $c = f$). Without loss of generality (check that you see why), we may suppose that $a = d = 1$. From the coefficients of x^3, we note that $(e + b) = 0$, so we have $e = -b$. Comparing coefficients of x^2, we have $0 = af + be + cd = f - b^2 + c = 2c - b^2$ so $2c = b^2$. Trying $c = 1$ gives the equation $2 = b^2$, which has no solution. Trying $c = 2$ we have $1 = b^2$, which does have a solution, try $b = 1$, so $e = -b = -1 = 2$. Finally, the coefficient of x gives $0 = ce + bf = -cb + bf = (-c + f)b$ which is consistent and gives no new information. So these equations do have a solution: $a = 1, b = 1, c = 2$ and $d = 1, e = 2, f = 2$ and we have the factorisation

$$x^4 + 1 = (x^2 + x + 2)(x^2 + 2x + 2) \qquad (= (x^2 + x - 1)(x^2 - x - 1))$$

(which you should check). Since neither of these quadratics has a zero (otherwise f would have a zero), our given quartic polynomial is a product of two irreducible quadratics.

The cases of polynomials over $\mathbb{R}$ and over $\mathbb{C}$ are unusual in that we can give an explicit description of the irreducibles. The situation for polynomials over $\mathbb{Z}_p$ is much more like that for integers. Just as we do not have a way to describe uniformly all prime numbers, so we cannot describe uniformly all irreducible polynomials over $\mathbb{Z}_p$.

Exercises 6.3

1. Prove, by induction on r, that if f is an irreducible polynomial and f divides the product $f_1 \cdots f_r$, then f divides one of $f_1, f_2, \ldots, f_r$.
2. Use Theorem 1.6.3 to factorise $x^{p-1} - 1$ over $\mathbb{Z}_p$.
3. Find all irreducible quadratic polynomials, with leading coefficient 1, over $\mathbb{Z}_p$ when p is 2, 3.
4. Find real numbers a, b such that the quartic polynomial $x^4 + 1$ has a decomposition as a product of two quadratics $(x^2 + ax + 1)(x^2 + bx + 1)$. Noting that $x^2 - y^2 = (x - y)(x + y)$, factorise $x^8 - 1$ over $\mathbb{R}$. Hence find a factorisation of $x^8 - 1$ over $\mathbb{C}$. Finally, factorise the polynomial $x^8 - 1$ as a product of irreducibles over $\mathbb{Z}_3$.
5. Find all irreducible cubic polynomials over $\mathbb{Z}_2$.
6. Give examples of polynomials f, g and h such that f divides gh, but f divides neither g nor h.

6.4 Polynomial congruence classes

To generalise the idea of congruence classes from Chapter 1, we take a fixed non-constant polynomial f and say that two polynomials r and s are **congruent modulo** f if $r - s$ is divisible by f. We then write

$$r \equiv s \mod f.$$

Example In $\mathbb{R}[x]$, we may take $f(x) = x^2 - 1$. Congruence classes modulo this polynomial have somewhat undesirable properties.

Let $r(x)$ be a polynomial of degree greater than 1. Then we can use the division algorithm for polynomials to write r in the form $qf + t$, with t being of degree strictly less than the degree of f (which is 2). That is, $t = ax + b$ for some, possibly zero, $a, b \in \mathbb{R}$. Thus every polynomial is congruent modulo $x^2 - 1$ to either a constant or a linear polynomial.

Consider the linear polynomials $x - 1$ and $x + 1$. Neither of these is congruent to the zero polynomial (because f does not divide either $x - 1$ or $x + 1$). However $(x + 1)(x - 1) = x^2 - 1$, so their product is congruent to the zero polynomial. (As in the case of integers modulo n we say that we have 'zero-divisors'.) The same complication will clearly arise whenever we take $f(x)$ to be a reducible polynomial, since if $f(x) = g(x)h(x)$ then the product of $g(x)$ and $h(x)$ will be congruent to zero. For this reason, we will sometimes restrict ourselves to the case when the polynomial $f(x)$ is irreducible. The situation is precisely analogous to that which arose when we considered $\mathbb{Z}_n$ when n is not a prime and so, for many purposes, we concentrated on $\mathbb{Z}_p$ only for p a prime.

Notation and terminology In line with the notation of Chapter 1, we will write $[r]_f$ to denote the set of all those polynomials s which are congruent to r modulo f. This set is referred to as the **polynomial congruence class** of r modulo f.

The argument used in the first part of the example above shows that each polynomial congruence class modulo f has a **standard representative** – the unique polynomial in the class which is of degree less than the degree of f. It can be found by taking any polynomial, r, in the class, applying the division algorithm to write $r = qf + t$ with $\deg(t) < \deg(f)$: then t is the standard representative of the class of r. To see that t is unique, suppose that $s \equiv r \mod f$: then $s = pf + r$ for some polynomial p. Since $r = qf + t$ this gives $s = (p + q)f + t$ and we see that r and s have the same remainder when divided by f.

We now consider operations on polynomial congruence classes.

Definition   Fix a non-constant polynomial f, and let r and s be any other polynomials. Then we define the **sum** and **product** of the polynomial congruence classes of r and s as follows

$$[r]_f + [s]_f = [r + s]_f \qquad \text{and} \qquad [r]_f[s]_f = [rs]_f.$$

As for the integers, there is a potential problem with this definition. We have defined the sum and the product of two congruence classes by reference to particular examples of polynomials in the classes. However, we need to be sure that if we chose to represent $[r]_f$ (or $[s]_f$) by some other polynomial in the class, then we would get the same congruence class for the sum (and for the product). We check this, in another proof which precisely generalises that of the corresponding result in Chapter 1 (Theorem 1.4.1)

Theorem 6.4.1   *Let f be a non-constant polynomial, and let r, s, t be any polynomials. Suppose that $[r]_f = [t]_f$. Then*

(i) $[r + s]_f = [t + s]_f$, *and*
(ii) $[rs]_f = [ts]_f$.

Proof   (i) Since $[r]_f = [t]_f$, f divides $r - t$, so we can write $r = t + kf$, for some polynomial k. Therefore

$$[r + s]_f = [t + kf + s]_f$$
$$= [s + t + kf]_f$$
$$= [s + t]_f \text{ (by definition of congruence classes)}$$

as required.

(ii) With the above notation,

$$[rs]_f = [(t + kf)s]_f$$
$$= [ts + kfs]_f$$
$$= [ts]_f$$

as required.   □

By applying this result twice, we obtain the following easy consequence.

Corollary 6.4.2   *If $[r]_f = [t]_f$ and $[s]_f = [u]_f$ then*

(i) $[r + s]_f = [t + u]_f$
(ii) $[rs]_f = [tu]_f$.

Note that nowhere in the above discussion did we need to specify whether the polynomials in question had coefficients which were real, complex or over $\mathbb{Z}_p$. We therefore see that these two results are independent of where the coefficients come from. We now consider some examples of polynomial congruence classes.

Example 1 In $\mathbb{R}[x]$, we take $f(x)$ to be the irreducible polynomial $x^2 + 1$. As in the first example of this section, because this is a polynomial of degree 2 each polynomial congruence class has, for its standard representative, a constant or a linear polynomial. If r and s are standard representatives of their polynomial classes we will have that $r + s$ is a standard representative of its class but this need not be so for the product rs. The formula for the product of two polynomial congruence classes is different from that in $\mathbb{R}[x]$ itself because we replace every polynomial by a standard representative. For instance, take $r(x) = x + 1$ and $s(x) = x + 2$. Then $[rs]_f = [x^2 + 3x + 2]_f = [(x^2 + 1) + 3x + 1]_f = [3x + 1]_f$. The general formula can be computed as follows:

$$[ax + b]_f[cx + d]_f = [acx^2 + (bc + ad)x + bd]_f$$
$$= [ac(x^2 + 1) + (bc + ad)x + bd - ac]_f$$
$$= [(bc + ad)x + bd - ac]_f.$$

Even though this is a somewhat suprising outcome, it may be familar to those who know the formula for the product of two complex numbers (see the Appendix for a discussion of multiplication of complex numbers). To see how this connection arises, identify the variable x (or, rather, its polynomial congruence class) with the complex number i (not an entirely unreasonable thing to do, since we are putting $x^2 + 1$ congruent to 0 so the class of x will satisfy the equation $[x]_f^2 = -[1]_f$). Then we can regard $[ax + b]_f$ as $ai + b$, $[cx + d]_f$ as $ci + d$, and we have 'rediscovered' the formula for determining the real and imaginary parts of the product of two complex numbers.

Example 2 Now we work over $\mathbb{Z}_2$, and take $f(x)$ to be the cubic equation $x^3 + x + 1$. Since $f(0) = 1$ and $f(1) = 1 + 1 + 1 = 1$, $f(x)$ has no linear factors and so, since it has degree 3, f must be irreducible. Every congruence class will have, as standard representative, a polynomial of degree less than or equal to 2. So in this example our polynomial congruence classes are represented by polynomials of degree less than 3. Since all coefficients are 0 or 1, there are only eight such polynomials:

$$0, 1, x, x + 1, x^2, x^2 + 1, x^2 + x, x^2 + x + 1.$$

Now, computing successive powers of x, we get $x, x^2, x^3 = (x^3 + x + 1) + (x + 1)$. Thus $[x^3]_f = [x + 1]_f$. Similarly $[x^4]_f = [x^3 x]_f = [x^2 + x]_f$. Then $[x^5]_f = [x^2 + x + 1]_f$, $[x^6]_f = [x^2 + 1]_f$ and $[x^7]_f = [1]_f$. Thus each of the seven non-zero polynomial congruence classes occurs as a power of $[x]_f$. This will not be the case in all examples but, in a case like this, where the set of non-zero polynomial congruence classes may be represented by powers of $[x]_f$, we say that $[x]_f$ is a **primitive** polynomial congruence class (meaning that every other polynomial congruence class is a power of this one).

It is clearly the case that the set of polynomial congruence classes is closed under addition. Also, the set of non-zero polynomial congruence classes is closed under multiplication. Furthermore, each non-zero polynomial congruence class has an inverse in the sense of the following definition.

Definition   A polynomial congruence class $[r]_f$ has an **inverse** if there is a polynomial congruence class $[s]_f$ such that $[r]_f [s]_f = [1]_f$ (then we write $[r]_f^{-1} = [s]_f$). Note that this equation means that $rs = tf + 1$ for some polynomial t. Clearly the congruence class of the zero polynomial cannot have an inverse.

Example   In the example above, where we work over $\mathbb{Z}_2$ and take $f(x)$ to be the cubic equation $x^3 + x + 1$, the inverses are as follows:

element	1	x	$x + 1$	x^2	$x^2 + 1$	$x^2 + x$	$x^2 + x + 1$
inverse	1	$x^2 + 1$	$x^2 + x$	$x^2 + x + 1$	x	$x + 1$	x^2

The set of non-zero congruence classes of polynomials modulo $f = x^3 + x + 1 \in \mathbb{Z}_2[x]$ is an example of a **field** in that the set is closed under addition, the non-zero elements are closed under multiplication and all have inverses (for the exact definition see Section 4.4). This is a field with a finite number of elements (namely 8). Fields with a finite number of elements were first discussed by our young French genius Galois. He proved that any finite field has a prime power number of elements, and that this number uniquely determines the structure of the field. For this reason the notation $GF(p^n)$ is often used for a finite (or 'Galois') field with p^n elements: so we would write this one as $GF(2^3)$ or just as $GF(8)$.

Next we see that we always have inverses of non-zero congruence classes provided we take f to be an irreducible polynomial.

Proposition 6.4.3   *Let f be an irreducible polynomial. Then every non-zero polynomial congruence class modulo f has an inverse.*

Proof   Let r be a polynomial not divisible by f (so $[r]_f$ is not equal to $[0]_f$). Consider a greatest common divisor of r and f. Such a polynomial is not a multiple of f (since r is not divisible by f), but must divide f. Since f is irreducible, this greatest common divisor must, therefore, be a non-zero constant polynomial c, say.

Then there are polynomials u and v, which we can find as in the previous section, such that

$$c = ur + vf.$$

Since c is a non-zero constant, it is valid to divide through by c to obtain

$$1 = u_1 r + v_1 f$$

where $u_1 = u/c$ and $v_1 = v/c$. This means that

$$[1]_f = [u_1 r]_f = [u_1]_f [r]_f,$$

so $[u_1]_f$ is an inverse for $[r]_f$.   □

This result is a generalisation of two results from Chapter 1, namely Corollary 1.4.5 and Corollary 1.4.6.

Remark   A general method for constructing a finite field with p^n elements now becomes clear. We search among the polynomials of degree n over $\mathbb{Z}_p$ for an irreducible one, f, say (we have stated, but not proved, that there will always be such a polynomial). Then we form the set of polynomial congruence classes modulo f. This carries the operations of addition and multiplication and we know that, since f is irreducible, each non-zero polynomial congruence class has an inverse. It is then not difficult to see that we have constructed our desired field with p^n elements. It can also be shown that, in this general situation, there will always be a primitive polynomial congruence class (not usually $[x]_f$, however).

Before finishing this section, we look at a further example where the polynomial $f(x)$ is not irreducible. This example will occur again in the next section.

Example   Let n be any integer and take $f(x) = x^n - 1$ over $\mathbb{Z}_p$. This is always reducible, since $f(1) = 0$. As we have seen, within the set of polynomial congruence classes, there will, therefore, be zero-divisors and such a class will not have an inverse. Nevertheless, we can still consider the set of polynomial congruence classes modulo f (even though they will not form a field). As before, standard representatives for these will be all polynomials of degree less than n.

Now, if we multiply the polynomial

$$g(x) = a_0 + a_1 x + a_2 x^2 + \cdots + a_{n-1} x^{n-1}$$

by the linear polynomial x, we see that

$$
\begin{aligned}
[xg(x)]_f &= [a_0 x + a_1 x^2 + \cdots + a_{n-2} x^{n-2} + a_{n-1} x^n]_f \\
&= [a_0 x + a_1 x^2 + \cdots + a_{n-2} x^{n-2} + a_{n-1} \cdot 1]_f \\
&= [a_{n-1} + a_0 x + a_1 x^2 + \cdots + a_{n-2} x^{n-2}]_f
\end{aligned}
$$

since $x^n - 1 = 0 \bmod f$. The important point to note here is that the coefficients of the powers of x have cycled round.

Exercises 6.4

1. Let f be the irreducible quadratic $x^2 + x + 2$ over $\mathbb{Z}_3$. Write down the (9) standard representatives of the congruence classes modulo f. Draw up the multiplication table of the (8) non-zero congruence classes. Find a polynomial g such that every congruence class modulo f is a power of the class $[g]_f$.

2. Let f be any non-constant polynomial and let r, s and t be any polynomials. Suppose that 1 is a greatest common divisor for f and t. Show that if $[rt]_f = [st]_f$ then $[r]_f = [s]_f$.

3. Find the inverses of the following polynomial congruence classes:
   (i) $[g]_f$ over $\mathbb{Z}_2$ when $f(x) = x^2 + x + 1$ and $g(x) = x + 1$;
   (ii) $[g]_f$ over $\mathbb{Z}_3$ when $f(x) = x^3 + x^2 + x + 2$ and $g(x) = x^2 + x$;
   (iii) $[g]_f$ over $\mathbb{R}$ when $f(x) = x^2 + 1$ and $g(x) = x + 1$.

6.5   Cyclic codes

Definition   A code C of **length** n (that is, whose words are of length n) over $\mathbf{B} = \mathbb{Z}_2$ is said to be **cyclic** if

(1) C is a linear code, and
(2) if $c_0 c_1 \ldots c_{n-1}$ is a codeword, then so is $c_{n-1} c_0 \ldots c_{n-2}$ (it will be clear later why it is more convenient to label the entries from 0 to $n-1$ rather than from 1 to n).

Examples   We consider two examples from Section 5.4.
   First, consider the code with codewords

   0000   0011   0101   1001   0110   1010   1100   1111.

When we cycle 1010, we obtain 0101 which is not a codeword. Thus this code is not cyclic.

For our second example take the 3-repetition code with codewords in $\mathbf{B}^6$. This has 4 codewords, namely

$$000000 \quad 101010 \quad 010101 \quad 111111.$$

It is clear that this is a cyclic code.

Our first aim is to give a fairly concrete description of all cyclic codes of length n. To do this we start by representing a general vector of length n, say $(a_0, a_1, \ldots, a_{n-1})$ by a polynomial over $\mathbf{B} = \mathbb{Z}_2$. This is done by using the individual entries in the vector as 'markers' for the appropriate power of x: thus $(a_0, a_1, \ldots, a_{n-1})$ corresponds to the polynomial of degree $n - 1$ which is

$$f(x) = a_0 + a_1 x + \cdots + a_{n-1} x^{n-1}.$$

Then multiplication of this polynomial by x gives a polynomial of degree n,

$$a_0 x + a_1 x^2 + \cdots + a_{n-2} x^{n-1} + a_{n-1} x^n.$$

If now we consider polynomial congruence classes modulo $f(x) = x^n - 1$ (so $x^n \equiv 1 \bmod f$), this is congruent to the polynomial

$$a_{n-1} + a_0 x + a_1 x^2 + \cdots + a_{n-2} x^{n-1},$$

which corresponds to the n-tuple $(a_{n-1}, a_0, a_1, \ldots, a_{n-2})$. Note that this corresponds to the 'cycling' operation which occurs in the second clause of the definition of cyclic code. We actually have a bijection with the set of polynomial congruence classes modulo $x^n - 1$ because each such class has a unique standard representative polynomial of degree less than n and so corresponds to a unique vector of length n. We will frequently move between these two interpretations of codeword as vectors or as polynomial congruence classes. It is not difficult from this point of view to establish our first result.

Proposition 6.5.1 *A code C of length n, regarded as a code of polynomial congruence classes with respect to $f(x) = x^n - 1$ is a cyclic code if and only if*

(i) *C is a linear code, and*
(ii) *for any polynomial t, if $[g]_f$ is a codeword in C, then so is $[tg]_f$.*

Proof Suppose first that C is a cyclic code. Then C is linear. Also, as we have seen, the word obtained from the cyclic permutation of a codeword is another codeword. Now, cyclic permutation corresponds precisely to multiplication by

$[x]_f$ in the set of polynomial congruence classes with respect to $f(x) = x^n - 1$. It follows (by induction) that multiplication by $[x^i]_f$ corresponds to this cyclic permutation being performed i times and hence, if the polynomial congruence class $[g]_f$ corresponds to a codeword, then so does $[x^i g]_f$. Since our code is linear any sum of such terms is a codeword. Therefore if $[g]_f$ is a codeword of a cyclic code, then for any polynomial t, the polynomial congruence class of the product tg is another codeword (remember that we are working over $\mathbb{Z}_2$ so every coefficient of a polynomial is either 0 or 1).

For the converse, suppose that C satisfies conditions (i) and (ii) of the proposition. Then C is linear and, since we can multiply a codeword by $[x]_f$ and obtain another codeword, we deduce that the cyclic condition is satisfied and so we do have a cyclic code. $\square$

Proposition 6.5.2 *Let C be a cyclic code. Then there is a polynomial congruence class $[g]_f$ such that every codeword in C is equivalent to the product of g and a polynomial.*

Proof Suppose that C is a cyclic code. Among all the non-zero codewords of C (regarded as polynomial congruence classes) choose one, g say, of smallest degree. Suppose that g has degree d. If now $[s]_f$ is any element in C, use the division algorithm to write $s = qg + r$ where either r is the zero polynomial or the degree of r is less than d. In this latter case, since $r = s - qg$ and since $[s]_f$ is a codeword, as is, by 6.5.1, $[qg]_f$, we deduce that $[r]_f$ is also in C. This contradicts the definition of d unless $r = 0$. We deduce that g divides s. Thus every codeword in C is equivalent to a product qg for some polynomial q. $\square$

A non-zero polynomial g of least degree in a cyclic code is called a **generator polynomial** for the code and we say that the code C is **generated** by g.

Example Our cyclic code above with codewords

$$000000 \quad 101010 \quad 010101 \quad 111111$$

corresponds to congruence classes modulo $f(x) = x^6 - 1$ of the polynomials, 0, $f_1(x) = 1 + x^2 + x^4$, $f_2(x) = x + x^3 + x^5$ and $f_3(x) = 1 + x + x^2 + x^3 + x^4 + x^5$. Clearly, f_1 is one choice of generator polynomial with

$$0 = 0 \cdot f(x), \qquad f_2(x) \equiv x f_1(x) \quad \text{and} \quad f_3(x) \equiv (1+x) f_1(x) \bmod f.$$

Note that, by 6.5.1 and 6.5.2, the codewords of a cyclic code are exactly those corresponding to congruence classes of the form $[tg]_f$ as t ranges over polynomials (of degree less than n), where g is a generator for the code. Conversely, if h is any non-zero polynomial of degree less than n, then the set of codewords corresponding to congruence classes of the form $[th]_f$ (as t ranges over polynomials of degree less than n) is easily checked to be a cyclic code. (It is linear since $[sh]_f + [th]_f = [(s+t)h]_f$ and it is cyclic because cycling corresponds to multiplying by x.) It need not be the case, however, that h is a generator polynomial for this code, because h might not satisfy the property of having least degree. But there will be some generator polynomial for the code defined in this way. In fact the next result tells us that any generator polynomial for a cyclic code of length n must be a factor of $x^n - 1$.

Corollary 6.5.3   *Let C be a cyclic code of length n. If g is a generator polynomial for C then g divides $x^n - 1$.*

Proof   Suppose that g does not divide $x^n - 1$, then we use the division algorithm to write $x^n - 1$ in the form $qg + r$ with the degree of r less than that of g. Since, note, $x^n - 1$ represents the zero word and we know that qg is a codeword, we deduce that r represents a codeword, contrary to the definition of g as having minimal degree, unless $r = 0$. Thus $x^n - 1 = q(x)g(x)$ and so g divides $x^n - 1$.   □

For instance, in our example above, we have $x^6 - 1 = (x^2 - 1)f_1(x)$.

Once we have a generator polynomial, g, of degree m for a cyclic code C of length n, we can write down an $n - m$ by m generator matrix by placing the coefficients of $g(x)$ in its first row, those of $xg(x)$ in its second, those of $x^2g(x)$ in the third row, and so on, with the coefficients of $x^{n-m-1}g(x)$ in the last row. Notice that this is a different type of generator matrix from those we used in Chapter 5, but it is nevertheless the case (it follows by 6.5.2) that we may obtain any codeword by taking linear combinations of the rows of our matrix.

As an example, the generator matrix associated with the generator polynomial $1 + x^2 + x^4$ in our previous example (so $n = 6, m = 4$) is the 2×6 matrix

$$\begin{pmatrix} 1\ 0\ 1\ 0\ 1\ 0 \\ 0\ 1\ 0\ 1\ 0\ 1 \end{pmatrix}.$$

Here is one further piece of terminology. We refer to a polynomial $p(x)$ which satisfies $x^n - 1 = p(x)g(x)$ as a **parity polynomial** for the code generated by g. The parity polynomial for our example above is $x^2 - 1$.

We now illustrate these ideas by discussing cyclic codes of length 6.

Example   The polynomial $x^6 - 1$ factorises as $(x^3 - 1)(x^3 + 1)$. Over $\mathbf{B} = \mathbb{Z}_2$ this is the same as $(x^3 + 1)^2$. Also, over $\mathbb{Z}_2$, $x + 1$ is a factor of $x^3 + 1$, so we have the factorisation $x^6 - 1 = x^6 + 1 = (x + 1)^2(x^2 + x + 1)^2$. Thus we can list the generator polynomials for the various cyclic codes of length 6. They are:

$$1, \quad x + 1, \quad x^2 + 1, \quad x^2 + x + 1, \quad x^3 + 1, \quad (x^2 + x + 1)^2,$$
$$(x + 1)^2(x^2 + x + 1), \quad (x + 1)(x^2 + x + 1)^2, \quad x^6 + 1.$$

We consider in turn the codes with these generator polynomials.

(1) The polynomial 1 clearly generates the whole of $\mathbf{B}^6$, so the set of code-words is all 64 vectors of length 6 over $\mathbf{B}$. Clearly this code has no error detecting or correcting properties.

(2) The polynomial $1 + x$ gives 32 codewords given as linear combinations of the rows of the generator matrix

$$\begin{pmatrix} 1 & 1 & 0 & 0 & 0 & 0 \\ 0 & 1 & 1 & 0 & 0 & 0 \\ 0 & 0 & 1 & 1 & 0 & 0 \\ 0 & 0 & 0 & 1 & 1 & 0 \\ 0 & 0 & 0 & 0 & 1 & 1 \end{pmatrix}.$$

With just a little thought one sees that the 32 codewords will be all those vectors in $\mathbf{B}^6$ with an even number of 1s. This means that the least weight of a non-zero codeword for this code is 2 so it detects one error. (With one error, we obtain a word with an odd number of 1s, but we have no way of knowing where the error lies.)

(3) The polynomial $1 + x^2$ gives the generator matrix

$$\begin{pmatrix} 1 & 0 & 1 & 0 & 0 & 0 \\ 0 & 1 & 0 & 1 & 0 & 0 \\ 0 & 0 & 1 & 0 & 1 & 0 \\ 0 & 0 & 0 & 1 & 0 & 1 \end{pmatrix}.$$

It is again clear that the least weight of a non-zero codeword for this code is 2. Any single error will give rise to an odd number of 1s and so will be immediately detected.

(4) We next consider the irreducible quadratic $1 + x + x^2$, with associated matrix

$$\begin{pmatrix} 1 & 1 & 1 & 0 & 0 & 0 \\ 0 & 1 & 1 & 1 & 0 & 0 \\ 0 & 0 & 1 & 1 & 1 & 0 \\ 0 & 0 & 0 & 1 & 1 & 1 \end{pmatrix}.$$

In this case, we write out the 16 codewords:

000000 111000 011100 100100 001110 110110 010101 101010
000111 111111 011011 100011 001001 110001 010010 101101

We note that for this code the least weight of a non-zero codeword is 2 so again the code detects a single error.

In fact we can always detect a single error in a code with a non-constant generator polynomial g; let p be the parity polynomial for C (so $[gp]_f = [0]_f$). Then if r is any codeword we have $r = tg$ for some t, so $[rp]_f = [tgp]_f = [t \cdot 0]_f = [0]_f$. In fact (as we shall see in Exercise 6.5.5) the converse of this result holds. Hence the only polynomial congruence classes $[r]_f$ with $[rp]_f$ being zero are codewords. Thus we can tell whether we have a codeword by multiplying by p.

(5) Now consider the cubic polynomial $(x + 1)(x^2 + x + 1) = x^3 + 1$. This gives the matrix

$$\begin{pmatrix} 1 & 0 & 0 & 1 & 0 & 0 \\ 0 & 1 & 0 & 0 & 1 & 0 \\ 0 & 0 & 1 & 0 & 0 & 1 \end{pmatrix}.$$

Thus there are 8 codewords. Clearly at least one of these (given by the first row of the matrix) has weight 2, so again the code detects (by multiplying by the parity polynomial) one error.

(6) For the polynomial $(x + 1)^2(x^2 + x + 1) = x^4 + x^3 + x + 1$ the corresponding matrix is

$$\begin{pmatrix} 1 & 1 & 0 & 1 & 1 & 0 \\ 0 & 1 & 1 & 0 & 1 & 1 \end{pmatrix}.$$

There are now just 4 codewords:

000000 110110 011011 101101.

Hence the least weight is 4. In fact this code is a two-word repetition code (every codeword has the form ww with w a word of even weight in $\mathbf{B}^3$). Hence this code detects up to 3 errors and it corrects one error, since with one error, one half of the word will be of odd weight so we can tell which is the correct half and hence recover the correct codeword.

(7) When the generator polynomial is $(x^2 + x + 1)^2 = x^4 + x^2 + 1$, the generator matrix is

$$\begin{pmatrix} 1 & 0 & 1 & 0 & 1 & 0 \\ 0 & 1 & 0 & 1 & 0 & 1 \end{pmatrix}$$

and so the codewords are

$$000000 \quad 101010 \quad 010101 \quad 111111.$$

This has least weight 3, so detects two errors, since a two-error word will have weight 1, 2, 4 or 5. It also corrects one error (indeed, this code is the three repetition code we have already met).

(8) For the generator polynomial $(x + 1)(x^2 + x + 1)^2 = x^5 + x^4 + x^3 + x^2 + x + 1$, the generator matrix is

$$\begin{pmatrix} 1 & 1 & 1 & 1 & 1 & 1 \end{pmatrix}.$$

The only codewords are 000000 and 111111, so the least distance is 6. This code detects up to 5 errors and corrects 2. With only two codewords of length 6 this code is certainly not efficient.

(9) In the final case, the generator polynomial is $x^n - 1$, so the only codeword is the zero vector (therefore this code cannot be used to transmit information).

Cyclic codes are extensively studied. Among the cyclic codes of special interest are the so-called quadratic residue codes. These have excellent error-correcting properties and so are of great practical use. More about cyclic codes can be found in [Hill].

Exercises 6.5

1. Let g be a polynomial over $\mathbb{Z}_2$. Show that if g is irreducible then the number of non-zero coefficients of powers of x (including the constant term) is an odd integer.

2. Factorise $x^5 - 1$ over $\mathbb{Z}_2$. Hence write down generator polynomials for all the cyclic codes of length 5 over **B**, and state how many errors each cyclic code detects and how many errors it corrects.

3. Let C be the cyclic code of length 7 with generator polynomial $x^3 + x^2 + 1$. List the codewords of C, and show that every 7-vector is within one error of a codeword of C.

4. Use the polynomial from Exercise 6.5.3 to determine all cyclic codes of length 7 over **B**, stating how many errors each cyclic code detects and how many errors it detects.

5. Let p be a parity polynomial for a cyclic code with generator polynomial g. Use the division algorithm to show that if c is a polynomial with $[cp]_f = [0]_f$ then c is a codeword.

Summary of Chapter 6

In this chapter, we defined polynomials and gave their basic 'arithmetic' properties. The exposition of our material was closely modelled on that of Chapter 1 for integers. This showed how essentially the same arguments from the earlier chapter can be used to produce a very similar theory for polynomials. In Section 6.1, we discussed the basic operations (addition, subtraction and multiplication) for polynomials. Section 6.2 was concerned with division for polynomials and included the Euclidean algorithm, one of the direct generalisations from Chapter 1. We considered factorisation of polynomials in Section 6.3. The results in this section depend on the ring of coefficients of our polynomials. Section 6.4 was a development of the idea of polynomial congruence classes (a direct generalization of congruence classes of integers). An important application of these was a method for, in principle, constructing any finite field. Finally, in Section 6.5, we showed how we could make use of polynomial congruence classes to produce an important special class of linear codes – the cyclic codes.

Appendix on complex numbers

The reader will be accustomed, from an early age, to the idea of extending number systems. The natural numbers are used for counting, but it soon becomes clear that questions involving natural numbers may not have answers which are natural numbers. For example: 'On a winter's day, the temperature is 6°C. At night the temperature falls by 10 degrees. What is the overnight temperature'? To deal with this problem, we extend the natural numbers to the set of integers, by including negatives. However, one soon meets integer equations with non-integer solutions. For example: 'Share 3 cookies between two people' (that is, 'Solve $2x - 3 = 0$'). Again, we extend our number system from integers to rationals (by including fractions). Even after extending to rationals, there are still unanswered questions. 'Find the ratio of the length of a diagonal of a square to the length of a side' (that is, 'Find x such that $x^2 = 2$'). This time we extend the rationals to the real numbers. It is usual to make do with the real number system for everyday life and for a good part of school life. As we have seen, however, a polynomial like $x^2 + 1$ cannot have real number zeros, since the square of a real number is never negative.

The way to meet this difficulty is to require the existence of a new number i for which $i^2 = -1$. Then the set of numbers of the form $z = a + ib$ (as a, b vary over the set of real numbers) is referred to as the set of **complex numbers**, with a being the **real part** of z and b being its **imaginary part**. We write $\mathbb{C}$ for the set of complex numbers:

$$\mathbb{C} = \{a + ib : a, b \in \mathbb{R} \text{ where } i^2 = -1\}.$$

Then we add two complex numbers by adding their real parts and their imaginary parts separately:

$$(a + ib) + (c + id) = (a + c) + i(b + d).$$

To multiply two complex numbers, we use the usual rules of algebra to simplify brackets. In addition, we regard the symbol i as commuting with any real number (so $a(\mathrm{i}d) = \mathrm{i}(ad)$) and recall that $\mathrm{i}^2 = -1$:

$$(a + \mathrm{i}b)(c + \mathrm{i}d) = ac + \mathrm{i}bc + \mathrm{i}ad + \mathrm{i}^2bd = (ac - bd) + \mathrm{i}(bc + ad).$$

It may occur to the reader that this process is, potentially, a never-ending one. Maybe, in order to solve equations involving complex numbers, we need to invent other number systems. However this is not the case. A result known as the Fundamental Theorem of Algebra says that every non-constant polynomial over the complex numbers has a complex number as a zero. As we saw in Chapter 6, this is sufficient information to be able to prove that every zero of a complex polynomial is a complex number. This fact is sometimes expressed by saying that the field of complex numbers is **algebraically closed**. This result is not without its controversy. It took more than two centuries for complex numbers to become widely accepted, even among mathematicians. One feature disliked by some, is that although the Fundamental Theorem shows that every complex polynomial has a complex zero, its proof does not tell us how to find such a zero.

Complex numbers are often illustrated by the Argand diagram. This is nothing more than the ordinary plane from Cartesian geometry, but with the usual x-axis now being thought of as the 'real'-axis (corresponding to the real part, a, of the complex number $z = a + \mathrm{i}b$) and the usual y-axis being thought of as the 'imaginary' axis (corresponding to $\mathrm{i}b$ where b is the imaginary part of $a + \mathrm{i}b$). For some purposes, it is convenient to think of points in the Argand diagram using polar coordinates. Thus a point is specified by giving its distance, d, from the origin (this distance is known as its **modulus**) and the angle, θ, between the positive real axis and the line joining the point to the origin (this angle is known as its **argument**). This gives the representation $z = d\mathrm{e}^{\mathrm{i}\theta}$ of a complex number, where $\mathrm{e} = 2.71828\ldots$ is the base of natural logarithms and where θ is measured in radians. For an explanation of this notation you should consult a book which deals with complex numbers. This way of considering complex numbers is particularly useful when one wishes to find powers and roots of a given complex number.

Given a complex number $z = a + \mathrm{i}b$, its **complex conjugate** is the complex number $a - \mathrm{i}b$ (so this is the reflection of the given complex number in the real axis of the Argand diagram). Thus a complex number which is equal to its complex conjugate has no (i.e. zero) imaginary part and is purely real. It is usual to denote the complex conjugate of z by $\bar{z}$. If we add a complex number to its complex conjugate, we clearly get a real number (twice the real part of the complex number). It is also true that if we multiply a complex number by its complex conjugate we obtain a real number. To see this, consider the product of a complex number z with its complex conjugate $\bar{z}$:

$$z\bar{z} = (a + ib)(a - ib) = a^2 + iba - iba + i^2(b)(-b) = a^2 + b^2$$

which is clearly a real number.

There are some very simple rules for dealing with conjugates.

Rules for complex conjugates   *Let z_1 and z_2 be complex numbers. Then*

(i) $\overline{z_1 + z_2} = \overline{z_1} + \overline{z_2}$ *and*
(ii) $\overline{z_1 z_2} = \overline{z_1}\,\overline{z_2}$.

Proof   (i) Consider the two sides of our claimed equality. First $\overline{z_1 + z_2}$ is the complex conjugate of $z_1 + z_2$. So, if $z_1 = a_1 + ib_1$ and $z_2 = a_2 + ib_2$, we have

$$\begin{aligned}
\overline{z_1 + z_2} &= \overline{(a_1 + ib_1) + (a_2 + ib_2)} \\
&= \overline{(a_1 + a_2) + i(b_1 + b_2)} \\
&= (a_1 + a_2) - i(b_1 + b_2).
\end{aligned}$$

The right-hand side is

$$\overline{z_1} + \overline{z_2} = (a_1 - ib_1) + (a_2 - ib_2) = (a_1 + a_2) - i(b_1 + b_2).$$

Since these two expressions are equal, we have proved (i). The proof of (ii) may now be safely left to the reader, although it is slightly more complicated because one needs to expand expressions for products of complex numbers, then gather together their real and imaginary parts. $\square$

Once these basic rules have been established, we can apply (ii) with $z_1 = z_2 = z$ to obtain $\overline{z^2} = \overline{z}^2$. It is then easy to prove (by mathematical induction) that, for all positive integers n, $\overline{z^n} = \overline{z}^n$. Suppose now that we have a complex polynomial

$$f(z) = a_0 + a_1 \cdot z + a_2 \cdot z^2 + \cdots + a_n \cdot z^n,$$

so the coefficients $a_0, a_1, \ldots, a_n$ are complex numbers. Using our rules for complex conjugates (and induction again), gives

$$\overline{f(z)} = \overline{a_0} + \overline{a_1} \cdot \overline{z} + \cdots + \overline{a_n} \cdot \overline{z^n}.$$

Now use the fact that $\overline{z^n} = \overline{z}^n$, to see that

$$\overline{f(z)} = \overline{a_0} + \overline{a_1} \cdot \overline{z} + \cdots + \overline{a_n} \cdot \overline{z}^n.$$

If in addition, each coefficient, a_i, is actually real (so that it is its own complex conjugate, $a_i = \overline{a_i}$), we obtain

$$\overline{f(z)} = a_0 + a_1\overline{z} + \cdots a_n\overline{z}^n \text{ which equals } f(\overline{z}).$$

Therefore if z is a zero of the polynomial $f(z)$ with real coefficients, then

$$0 = \overline{0} = \overline{f(z)} = f(\overline{z}).$$

We have therefore proved the statement used in Chapter 6: if $f(z)$ is a complex polynomial with real coefficients, then whenever z is a zero of f, so is $\overline{z}$.

Answers

Chapter 1

Exercises 1.1

1. (i) $2 \cdot 11 + (-3) \cdot 7 = 1$; (ii) $2 \cdot (-28) + (-1)(-63) = 7$;
   (iii) $7 \cdot 91 + (-5) \cdot 126 = 7$; (iv) $(-9)630 + 43 \cdot 132 = 6$;
   (v) $35 \cdot 7245 + (-53)4784 = 23$; (vi) $(-31)6499 + 47 \cdot 4288 = 67$.
   Note that there are many ways of expressing the gcd.
2. $20 \cdot 6 + (-10) \cdot 14 + 21 = 1$.
4. One example occurs when a is 4, b is 2 and c is 6.
5. Take $a = 2$, $b = 4$ and $c = 12$ for an example.
7. We start with both jugs empty, which we write as $(0, 0)$; fill the larger jug, $(0, 17)$; fill the smaller from the larger, $(12, 5)$; empty the smaller, $(0, 5)$; pour the remains into the smaller, $(5, 0)$; then continue as $(5, 17)$, $(12, 10)$, $(0, 10)$, $(10, 0)$, $(10, 17)$, $(12, 15)$, $(0, 15)$, $(12, 3)$, $(0, 3)$, $(3, 0)$, $(3, 17)$, $(12, 8)$. Note that we add units of 17 and subtract units of 12 and, in effect, produce 8 as the linear combination $4 \cdot 17 - 5 \cdot 12$ of 17 and 12.

Exercises 1.2

1. $a_1 = 1, a_2 = 3, a_3 = 7, a_4 = 15$ and $a_5 = 31$. Now to prove, using induction, that a_{n+1} is a power of 2, first notice (for the base case) that when $n = 1$, $a_1 + 1 = 2$ which is a power of 2. Next suppose that $a_k + 1$ is a power of 2, say $a_k + 1 = 2^n$. Then

$$a_{k+1} + 1 = (2a_k + 1) + 1 = 2a_k + 2 = 2(a_k + 1) = 2 \cdot 2^n = 2^{n+1}$$

which is also a power of 2 as required.

4. $1^3 + 2^3 + \cdots + n^3 = \{n(n+1)/2\}^2 = \frac{1}{4}n^4 + \frac{1}{2}n^3 + \frac{1}{4}n^2.$
6. $1 + 3 + \cdots + (2n-1) = n^2.$
11. $2^{11} - 1 = 23 \times 89.$
12. (a) False, the given argument breaks down on a set with 2 elements.
    (b) False, the base case $n = 1$ is untrue.

Exercises 1.3

1. The primes less than 250 are
   2, 3, 5, 7, 11, 13, 17, 19, 23, 29, 31, 37, 41, 43, 47, 53, 59, 61, 67, 71, 73, 79,
   83, 89, 97, 101, 103, 107, 109, 113, 127, 131, 137, 139, 149, 151, 157, 163,
   167, 173, 179, 181, 191, 193, 197, 199, 211, 223, 227, 229, 233, 239, 241.
3. (a)
$$136 = 2 \cdot 68 = 2^2 \cdot 34 = 2^3 \cdot 17$$
$$150 = 2 \cdot 75 = 2 \cdot 5 \cdot 15 = 2 \cdot 3 \cdot 5^2$$
$$255 = 5 \cdot 51 = 3 \cdot 5 \cdot 17.$$

After trying small primes, we see that
$$713 = 23 \cdot 31$$
$$3549 = 3 \cdot 1183 = 3 \cdot 7 \cdot 169 = 3 \cdot 7 \cdot 13^2.$$

Checking divisibility for all primes less than 70 shows that 4591 is prime.
(b) Thus $(136, 150) = 2$, lcm $(136, 150) = 2^3 \times 3 \times 5^2 \times 17 = 10200$
$(255, 3549) = 3$, lcm $(255, 3549) = 3 \times 5 \times 7 \times 13^2 \times 17 = 301665.$
4. If $c_n = p_1 \times \cdots \times p_n + 1$ then $c_1 = 3, c_2 = 7, c_3 = 31, c_4 = 211$ and
   $c_5 = 2311$ are all, as you may check, prime. But $c_6 = 30031 = 59 \cdot 509$ is
   not prime.

Exercises 1.4

1. (i) $48 - 8 = 40$ which is not divisible by 14 so the assertion is false.
   (ii) $48 - (-8) = 56$ which is divisible by 14, so $-8 \equiv 48 \bmod 14.$
   (iii) $10 - 0 = 10$ which is not divisible by 100, so the assertion is false.
   (iv) $357482 - 7754 = 349728$ which is divisible by 3643, so $357482 \equiv$
         7754 mod 3643.
   (v) $135227 - 16023 = 1309204$ which is divisible by 25177 so the
        congruence holds.
   (vi) $33303 - 4015 = 29288.$ Since 1295 does not divide 29288, the
         congruence does not hold.

2. When n is 6, we obtain

+	0	1	2	3	4	5
0	0	1	2	3	4	5
1	1	2	3	4	5	0
2	2	3	4	5	0	1
3	3	4	5	0	1	2
4	4	5	0	1	2	3
5	5	0	1	2	3	4

×	0	1	2	3	4	5
0	0	0	0	0	0	0
1	0	1	2	3	4	5
2	0	2	4	0	2	4
3	0	3	0	3	0	3
4	0	4	2	0	4	2
5	0	5	4	3	2	1

When n is 7, we obtain

+	0	1	2	3	4	5	6
0	0	1	2	3	4	5	6
1	1	2	3	4	5	6	0
2	2	3	4	5	6	0	1
3	3	4	5	6	0	1	2
4	4	5	6	0	1	2	3
5	5	6	0	1	2	3	4
6	6	0	1	2	3	4	5

×	0	1	2	3	4	5	6
0	0	0	0	0	0	0	0
1	0	1	2	3	4	5	6
2	0	2	4	6	1	3	5
3	0	3	6	2	5	1	4
4	0	4	1	5	2	6	3
5	0	5	3	1	6	4	2
6	0	6	5	4	3	2	1

3. (i)   The inverse of 7 modulo 11 is 8;
   (ii) the inverse of 10 modulo 26 does not exist;
   (iii) the inverse of 11 modulo 31 is 17;
   (iv) the inverse of 23 modulo 31 is 27; and
   (v)   the inverse of 91 modulo 237 is 112.

4. When n is 16, G_n has 8 elements, 1, 3, 5, 7, 9, 11, 13 and 15. The multiplication table is

	1	3	5	7	9	11	13	15
1	1	3	5	7	9	11	13	15
3	3	9	15	5	11	1	7	13
5	5	15	9	3	13	7	1	11
7	7	5	3	1	15	13	11	9
9	9	11	13	15	1	3	5	7
11	11	1	7	13	3	9	15	5
13	13	7	1	11	5	15	9	3
15	15	13	11	9	7	5	3	1

When n is 15 the elements of G_n are 1, 2, 4, 7, 8, 11, 13 and 14. The table is

	1	2	4	7	8	11	13	14
1	1	2	4	7	8	11	13	14
2	2	4	8	14	1	7	11	13
4	4	8	1	13	2	14	7	11
7	7	14	13	4	11	2	1	8
8	8	1	2	11	4	13	14	7
11	11	7	14	2	13	1	8	4
13	13	11	7	1	14	8	4	2
14	14	13	11	8	7	4	2	1

9. (i) The first calculation is in error; we can draw no conclusions about the second or third (in fact the second is wrong, but the third is correct).

   (ii) The underlined digit should be 3.

Exercises 1.5

1. (i)   no solution;

   (ii) $[4]_{11}$;

   (iii) $[11]_{21}$ or $[11]_{84}$, $[32]_{84}$, $[53]_{84}$, $[74]_{84}$;

   (iv) $[6]_{17}$;

   (v)   no solution;

   (vi) $[7]_{20}$ or $[7]_{100}$, $[27]_{100}$, $[47]_{100}$, $[67]_{100}$, and $[87]_{100}$;

   (vii) $[10]_{107}$.

2. (i) $[172]_{264}$;

   (ii) $[7]_{20}$;

   (iii) $[123]_{280}$.

3. 1944

4. (i) $x^4 + x^2 + 1 = (x^2 + 2)(x^2 + 2) = (x + 1)(x + 1)(x + 2)(x + 2)$.

   (ii) Reduce modulo 3.

5. The minimum number of gold pieces was 408.

Exercises 1.6

1. (i) 5; (ii) 6; (iii) 16; (iv) 20.

2. (i) $5^{20} \equiv 4 \bmod 7$; (ii) $2^{16} \equiv 0 \bmod 8$; (iii) $7^{1001} \equiv 7 \bmod 11$;

   (iv) $6^{76} \equiv 9 \bmod 13$.

5. $\phi(32) = 16$; $\phi(21) = 12$; $\phi(120) = 32$; $\phi(384) = 128$.

6. (i) $2^{25} \equiv 2 \bmod 21$;

   (ii) $7^{66} \equiv 7^2 \equiv 49 \bmod 120$.

   (iii) $\phi(100) = 40$. So the last two digits of 7^{162} are 49. Note that, since $(5, 100) \neq 1$, Euler's Theorem cannot be applied to calculate the last

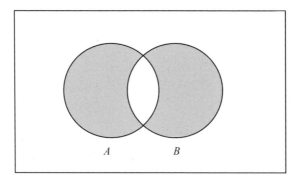

A B

Fig. A1

two digits of 5^{121}. It can be seen directly that $5^k \equiv 25 \bmod 100$ for $k \geq 2$. Using this plus $3^{40} \equiv 1 \bmod 100$ and, say, $5^2 \cdot 3^4 \equiv 25 \cdot 81 \equiv 25 \bmod 100$, it follows that the last two digits of $5^{143} \cdot 3^{312}$ are 25. So the answer is 75.

10. $2^{37} - 1 = 223 \cdot 616318177$.

11. $2^{32} + 1 = 4294967297 = 641 \cdot 6700417$.

12. The message is FOOD.

13. The message is JOHN.

Chapter 2

Exercises 2.1

1. $X = W = Z; Y = V$.

2. Subsets of $\{a, b, c\}$ are Ø $\{a\}$, $\{b\}$, $\{c\}$, $\{a, b\}$, $\{a, c\}$, $\{b, c\}$ and $\{a, b, c\}$.
   Subsets of $\{a, b, c, d\}$ are Ø, $\{a\}$, $\{b\}$, $\{c\}$, $\{d\}$, $\{a, b\}$, $\{a, c\}$, $\{a, d\}$, $\{b, c\}$, $\{b, d\}$, $\{c, d\}$, $\{a, b, c\}$, $\{a, b, d\}$, $\{a, c, d\}$, $\{b, c, d\}$ and $\{a, b, c, d\}$.
   If X has n elements, the set of subsets of X has 2^n elements.

4. The symmetric difference is shaded in Fig. A1.

6. $X \times Y = \{(0,2), (0,3), (1,2), (1,3)\}$. This means that the set has $2^4 = 16$ subsets: Ø, $\{(0,2)\}$, $\{(0,3)\}$, $\{(1,2)\}$, $\{(1,3)\}$, $\{(0,2), (0,3)\}$, $\{(0,2), (1,2)\}$, $\{(0,2), (1,3)\}$, $\{(0,3), (1,2)\}$, $\{(0,3), (1,3)\}$, $\{(1,2), (1,3)\}$, $\{(0,2), (0,3), (1,2)\}$, $\{(0,2) (0,3), (1,3)\}$ $\{(0,2), (1,2), (1,3)\}$, $\{(0,3), (1,2), (1,3)\}$ and $\{(0,2), (0,3), (1,2), (1,3)\}$.

7. (i) True.
   (ii) False. One possible counterexample is given by taking A to be $\{1\}$, B to be $\{2\}$, $C = \{a\}$ and $D = \{b\}$. Then $(1, b)$ is in the right-hand term but not in the left-hand term.

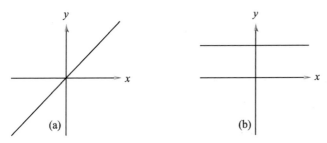

Fig. A2 (a) The graph of the identity function $y = f(x) = x$. (b) The graph of the constant function $y = f(x) = 1$.

8. $X \times Y$ has mn elements.
9. Take $A = B = \{1, 2\}$ and $X = \{(1, 1), (2, 2)\}$.

Exercises 2.2

1. A function is given by specifying $f(0)$ (two possibilities 0 or 5 for this), $f(1)$ (again 0 or 5) and $f(2)$ (again 0 or 5). There are $2 \times 2 \times 2 = 8$ such functions.
2. (i) bijective; (ii) not injective but surjective; (iii) neither injective nor surjective; (iv) surjective, not injective; (v) injective, not surjective.
3. See Fig. A2.
4. $fg(x) = x^2 - 1$; $gf(x) = x^2 + 2x - 1$; $f^2(x) = x + 2$; $g^2(x) = x^4 - 4x^2 + 2$.
5. For instance:
   (i) $f(x) = \log(x)$; (ii) $f(x) = \tan(x)$; (iii) $f(2k) = k$ and $f(2k - 1) = -k$.
6. (Compare Theorem 4.1.1.) There are 6 bijections.
7. (i) The inverse is $(4 - 3x)$;
   (ii) the inverse of $f(x) = (x - 1)^3$ is $1 + (x)^{1/3}$.

Exercises 2.3

1. (a) is R (reflexive); not S (symmetric); not WA (weakly antisymmetric); is not T (transitive) – consider $a = 2, b = 1, c = 0$.
   (b) R, S, not WA, T.
   (c) not R, S, not WA, not T.
   (d) not R, S, not WA, not T.
   (e) R, not S, not WA, T.
   (f) R, S, not WA, T.
   (g) R, not S, WA, not antisymmetric, T.

Fig. A3

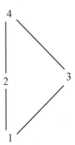

Fig. A4

2. (c) The relation of equality on any non-empty set is an example.
   (f) For instance, take
   $$X = \{a, b,\} \text{ and}$$
   $$R = \{(b, b)\}.$$
3. There is an unjustified hidden assumption that for each $x \in X$ there exists some $y \in X$ with $x R y$.
5. The Hasse diagram is as shown in Fig. A3.
6. The adjacency matrix is given below and Hasse diagram is as shown in Fig. A4.

	1	2	3	4
1	1	1	1	1
2	0	1	0	1
3	0	0	1	1
4	0	0	0	1

7. The equivalence classes are $\{1,2\}$ and $\{3,4\}$.

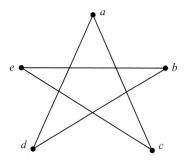

Fig. A5

8. The equivalence classes are {(1,1)}, {(1,2), (2,1)}, {(1,3), (2,2), (3,1)}, {(1,4), (2,3), (3,2), (4,1)}, {(2,4), (3,3), (4,2)} {(3,4), (4,3)}, {(4,4)}.
9. Not transitive (e.g. aRd, dRb but not aRb). The digraph requires no arrows since the relation is symmetric. See Fig. A5.

Exercises 2.4

1. The state diagrams are as shown in Fig. A6.
2. The tables are:

(1)

	a	b
0	1	1
1	1	2
2	0	2

(2)

	a	b
0	1	0
1	1	1
2	1	2

(3)

	a	b
0	0	1
1	2	1
2	2	3
3	3	3

3. (a) The words accepted by the machine are those in which the number of b's is of the form $1 + 3k$;
   (b) words with at least one a;
   (c) no words are accepted;
   (d) words of the form *b*a*b* where each * can denote any sequence (possibly empty) of letters.
4. The words accepted are those with an odd number of letters. The state diagram is as shown in Fig. A7 with $F = \{1\}$;
5. See Fig. A8.
6. The state diagram is as shown in Fig. A9 with $F = \{4\}$.

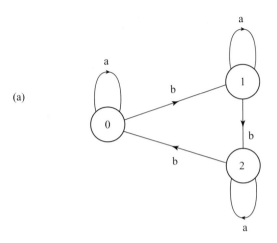

(a)

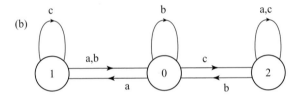

(b)

Fig. A6

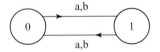

Fig. A7

Chapter 3

Exercises 3.1

1. (a) (i) $(p \wedge q) \to r$; (ii) $(\neg t \wedge p) \to ((s \vee q) \wedge \neg (s \wedge q))$.
   (b) (i)   Either it is raining on Venus and the Margrave of Brandenburg
             carries his umbrella, or the umbrella will dissolve;
       (ii) It is raining on Venus, and either the Margrave of Brandernburg
             carries his umbrella or the umbrella will dissolve;
       (iii) The fact that it is not raining on Venus implies both that X loves Y
             and also that if the umbrella will dissolve then Y does not love Z;

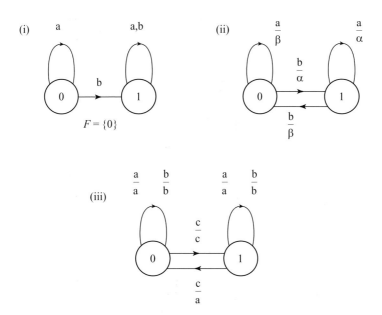

Fig. A8

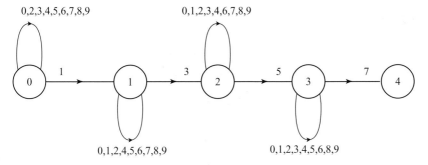

Fig. A9

(iv) If neither X does not love Y nor Y does not love Z then it is raining on Venus. (Equivalently: if X loves Y and Y loves Z then it is raining on Venus.)

2. (i) neither tautology nor contradiction; (ii) neither; (iii) contradiction; (iv) tautology; (v) neither; (vi) tautology.

3. The statement $p \wedge q$ is logically equivalent to $p \wedge (p \rightarrow q)$. Also $(p \wedge q) \leftrightarrow p$ is logically equivalent to $p \rightarrow q$.

Exercises 3.2

1. (a) (There are other, equivalent, ways of saying these.)
    (i) Everyone who is Scottish likes whisky.
    (ii) Everyone who likes whisky is Scottish.
    (iii) There is someone who is Scottish and does not like whisky.
    (iv) Not everyone who is Scottish likes whisky.
    (v) Not everyone is Scottish and likes whisky.
    (vi) There are at least two people who like whisky.
   (b) (There are other correct solutions.)
    (i) $\exists x\, (\neg S(x) \wedge W(x))$
    (ii) $(\exists x(S(x))) \rightarrow (\exists y(S(y) \wedge W(y)))$
    (iii) $\forall x\, (\neg S(x) \rightarrow \neg W(x))$
    (iv) $\exists x \exists y\, (x \neq y \wedge \neg S(x) \wedge \neg S(y) \wedge W(x) \wedge W(y))$.
2. (i) True, (ii) False, (iii) True, (iv) False.

Exercises 3.3

1. Probably we could construct a proof of this fact using almost any of our methods. Here are just two proofs.
   (i) Proof by induction: when $n = 1$, $n^2 + n + 1$ is 3 which is odd. Now suppose that $n^2 + n + 1$ is odd, then
   $(n + 1)^2 + (n+ 1) + 1 = n^2 + 2n + 1 + n + 1 + 1 = (n^2 + n + 1) + 2n + 2$.
   Since $n^2 + n + 1$ is odd and $2n + 2$ is even (being divisible by 2), we see that $(n + 1)^2 + (n + 1) + 1$ is odd as required.
   (ii) Proof by cases: if n is even then n^2 is also even and so $n^2 + n$ is even, so $n^2 + n + 1$ is odd. If n is odd (say $n = 2k + 1$), then n^2 is odd (since it would be $4k^2 + 2k + 1$) so $n^2 + n$ is even and then $n^2 + n + 1$ is odd.
   Thus in either case $n^2 + n + 1$ is odd.
2. Here again methods like argument by cases, contrapositive and contradiction, all lead to fairly easy proofs. Again we give two proofs.
   (i) Proof by cases: if $a + b$ is odd, we consider four cases.
      (a) a, b are both even. In that case $a + b$ is also even, so this case cannot arise.
      (b) If a is even and b is odd, then $a + b$ is odd, so this case can arise.
      (c) If b is even and a is odd, then $a + b$ will be odd and this case can also arise.
      (d) If both a, b are odd then $a + b$ is even so this case does not arise. These four cases show that if $a + b$ is odd then precisely one of a, b is odd.

(ii) Proof by contradiction: suppose that $a + b$ is odd but either both or neither of a, b are odd. In either of these cases $a + b$ would be even. (Note that this argument also needs a slight recourse to cases.)

3. Suppose that a, b are integers with $a + b$ even. We want to show that $a - b$ is even. In this case induction does not seem appropriate, but again most other methods could work. We demonstrate two methods.

   (i) Contrapositive: we will show that if $a - b$ is odd then $a + b$ must be odd. If $a - b$ is odd one of a, b must be odd, the other being even (for if both were even (or odd) then $a + b$ would be even). But then $a + b$ is odd.

   (ii) Proof by contradiction: suppose that $a + b$ is even but $a - b$ is odd. Adding these gives

$$(a + b) + (a - b) = 2a.$$

However $2a$ is even whereas the sum of an even and an odd integer must be odd, contradiction.

For the last part, to give a counterexample to the claim that if $a + b$ is even then ab is even, take $a = b = 1$. Then $a + b = 2$ which is even, but $ab = 1$ which is odd.

Chapter 4

Exercises 4.1

1. $\pi_1\pi_2 = \begin{pmatrix} 1 & 2 & 3 & 4 & 5 & 6 & 7 & 8 & 9 \\ 7 & 8 & 9 & 4 & 5 & 6 & 1 & 2 & 3 \end{pmatrix}$; $\pi_2\pi_3 = \begin{pmatrix} 1 & 2 & 3 & 4 & 5 & 6 & 7 & 8 & 9 \\ 6 & 5 & 4 & 3 & 2 & 1 & 9 & 8 & 7 \end{pmatrix}$;

$\pi_3\pi_1 = \begin{pmatrix} 1 & 2 & 3 & 4 & 5 & 6 & 7 & 8 & 9 \\ 6 & 5 & 4 & 9 & 8 & 7 & 3 & 2 & 1 \end{pmatrix}$; $\pi_3\pi_2 = \begin{pmatrix} 1 & 2 & 3 & 4 & 5 & 6 & 7 & 8 & 9 \\ 3 & 2 & 1 & 9 & 8 & 7 & 6 & 5 & 4 \end{pmatrix}$;

$\pi_2\pi_1\pi_3 = \begin{pmatrix} 1 & 2 & 3 & 4 & 5 & 6 & 7 & 8 & 9 \\ 4 & 5 & 6 & 1 & 2 & 3 & 7 & 8 & 9 \end{pmatrix}$;

$\pi_2\pi_2\pi_2 = \begin{pmatrix} 1 & 2 & 3 & 4 & 5 & 6 & 7 & 8 & 9 \\ 9 & 8 & 7 & 6 & 5 & 4 & 3 & 2 & 1 \end{pmatrix}$;

$\pi_4\pi_5 = \begin{pmatrix} 1 & 2 & 3 & 4 & 5 & 6 & 7 & 8 & 9 & 10 & 11 & 12 \\ 10 & 6 & 1 & 7 & 3 & 5 & 2 & 8 & 4 & 9 & 12 & 11 \end{pmatrix}$;

$\pi_5\pi_4 = \begin{pmatrix} 1 & 2 & 3 & 4 & 5 & 6 & 7 & 8 & 9 & 10 & 11 & 12 \\ 5 & 12 & 7 & 2 & 8 & 4 & 6 & 3 & 9 & 11 & 10 & 1 \end{pmatrix}$;

$$\pi_1\pi_3 = \begin{pmatrix} 1 & 2 & 3 & 4 & 5 & 6 & 7 & 8 & 9 \\ 6 & 5 & 4 & 9 & 8 & 7 & 3 & 2 & 1 \end{pmatrix}; \pi_2\pi_2 = \begin{pmatrix} 1 & 2 & 3 & 4 & 5 & 6 & 7 & 8 & 9 \\ 1 & 2 & 3 & 4 & 5 & 6 & 7 & 8 & 9 \end{pmatrix};$$

$$\pi_2\pi_1 = \begin{pmatrix} 1 & 2 & 3 & 4 & 5 & 6 & 7 & 8 & 9 \\ 7 & 8 & 9 & 4 & 5 & 6 & 1 & 2 & 3 \end{pmatrix}; \pi_3\pi_3 = \begin{pmatrix} 1 & 2 & 3 & 4 & 5 & 6 & 7 & 8 & 9 \\ 7 & 8 & 9 & 1 & 2 & 3 & 4 & 5 & 6 \end{pmatrix};$$

$$\pi_2\pi_1\pi_2 = \begin{pmatrix} 1 & 2 & 3 & 4 & 5 & 6 & 7 & 8 & 9 \\ 3 & 2 & 1 & 6 & 5 & 4 & 9 & 8 & 7 \end{pmatrix};$$

$$\pi_2\pi_3\pi_2 = \begin{pmatrix} 1 & 2 & 3 & 4 & 5 & 6 & 7 & 8 & 9 \\ 7 & 8 & 9 & 1 & 2 & 3 & 4 & 5 & 6 \end{pmatrix};$$

$$\pi_4\pi_4 = \begin{pmatrix} 1 & 2 & 3 & 4 & 5 & 6 & 7 & 8 & 9 & 10 & 11 & 12 \\ 8 & 9 & 1 & 2 & 3 & 4 & 5 & 6 & 7 & 12 & 10 & 11 \end{pmatrix};$$

$$\pi_5\pi_5 = \begin{pmatrix} 1 & 2 & 3 & 4 & 5 & 6 & 7 & 8 & 9 & 10 & 11 & 12 \\ 10 & 3 & 7 & 9 & 8 & 6 & 2 & 5 & 4 & 12 & 11 & 1 \end{pmatrix}.$$

2. $\pi_1^{-1} = \begin{pmatrix} 1 & 2 & 3 & 4 & 5 & 6 & 7 & 8 & 9 \\ 3 & 2 & 1 & 6 & 5 & 4 & 9 & 8 & 7 \end{pmatrix}; \pi_2^{-1} = \begin{pmatrix} 1 & 2 & 3 & 4 & 5 & 6 & 7 & 8 & 9 \\ 9 & 8 & 7 & 6 & 5 & 4 & 3 & 2 & 1 \end{pmatrix};$

$$\pi_3^{-1} = \begin{pmatrix} 1 & 2 & 3 & 4 & 5 & 6 & 7 & 8 & 9 \\ 7 & 8 & 9 & 1 & 2 & 3 & 4 & 5 & 6 \end{pmatrix};$$

$$\pi_4^{-1} = \begin{pmatrix} 1 & 2 & 3 & 4 & 5 & 6 & 7 & 8 & 9 & 10 & 11 & 12 \\ 2 & 3 & 4 & 5 & 6 & 7 & 8 & 9 & 1 & 12 & 10 & 11 \end{pmatrix};$$

$$\pi_5^{-1} = \begin{pmatrix} 1 & 2 & 3 & 4 & 5 & 6 & 7 & 8 & 9 & 10 & 11 & 12 \\ 10 & 3 & 7 & 5 & 9 & 6 & 2 & 4 & 8 & 12 & 11 & 1 \end{pmatrix}.$$

3. $\pi_1\pi_2 = (1\ 7)(2\ 8)(3\ 9);$ $\pi_2\pi_3 = (1\ 6)(2\ 5)(3\ 4)(7\ 9);$

   $\pi_3\pi_1 = (1\ 6\ 7\ 3\ 4\ 9)(2\ 5\ 8);$ $\pi_2\pi_3 = (1\ 3)(4\ 9)(5\ 8)(6\ 7);$

   $\pi_2\pi_1\pi_3 = (1\ 4)(2\ 5)(3\ 6);$ $\pi_2\pi_2\pi_2 = (1\ 9)(2\ 8)(3\ 7)(4\ 6);$

   $\pi_4\pi_5 = (1\ 10\ 9\ 4\ 7\ 2\ 6\ 5\ 3)(11\ 12);$

   $\pi_5\pi_4 = (1\ 5\ 8\ 3\ 7\ 6\ 4\ 2\ 12)(10\ 11);$

   $\pi_1\pi_3 = (1\ 6\ 7\ 3\ 4\ 9)(2\ 5\ 8);$ $\pi_2\pi_2 = \text{id};$

   $\pi_2\pi_1 = (1\ 7)(2\ 8)(3\ 9);$ $\pi_3\pi_3 = (1\ 7\ 4)(2\ 8\ 5)(3\ 9\ 6);$

   $\pi_2\pi_1\pi_2 = (1\ 3)(4\ 6)(7\ 9);$ $\pi_2\pi_3\pi_2 = (1\ 7\ 4)(2\ 8\ 5)(3\ 9\ 6);$

   $\pi_4\pi_4 = (1\ 8\ 6\ 4\ 2\ 9\ 7\ 5\ 3)(10\ 12\ 11);$

   $\pi_5\pi_5 = (1\ 10\ 12)(2\ 3\ 7)(4\ 9)(5\ 8);$

4. (i) $(1\ 8\ 4\ 6\ 2\ 3);$ (ii) $(1\ 2\ 7\ 5\ 4\ 9\ 3\ 12\ 10);$

   (iii) $(1\ 5\ 9\ 4\ 8\ 3\ 7\ 2\ 6)\,(10\ 11).$

5. The table is

	id	(1234)	(13)(24)	(1432)	(13)	(24)	(12)(34)	(14)(23)
id	id	(1234)	(13)(24)	(1432)	(13)	(24)	(12)(34)	(14)(23)
(1234)	(1234)	(13)(24)	(1432)	id	(14)(23)	(12)(34)	(13)	(24)
(13)(24)	(13)(24)	(1432)	id	(1234)	(24)	(13)	(14)(23)	(12)(34)
(1432)	(1432)	id	(1234)	(13)(24)	(12)(34)	(14)(23)	(24)	(13)
(13)	(13)	(12)(34)	(24)	(14)(23)	id	(13)(24)	(1234)	(1432)
(24)	(24)	(14)(23)	(13)	(12)(34)	(13)(24)	id	(1432)	(1234)
(12)(34)	(12)(34)	(24)	(14)(23)	(13)	(1432)	(1234)	id	(13)(24)
(14)(23)	(14)(23)	(13)	(12)(34)	(24)	(1234)	(1432)	(13)(24)	id

6. $s = (1\ 2\ 4\ 8\ 5\ 10\ 9\ 7\ 3\ 6)$; $t = (2\ 3\ 5\ 9\ 8\ 6)(4\ 7)$;
   $c = (1\ 6)(2\ 7)(3\ 8)(4\ 9)(5\ 10)$; $cs = (1\ 7\ 8\ 10\ 4\ 3)(2\ 9)$;
   $scs = (1\ 3\ 2\ 7\ 5\ 10\ 8\ 9\ 4\ 6)$;
   s, 10 times; t, 6 times; cs, 6 times; scs, 10 times.

Exercises 4.2

1. (i) The permutation has order 30 and is odd; (ii) order 30, odd; (iii) order 4, even; (iv) order 1, even.
2. An example is given by the transpositions (1 2) and (2 3).
3. An example is (1 2)(3 4).
5. An example is provided by the transpositions in 2 above.
6. The orders are 5, 6 and 2 respectively.
7. The orders are 2, 3 and 5.
9. The identity element has order 1, the elements (1 2)(3 4), (1 3)(2 4) and (1 4)(2 3) have order 2 and the remaining 8 elements all have order 3: (1 2 3), (1 2 4), (1 3 4), (2 3 4), (1 3 2), (1 4 2), (1 4 3), and (2 4 3).
11. The highest possible order of an element of $S(8)$ is 15, of $S(12)$ is 60 and of $S(15)$ is 105.
12. $o(s) = 10$, $\text{sgn}(s) = -1$, $o(t) = 6$, $\text{sgn}(t) = 1$, $o(c) = 2$, $\text{sgn}(c) = -1$, $o(cs) = 6$, $\text{sgn}(cs) = 1$, $o(scs) = 10$, $\text{sgn}(scs) = -1$.

Exercises 4.3

1. (i) No; 0 has no inverse.   (ii) This is a group.
   (iii) No: 2 has no inverse.   (iv) This is not a group: not all the functions have inverses.
   (v) This is a group.   (vi) This is a group.
   (vii) No: non-associative.   (viii) This is a group.
2. Take G to be $S(3)$, a to be (1 2) and b to be (1 3).

5. The required matrix is

$$\begin{pmatrix} \frac{1}{3} & \frac{1}{3} & \frac{1}{3} \\ \frac{1}{3} & \frac{1}{3} & \frac{1}{3} \\ \frac{1}{3} & \frac{1}{3} & \frac{1}{3} \end{pmatrix}$$

7. The table for $D(4)$ is as shown:

	e	ρ	ρ^2	ρ^3	R	ρR	$\rho^2 R$	$\rho^3 R$
e	e	ρ	ρ^2	ρ^3	R	ρR	$\rho^2 R$	$\rho^3 R$
ρ	ρ	ρ^2	ρ^3	e	ρR	$\rho^2 R$	$\rho^3 R$	R
ρ^2	ρ^2	ρ^3	e	ρ	$\rho^2 R$	$\rho^3 R$	R	ρR
ρ^3	ρ^3	e	ρ	ρ^2	$\rho^3 R$	R	ρR	$\rho^2 R$
R	R	$\rho^3 R$	$\rho^2 R$	ρR	e	ρ^3	ρ^2	ρ
ρR	ρR	R	$\rho^3 R$	$\rho^2 R$	ρ	e	ρ^3	ρ^2
$\rho^2 R$	$\rho^2 R$	ρR	R	$\rho^3 R$	ρ^2	ρ	e	ρ^3
$\rho^3 R$	$\rho^3 R$	$\rho^2 R$	ρR	R	ρ^3	ρ^2	ρ	e

8. The completed table is

	a	b	c	d	f	g
a	c	g	a	f	d	b
b	d	f	b	g	c	a
c	a	b	c	d	f	g
d	b	a	d	c	g	f
f	g	c	f	a	b	d
g	f	d	g	b	a	c

Note that c is the identity element.
Thus $ax = b$ has one solution (g); $xa = b$ also has one (d); $x^2 = c$ has four solutions $(c, a, d,$ and $g)$ and $x^3 = d$ has one solution (d).

Exercises 4.4

1. (i), (ii) and (iii) are semigroups, (iv) and (v) are not.
3. (i) A non-commutative ring with identity and zero-divisors;
   (ii)   not a ring (not closed under addition);
   (iii) not a ring (additive inverses missing);
   (iv)   commutative ring with identity and zero-divisors;
   (v)   commutative ring with no identity and no zero-divisors;
   (vi)   commutative ring with no identity but zero-divisors;
   (vii) commutative ring with no identity and no zero-divisors;
   (viii) commutative ring with identity and no zero-divisors.

7. Take, for example, R to be the set of all 2×2 matrices with

$$x = \begin{pmatrix} 1 & 0 \\ 0 & -1 \end{pmatrix}, \quad y = \begin{pmatrix} 0 & 1 \\ 1 & 0 \end{pmatrix}.$$

8. Take, for example, R to be $\mathbb{Z}_2$ and $x = y = [1]_2$.
9. (i) Is a vector space; the other two fail the distributivity axiom: $(\lambda + \mu)A$ is not equal to $\lambda A + \mu A$.

Chapter 5

Exercises 5.1

2. If $axba^{-1} = b$, multiply on the right first by a then by b^{-1} to obtain $ax = bab^{-1}$. Now multiply by a^{-1} on the left to obtain $x = a^{-1}bab^{-1}$.
3. Let G be the cyclic group with 12 elements and square each of the 12 to get

element	e	x	x^2	x^3	x^4	x^5	x^6	x^7	x^8	x^9	x^{10}	x^{11}
square	e	x^2	x^4	x^6	x^8	x^{10}	e	x^2	x^4	x^6	x^8	x^{10}

(remembering that $x^{12} = e$). It is now clear that several elements of G are not squares of other elements, for example, there is no element g with $g^2 = x^3$.

4. (i) A subgroup; (ii) not a subgroup; (iii) not closed; (iv) a subgroup.
5. Take G to be $S(3)$, a to be $(1\ 2)$, b to be $(1\ 3)$ and c to be $(2\ 3)$.
6. Since the number of elements in $\langle x^d \rangle$ is the order of x^d, we first calculate these orders

element	e	x	x^2	x^3	x^4	x^5	x^6	x^7	x^8	x^9	x^{10}	x^{11}
order	1	12	6	4	3	12	2	12	3	4	6	12

It is clear that the subgroup generated by x is the whole group G. From the table of orders we also see that G can be generated by x^5, x^7 and x^{11}. Each of these powers (1, 5, 7 or 11), has greatest common divisor 1 with 12, confirming that $\langle x^d \rangle$ has 12 (= 12/1) elements in these cases. Next consider x^2. We see that, since x^2 has order 6, the subgroup has 6 elements in this case. The only other element of order 6 is x^{10}. It is clear that x^2 and x^{10} generate the same subgroup with 6 elements and that 6 is 12/2 where 2 is the greatest common divisor of 12 with both 2 and with 10. The next element in the list is x^3 which generates a subgroup with 4 elements. The other element of order 4 is x^9. Again these two elements actually generate the same subgroup and $(12, 3) = (12, 9) = 3 \ (= 12/4)$. Next x^4 and x^8

have order 3 and x^8 is the square of x^4, so they generate the same subgroup with 3 elements and $(12, 4) = (12, 8) = 4 \ (= 12/3)$. The only non-identity element we have not discussed is x^6 and this is the unique element of order 2 so the subgroup it generates has 2 $(= 12/6)$ elements.

7. Let m be minimal such that x^m is in H and let x^k be any other element in H (we know that any element of H is a power of x because G is cyclic). Use the division algorithm to write $k = qm + r$ with r less than m. Then x^k is in H (given) and x^{qm} is in H (because x^m is), so since H is a subgroup,

$$x^k(x^{qm})^{-1} = x^{qm+r}x^{-qm} = x^r$$

is an element of H. This contradicts the minimality of m, unless $r = 0$. We have therefore shown that every element of H is a power of x^m and so H is cyclic.

8. We first use induction to show that $(g^{-1}xg)^k = g^{-1}x^kg$. The base case is clear, so suppose that $(g^{-1}xg)^k = g^{-1}x^kg$ for some $k \geq 1$. Then

$$(g^{-1}xg)^{k+1} = (g^{-1}xg)^k(g^{-1}xg) = g^{-1}x^kgg^{-1}xg$$
$$= g^{-1}x^kxg = g^{-1}x^{k+1}g$$

as required. Now suppose that x has order 3. Then $x^3 = e$ and so, for all g in G,

$$(g^{-1}xg)^3 = g^{-1}x^3g = g^{-1}eg = e$$

so the order of $g^{-1}xg$ divides 3. Since $g^{-1}xg$ does not have order 1 (otherwise x would be e and would not have order 3), we have shown that if x has order 3 then so does $g^{-1}xg$. For the converse, suppose that $g^{-1}xg$ has order 3, then $e = (g^{-1}xg)^3 = g^{-1}x^3g$ (by the first part). It then follows that $x^3 = e$, so the order of x divides 3. However, x does not have order 1 (otherwise $x = e$ and then $g^{-1}xg = e$ therefore does not have order 3), so g has order 3.

10. One generator for G_{23} is $[5]_{23}$. A generator for G_{26} is $[7]_{26}$. However, G_8 is not cyclic.

Exercises 5.2

1. The left cosets are
   $\{[1]_{14}, [13]_{14}\}, \{[3]_{14}, [11]_{14}\}, \text{ and } \{[5]_{14}, [9]_{14}\}$.
3. The left cosets are
   $\{1, \tau\}, \{r, r\tau\}, \{r^2, r^2\tau\} \text{ and } \{r^3, r^3\tau\}$
   where r represents rotation through $\pi/4$.

5. Since $\phi(20)$ is 8, the possible orders of elements of G_{20} are 1, 2, 4 or 8. The actual order of $[1]_{20}$ is 1, of $[3]_{20}$ is 4, of $[7]_{20}$ is 4, of $[9]_{20}$ is 2, of $[11]_{20}$ is 2, of $[13]_{20}$ is 4, of $[17]_{20}$ is 4 and of $[19]_{20}$ is 2.

Exercises 5.3

1. (i) $\mathbb{Z}_2 \times \mathbb{Z}_2$; (ii) $\mathbb{Z}_4$; (iii) $\mathbb{Z}_4$ and $\mathbb{Z}_2 \times \mathbb{Z}_2$ respectively.
3. Take G to be $S(3)$ and g to be $(1\ 2\ 3)$ to see that f need not be the identity function.
4. The group $\mathbb{Z}_2 \times \mathbb{Z}_2$ is not cyclic.
7. The tables are as shown:

(i)

$\mathbb{Z}_4 \times \mathbb{Z}_2$	(0, 0)	(1, 0)	(2, 0)	(3, 0)	(0, 1)	(1, 1)	(2, 1)	(3, 1)
(0, 0)	(0, 0)	(1, 0)	(2, 0)	(3, 0)	(0, 1)	(1, 1)	(2, 1)	(3, 1)
(1, 0)	(1, 0)	(2, 0)	(3, 0)	(0, 0)	(1, 1)	(2, 1)	(3, 1)	(0, 1)
(2, 0)	(2, 0)	(3, 0)	(0, 0)	(1, 0)	(2, 1)	(3, 1)	(0, 1)	(1, 1)
(3, 0)	(3, 0)	(0, 0)	(1, 0)	(2, 0)	(3, 1)	(0, 1)	(1, 1)	(2, 1)
(0, 1)	(0, 1)	(1, 1)	(2, 1)	(3, 1)	(0, 0)	(1, 0)	(2, 0)	(3, 0)
(1, 1)	(1, 1)	(2, 1)	(3, 1)	(0, 1)	(1, 0)	(2, 0)	(3, 0)	(0, 0)
(2, 1)	(2, 1)	(3, 1)	(0, 1)	(1, 1)	(2, 0)	(3, 0)	(0, 0)	(1, 0)
(3, 1)	(3, 1)	(0, 1)	(1, 1)	(2, 1)	(3, 0)	(0, 0)	(1, 0)	(2, 0)

(ii)

$G_5 \times G_3$	(1, 1)	(2, 1)	(4, 1)	(3, 1)	(1, 2)	(2, 2)	(4, 2)	(3, 2)
(1, 1)	(1, 1)	(2, 1)	(4, 1)	(3, 1)	(1, 2)	(2, 2)	(4, 2)	(3, 2)
(2, 1)	(2, 1)	(4, 1)	(3, 1)	(1, 1)	(2, 2)	(4, 2)	(3, 2)	(1, 2)
(4, 1)	(4, 1)	(3, 1)	(1, 1)	(2, 1)	(4, 2)	(3, 2)	(1, 2)	(2, 2)
(3, 1)	(3, 1)	(1, 1)	(2, 1)	(4, 1)	(3, 2)	(1, 2)	(2, 2)	(4, 2)
(1, 2)	(1, 2)	(2, 2)	(4, 2)	(3, 2)	(1, 1)	(2, 1)	(4, 1)	(3, 1)
(2, 2)	(2, 2)	(4, 2)	(3, 2)	(1, 2)	(2, 1)	(4, 1)	(3, 1)	(1, 1)
(4, 2)	(4, 2)	(3, 2)	(1, 2)	(2, 2)	(4, 1)	(3, 1)	(1, 1)	(2, 1)
(3, 2)	(3, 2)	(1, 2)	(2, 2)	(4, 2)	(3, 1)	(1, 1)	(2, 1)	(4, 1)

(iii)

$\mathbb{Z}_2 \times \mathbb{Z}_2 \times \mathbb{Z}_2$	(0, 0, 0)	(1, 0, 0)	(0, 1, 0)	(1, 1, 0)	(0, 0, 1)	(1, 0, 1)	(0, 1, 1)	(1, 1, 1)
(0, 0, 0)	(0, 0, 0)	(1, 0, 0)	(0, 1, 0)	(1, 1, 0)	(0, 0, 1)	(1, 0, 1)	(0, 1, 1)	(1, 1, 1)
(1, 0, 0)	(1, 0, 0)	(0, 0, 0)	(1, 1, 0)	(0, 1, 0)	(1, 0, 1)	(0, 0, 1)	(1, 1, 1)	(0, 1, 1)
(0, 1, 0)	(0, 1, 0)	(1, 1, 0)	(0, 0, 0)	(1, 0, 0)	(0, 1, 1)	(1, 1, 1)	(0, 0, 1)	(1, 0, 1)
(1, 1, 0)	(1, 1, 0)	(0, 1, 0)	(1, 0, 0)	(0, 0, 0)	(1, 1, 1)	(0, 1, 1)	(1, 0, 1)	(0, 0, 1)
(0, 0, 1)	(0, 0, 1)	(1, 0, 1)	(0, 1, 1)	(1, 1, 1)	(0, 0, 0)	(1, 0, 0)	(0, 1, 0)	(1, 1, 0)
(1, 0, 1)	(1, 0, 1)	(0, 0, 1)	(1, 1, 1)	(0, 1, 1)	(1, 0, 0)	(0, 0, 0)	(1, 1, 0)	(0, 1, 0)
(0, 1, 1)	(0, 1, 1)	(1, 1, 1)	(0, 0, 1)	(1, 0, 1)	(0, 1, 0)	(1, 1, 0)	(0, 0, 0)	(1, 0, 0)
(1, 1, 1)	(1, 1, 1)	(0, 1, 1)	(1, 0, 1)	(0, 0, 1)	(1, 1, 0)	(0, 1, 0)	(1, 0, 0)	(0, 0, 0)

(iv)

$G_{12} \times G_4$	(1, 1)	(5, 1)	(7, 1)	(11, 1)	(1, 3)	(5, 3)	(7, 3)	(11, 3)
(1, 1)	(1, 1)	(5, 1)	(7, 1)	(11, 1)	(1, 3)	(5, 3)	(7, 3)	(11, 3)
(5, 1)	(5, 1)	(1, 1)	(11, 1)	(7, 1)	(5, 3)	(1, 3)	(11, 3)	(7, 3)
(7, 1)	(7, 1)	(11, 1)	(1, 1)	(5, 1)	(7, 3)	(11, 3)	(1, 3)	(5, 3)
(11, 1)	(11, 1)	(7, 1)	(5, 1)	(1, 1)	(11, 3)	(7, 3)	(5, 3)	(1, 3)
(1, 3)	(1, 3)	(5, 3)	(7, 3)	(11, 3)	(1, 1)	(5, 1)	(7, 1)	(11, 1)
(5, 3)	(5, 3)	(1, 3)	(11, 3)	(7, 3)	(5, 1)	(1, 1)	(11, 1)	(7, 1)
(7, 3)	(7, 3)	(11, 3)	(1, 3)	(5, 3)	(7, 1)	(11, 1)	(1, 1)	(5, 1)
(11, 3)	(11, 3)	(7, 3)	(5, 3)	(1, 3)	(11, 1)	(7, 1)	(5, 1)	(1, 1)

Of these, the first two are isomorphic to each other and also $\mathbb{Z}_2 \times \mathbb{Z}_2 \times \mathbb{Z}_2$ is isomorphic to $G_{12} \times G_4$.

8. The possible orders of the elements in $G \times H$ are the integers of the form $\mathrm{lcm}\{a, b\}$ where a divides 6 and b divides 14. Namely: 1, 2, 3, 6, 7, 14, 21, 42.

Exercises 5.4

2. The first and second detect one error and correct none; the third detects two and corrects one and the fourth detects none and corrects none.
3. The codewords are

000000111   001001110   010010101   011011100
100100011   101101010   110110001   111111000

The code detects two errors and corrects one error.
4. The decoding table is

000000   100110   010101   001011   110011   101101   011110   111000
000001   100111   010100   001010   110010   101100   011111   111001
000010   100100   010111   001001   110001   101111   011100   111010
000100   100010   010001   001111   110111   101001   011010   111100
001000   101110   011101   000011   111011   100101   010110   110000
010000   110110   000101   011011   100011   111101   001110   101000
100000   000110   110101   101011   010011   001101   111110   011000
001100   101010   011001   000111   111111   100001   010010   110100

5. Corrected words are

101110100010   111111111100   000000000000
001000100011   001110101100

6. The two-column decoding table is

Syndrome	Coset leader
0000	0000000
1101	1000000
1110	0100000
1011	0010000
1000	0001000
0100	0000100
0010	0000010
0001	0000001
0011	0000011
0101	0000101
1001	0001001
0110	0000110
1010	0001010
1100	0001100
1111	1000010
0111	0000111

The syndrome of 1100011 is 0000 so this is a codeword;
the syndrome of 1011000 is 1110 so we correct to 1111000;
the syndrome of 0101110 is 0000 so this is a codeword;
the syndrome of 0110001 is 0100 so corrected word is 0110101;
the syndrome of 1010110 is 0000 so this is a codeword.

7. The two-column decoding table is

Syndrome	Coset leader
000	000000
101	100000
110	010000
011	001000
100	000100
010	000010
001	000001
111	001100

The message is THE END.

Chapter 6

Exercises 6.1

1. (i) $2x^2 + 2x$,
    (ii) $-3x^2 + 2x$,
    (iii) $2x^2 + (7 - 5i)x + (3 - 3i)$,
    (iv) $-3ix^2 + 2ix$,
    (v) $2x^2 + x$,
    (vi) $x^2 + 2x$.
2. (i) $x^3 + 8x^2 + 10x + 3$,
    (ii) $x^5 - x^4 - 2x^2 - 1$,
    (iii) $ix^3 + (3 + 7i)x^2 + (21 + 3i)x + 9$,
    (iv) $-x^5 - (1 + 2i)x^4 + (1 - i)x^3 + (1 + 3i)x^2 + (1 - i)x - 1$,
    (v) $x^4 + x^2 + 1$,
    (vi) $x^5 + x^3 + x^2 + 1$.
3. In the three cases the zeros are: (i) $x = 1, 1 + i$ or $1 - i$, (ii) $x = 7i$ or $-i$,
    (iii) $x = [4]_5$ is the only zero.

Exercises 6.2

1. (i) $f(x) = (x^2 + 3x + 6)g(x) + (10x - 5)$,
    (ii) $f(x) = (x + 6)g(x) + (24x - 35)$,
    (iii) $f(x) = (x + 6)g(x) + 3x$.
2. (i) Experiment with small values for x to see that $x = 1$ is a zero. Thus
        $x - 1$ divides the polynomial, and

    $$x^3 - x^2 - 4x + 4 = (x - 1)(x^2 - 4) = (x - 1)(x - 2)(x + 2).$$

    (ii) In this case, we see that $x = 2$ is a zero and

    $$x^3 - 3x^2 + 3x - 2 = (x - 2)(x^2 - x + 1).$$

    Using the formula to find the zeros of the quadratic $x^2 - x + 1$, we
    see at once that this quadratic has no real roots, so we already have a
    decomposition into irreducible real polynomials.

    (iii) If we continue the factorisation over $\mathbb{C}$, we see that

    $$x^3 - 3x^2 + 3x - 2 = (x - 2)(x - w)(x - \overline{w}),$$

    where $w = \frac{1 + i\sqrt{3}}{2}$.

    (iv) Over $\mathbb{Z}_7$, we clearly only need to seek for roots of $g(x) = x^2 - x + 1$
    which is done by substituting the seven possible values for x. Then

$g(0) = 1$, $g(1) = 1$, $g(2) = 3$. However $g(3) = 9 - 3 + 1 = 7 = 0$, so 3 is a zero and so $x - 3$ divides $g(x)$. This completes the factorisation as

$$x^3 - 3x^2 + 3x - 2 = (x - 2)(x - 3)(x - 5).$$

(v) It is clear that $x = -1$ is a root of the given polynomial and

$$x^3 + x^2 + x + 1 = (x + 1)(x^2 + 1) = (x + 1)(x + 1)^2 = (x + 1)^3.$$

3. (i) We first see that

$$x^3 + 1 = (x - 1)(x^2 + x - 1) + 2x.$$

Then since $x^2 + x - 1 = 2x(\frac{1}{2}x + \frac{1}{2}) - 1$, a greatest common divisor for the given polynomials is $(-)1$. Then

$$
\begin{aligned}
-1 &= (x^2 + x - 1) - 2x(\tfrac{1}{2}x + \tfrac{1}{2}) \\
&= (x^2 + x - 1) - \left((x^3 + 1) - (x - 1)(x^2 + x - 1)\right)(\tfrac{1}{2}x + \tfrac{1}{2}) \\
&= -\tfrac{1}{2}(x + 1)(x^3 + 1) + \tfrac{1}{2}(x^2 + 1)(x^2 + x - 1).
\end{aligned}
$$

(ii) The first step is to note that

$$x^4 + x + 1 = (x)(x^3 + x + 1) + x^2 + 1.$$

Then we find that

$$x^3 + x + 1 = (x)(x^2 + 1) + 1.$$

It follows that 1 is a gcd for the two given polynomials and that

$$
\begin{aligned}
1 &= (x^3 + x + 1) - (x)(x^2 + 1) \\
&= (x^3 + x + 1) + (x)\left((x^4 + x + 1) - (x)(x^3 + x + 1)\right) \\
&= (x^3 + x + 1)(x^2 + 1) + x(x^4 + x + 1).
\end{aligned}
$$

(iii) The first step is to note that

$$x^3 - ix^2 + 2x - 2i = (x - i)(x^2 + 1) + x - i.$$

Then, since $x^2 + 1 = (x + i)(x - i)$, a greatest common divisor is $x - i$. Also $x - i = x^3 - ix^2 + 2x - 2i - (x - i)(x^2 + 1)$.

4. We are given that $f(x) = (x - \alpha)g(x) + r(x)$, so substitute $x = \alpha$, to obtain $f(\alpha) = (\alpha - \alpha)g(\alpha) + r(\alpha)$. Since $(\alpha - \alpha)$ is the zero polynomial, and multiplying any polynomial by the zero polynomial gives the zero polynomial, we see that $f(\alpha) = r(\alpha)$, as required.

Exercises 6.3

1. The base case for the induction may be taken for granted (the result is clear when $n = 1$). Now suppose that the result holds when $r = k$ and suppose that f divides the product $f_1(x) \ldots f_{k+1}(x)$. Write $g(x)$ for the product $f_1(x) \ldots f_k(x)$, so we know that f divides $g(x)f_{k+1}(x)$. By the results in this section, we deduce that f divides at least one of $g(x)$ or $f_{k+1}(x)$. Using induction on $g(x)$, we deduce that f divides one of
$$f_1(x), f_2(x), \ldots, f_{k+1}(x).$$

2. Fermat's Theorem implies that each of the non-zero elements of $\mathbb{Z}_p$ is a zero of the polynomial $x^{p-1} - 1$, and so for each of these $p - 1$ elements i, say, $(x - i)$ divides $x^{p-1} - 1$. Since this polynomial of degree $p - 1$ is divisible by $p - 1$ linear factors, we see that this must be the factorisation of the polynomial.

3. Any quadratic over $\mathbb{Z}_2$ with leading coefficient 1 has to be of the form $x^2 + ax + b$. If $b = 0$, then $x = 0$ would be a root. Therefore we may take our quadratic to be $x^2 + ax + 1$ (since 1 is the only non-zero element in $\mathbb{Z}_2$). Substituting $x = 1$ gives $1 + a + 1$, so if the quadratic is irreducible, this must be non-zero and so the only irreducible quadratic over $\mathbb{Z}_2$ is $x^2 + x + 1$.

   Over $\mathbb{Z}_3$ our irreducible quadratic will have the form $x^2 + ax + b$ where b is 1 or -1. If $b = 1$, the condition that 1 is not a root is that $a - 1$ is non-zero, and the condition that -1 is not a root is that $-a - 1$ is non-zero. The only value of a satisfying both these conditions is $a = 0$. It follows that in this case $x^2 + 1$ is the only irreducible. When $b = -1$, we see that $f(1) = a$ and $f(-1) = -a$, so both $x^2 + x - 1$ and $x^2 - x - 1$ are irreducible. This gives three irreducible quadratics, namely $x^2 + 1$, $x^2 + x - 1$ and $x^2 - x - 1$.

4. If $x^4 + 1 = (x^2 + ax + 1)(x^2 + bx + 1)$, equating coefficients of x^3 (or of x) gives $a + b = 0$ so $a = -b$. Now equate coefficients of x^2 to see that $0 = 2 + ab$, so $ab = -2$ and hence $a^2 = 2$. Thus we may take a to be $\sqrt{2}$ and b to be $-\sqrt{2}$. Then $x^8 - 1 = (x^4 - 1)(x^4 + 1)$. Also $x^4 - 1 = (x^2 + 1)(x^2 - 1)$. Now, $x^2 + 1$ does not factorise over $\mathbb{R}$ whereas $x^2 - 1 = (x + 1)(x - 1)$. Since $x^2 + \sqrt{2}x + 1$ and $x^2 - \sqrt{2}x + 1$ have no real roots, the factorisation of $x^8 - 1$ over $\mathbb{R}$ is

$$x^8 - 1 = (x - 1)(x + 1)(x^2 + 1)(x^2 + \sqrt{2}x + 1)(x^2 - \sqrt{2}x + 1).$$

   Then, using the quadratic formula, we see that over $\mathbb{C}$ the quadratic $x^2 + \sqrt{2}x + 1$ has zeros $\omega = \frac{-\sqrt{2}+i\sqrt{2}}{2}$ and $\overline{\omega} = \frac{-\sqrt{2}-i\sqrt{2}}{2}$. Similarly, we can find the real and imaginary parts of the zeros of the quadratic

$x^2 - \sqrt{2}x + 1$ (these turn out to be ω^3 and $\overline{\omega^3}$.) The factorisation of $x^8 - 1$ as a product of 8 linear terms is then

$$x^8 - 1 = (x + 1)(x - 1)(x + i)(x - i)(x - \omega)(x - \overline{\omega})(x - \omega^3)(x - \overline{\omega^3}).$$

When we come to factorise this polynomial over $\mathbb{Z}_3$, we need to find the factorisations of $x^2 + 1$ and $x^4 + 1$. The quadratic is irreducible. The quartic has no linear factors, since neither 1 nor 2 is a root of the polynomial. Since we know (from Exercise 6.3.3) the irreducible quadratics over $\mathbb{Z}_3$, it only remains to see if two of the three can multiply together to give $x^4 + 1$. Since the constant term is 1, the only candidates are $x^2 + x - 1$ and $x^2 - x - 1$. A simple calculation shows that the product of these is indeed $x^4 + 1$, so the complete factorisation of $x^8 - 1$ over $\mathbb{Z}_3$ is

$$x^8 - 1 = (x - 1)(x + 1)(x^2 + 1)(x^2 + x - 1)(x^2 - x - 1).$$

5. For cubics over $\mathbb{Z}_2$, we again can take the coefficient of x^3 to be 1 and the constant term to be 1, so we consider $f(x) = x^3 + ax^2 + bx + 1$. Putting $x = 1$, we obtain $a + b$, so provided that $a + b$ is non-zero (i.e. a is not equal to b), f will have no linear factor, so will be irreducible. The irreducible cubics are therefore $x^3 + x^2 + 1$ and $x^3 + x + 1$.

6. A general example may be made by taking g and h to be different irreducibles and f to be any scalar multiple of gh, for example, $g(x) = x - 1, h(x) = x + 1$ and $f(x) = x^2 - 1$.

Exercises 6.4

1. It follows from our general theory that the polynomial congruence classes are:

$$[0]_f, [1]_f, [2]_f, [x]_f, [1 + x]_f, [2 + x]_f, [2x]_f, [1 + 2x]_f, [2 + 2x]_f.$$

Now using the fact that $f = x^2 + x + 2$, we obtain the following table for the non-zero representatives (we have omitted the brackets and subscripts):

	1	2	x	$1 + x$	$2 + x$	$2x$	$2x + 1$	$2x + 2$
1	1	2	x	$1 + x$	$2 + x$	$2x$	$2x + 1$	$2x + 2$
2	2	1	$2x$	$2 + 2x$	$1 + 2x$	x	$x + 2$	$x + 1$
x	x	$2x$	$2x + 1$	1	$1 + x$	$x + 2$	$2 + 2x$	2
$1 + x$	$1 + x$	$2 + 2x$	1	$x + 2$	$2x$	2	x	$2x + 1$
$2 + x$	$2 + x$	$1 + 2x$	$1 + x$	$2x$	1	$2 + 2x$	1	$2x$
$2x$	$2x$	x	$x + 2$	2	$2 + 2x$	$1 + 2x$	$x + 1$	1
$1 + 2x$	$1 + 2x$	$x + 2$	$2x + 2$	x	1	$x + 1$	2	$2x$
$2 + 2x$	$2 + 2x$	$x + 1$	2	$1 + 2x$	$2x$	1	$2x$	$x + 2$

Now to find a representative whose powers give all the others, first consider x. Its square is $2x + 1$ whose square is 2 so the eighth power of x is 1. In fact it follows from this that x has eight distinct powers and so these must be all the non-zero polynomial congruence classes.

2. Since 1 is a greatest common divisor for f and t, we know that there exist polynomials u, v such that $1 = uf + vt$. Multiply both sides of this equation by $r - s$ to get $r - s = u(r - s)f + v(r - s)t$. Now suppose that $[rt]_f = [st]_f$, so f divides $rt - st = (r - s)t$. In that case f divides the right-hand side of the above equation, so f divides $r - s$ and $[r]_f = [s]_f$.

3. (i) Since $f(x) = x^2 + x + 1$ is irreducible, our given polynomials, f, g have 1 as a greatest common divisor. Also $x^2 + x + 1 = (x)(x + 1) + 1$ and so $1 = (x^2 + x + 1) - x(x + 1)$. Thus an inverse for $x + 1$ is x.

   (ii) Now consider $x^3 + x^2 + x + 2$ and $x^2 + x$. We have that $x^3 + x^2 + x + 2 = (x)(x^2 + x) + x + 2$, and $x^2 + x = (x + 2)(x + 2) + 2$ (remember $p = 3!$). Finally 2 divides $x + 2$, so 2 (or 1) is a greatest common divisor for our given polynomials. This means that

   $$1 = 2(x^2 + x) - 2(x + 2)(x + 2)$$
   $$= -(x^2 + x) + (x + 2)(x + 2)$$
   $$= -(x^2 + x) + (x + 2)((x^3 + x^2 + x + 2) - (x)(x^2 + x)).$$

   After rearranging, this means that an inverse for $x^2 + x$ modulo $x^3 + x^2 + x + 2$ is $2x^2 + x + 2$.

   (iii) Since $x^2 + 1 = (x + 1)(x - 1) + 2$, a greatest common divisor is 2 and $2 = (x^2 + 1) - (x - 1)(x + 1)$, so $1 = (x^2 + 1)/2 - (x - 1)(x + 1)/2$. Thus an inverse for $x + 1$ is $-(x - 1)/2$.

Exercises 6.5

1. Let g be a polynomial over **B**. If g is irreducible, then 1 is not a zero of g so $g(1)$ is equal to 1. However, since every power of 1 is 1 itself, $g(1)$ is simply the sum of the coefficients of g (including the constant term). Since those coefficients which are zero do not contribute to this sum, we deduce that the number of powers of x with non-zero coefficient must be an odd integer.

2. Clearly $x = 1$ is a zero of $x^5 - 1$, and $x^5 - 1 = (x - 1)(x^4 + x^3 + x^2 + x + 1)$. Now the above quartic has no zeros, so the only possible factorisation would be as a product of irreducible quadratics. However, we

saw in Exercise 6.3.3, that the only irreducible quadratic over $\mathbf{B}$ is $x^2 + x + 1$. Since the square of $x^2 + x + 1$ is $x^4 + x^2 + 1$, we deduce that $x^4 + x^3 + x^2 + x + 1$ is irreducible. Thus the only possible generator polynomials for cyclic codes are

$$1, \qquad x + 1, \qquad x^4 + x^3 + x^2 + x + 1, \qquad \text{and} \qquad x^5 - 1.$$

The first gives all vectors of length 5 as codewords (and so detects and corrects zero errors), the last has no non-zero codewords. The generator matrices corresponding to $x + 1$ and $x^4 + x^3 + x^2 + x + 1$ are, respectively,

$$\begin{pmatrix} 1 & 1 & 0 & 0 & 0 \\ 0 & 1 & 1 & 0 & 0 \\ 0 & 0 & 1 & 1 & 0 \\ 0 & 0 & 0 & 1 & 1 \end{pmatrix}; \qquad (1\ 1\ 1\ 1\ 1).$$

It is clear that the first of these produces the code consisting of the 16 words of even length in $\mathbf{B}^5$ and so detects an error, but cannot correct any error. The second gives a code with 2 words, and so detects up to 4 errors with 2, or fewer errors, being corrected.

3. The matrix associated with the given code is

$$\begin{pmatrix} 1 & 0 & 1 & 1 & 0 & 0 & 0 \\ 0 & 1 & 0 & 1 & 1 & 0 & 0 \\ 0 & 0 & 1 & 0 & 1 & 1 & 0 \\ 0 & 0 & 0 & 1 & 0 & 1 & 1 \end{pmatrix}$$

This code has 16 codewords

0000000	1011000	0101100	1110100
0010110	1001110	0111010	1100010
0011101	1000101	0110001	1101001
0001011	1010011	0100111	1111111

It is clear that the minimum distance between codewords is 3. If, therefore, we add any vector with six zeros and a single 1 to a codeword, we cannot obtain another codeword. It follows that each of the 16 codewords is a distance of 1 away from seven non-codewords, so there are $8 \times 16 = 2^3 \times 2^4 = 2^7$ codewords in these (disjoint, note) 'spheres of radius one' around codewords. As we remarked in the text, this is precisely one of the basic properties of the Hamming code. (In fact looking at the generator matrix for the Hamming code on page 249 in the text, we can see each row is one of the above codewords.)

4. To determine all cyclic codes of length 7, we needed to factorise $x^7 - 1$. Clearly $x - 1$ is a factor. By now the codes associated with $1, x + 1$ and $x^7 - 1$ are familiar, so we are only left with those polynomials which divide $x^6 + x^5 + x^4 + x^3 + x^2 + x + 1$. However, we are given one of these in Exercise 6.5.3, so it is only a matter of working out what happens when we divide $x^6 + x^5 + x^4 + x^3 + x^2 + x + 1$ by $x^3 + x^2 + 1$. The answer turns out to be $g(x) = x^3 + x + 1$ and so we have a complete list of cyclic codes once we know the code associated with $g(x)$. As in Exercise 6.5.3, we can easily write now the generator matrix for this code and hence its codewords. It then turns out that the minimum distance is again 3 and so this code detects up to 2 errors and corrects up to 1 error.

5. Let $p(x)$ be a parity polynomial for a cyclic code of length n and generator polynomial $g(x)$. This means that $p(x)g(x) = x^n - 1 = f(x)$. Thus if g has degree k, then p has degree $n - k$. Now suppose that $c(x)$ is a polynomial with $[c(x)p(x)]_f = [0]_f$, so $f(x)$ divides $c(x)p(x)$. Write $c(x)$ in the form $q(x)g(x) + r(x)$ (with r either zero or of degree less than k) and multiply throughout by $p(x)$ to get $[0]_f = [q(x)p(x)g(x) + r(x)p(x)]_f$. Thus $[0]_f = [r(x)p(x)]_f$ which is impossible unless $r(x)$ is zero, otherwise $r(x)p(x)$ would have degree less than n. We deduce that $r(x)$ is the zero polynomial and so $g(x)$ divides $c(x)$ and, therefore, $c(x)$ is a codeword.

References and further reading

Allenby, R.B.J.T., *Rings, Fields and Groups*, Edward Arnold, London, 1983.
[Further reading in algebra.]

Bell, E.T., *Men of Mathematics*, Simon and Schuster, New York, 1937. Pelican edition (2 vols.), 1953.
[Anecdotal, and not very reliable: but probably the best known biographical/historical work.]

Biggs, N.L., *Discrete Mathematics*, Clarendon Press, Oxford, 1985.
[Comprehensive and readable.]

Boole, G., *An Investigation of the Laws of Thought*, Dover, New York, 1957 (reprint of the 1854 edition).

Boyer, C.B., *A History of Mathematics*, Wiley, New York, 1968.
[From ancient times to the twentieth century; readable and recommended. Contains an extensive annotated bibliography.]

Bühler, W.K., *Gauss*, Springer-Verlag, Berlin, 1981.
[Biography of Gauss.]

Carroll, L., *Symbolic Logic and the Game of Logic*, Dover, New York, 1958 (reprint of the 1896 original).

Dauben, J.W., *Georg Cantor*, Harvard, Cambridge, MA, 1979.
[Engrossing account of a radical shift in mathematics.]

Davenport, H., *The Higher Arithmetic*, 5th edn. Cambridge University Press, Cambridge, 1982.
[Readable account of elementary number theory.]

Diffie, W. and Hellman, M.E., New directions in cryptography, *IEEE Transactions on Information Theory*, **22** (1976), 644–654.

Enderton, H.B., *A Mathematical Introduction to Logic*, 2nd edn., Academic Press, New York, 2001.
[Readable, quite advanced.]

Enderton, H.B., *Elements of Set Theory*, Academic Press, New York, 1977.
[Readable.]

Eves, H., *An Introduction to the History of Mathematics*, 5th edn., Holt, Rinehart and Winston, New York, 1983.
[A popular textbook.]

Fauvel, J. and Gray, J. (eds.), *The History of Mathematics: A Reader*, Macmillan/Open University, London and Milton Keynes, 1988.
[Contains excerpts from original sources.]

Flegg, H.G., *Boolean Algebra*, Macdonald, London, 1971.

Fraenkel, A.A., *Set Theory and Logic*, Addison-Wesley, Reading, MA, 1966.
[Further reading, especially on infinite arithmetic.]

Fraleigh, J.B., *A First Course in Abstract Algebra*, 6th edn., Addison-Wesley, Reading, MA, 1999.
[Further reading in algebra.]

Gauss, C.F., *Disquisitiones Arithmeticae*, translated by A.A. Clarke, Yale University Press, New Haven, CT, 1966, revised by W.C. Waterhouse, Springer-Verlag, New York, 1986.

Grattan-Guinness, I., *The Development of the Foundations of Mathematical Analysis from Euler to Riemann*, MIT Press, Cambridge, MA, 1970.
[Contains much more on the development of the notion of function.]

Hankins, T.L., *Sir William Rowan Hamilton*, Johns Hopkins University Press, Baltimore, MD, 1980.
[Hamilton's life and work.]

Heath, T.L., *The Thirteen Books of Euclid's Elements* (3 vols.), Cambridge University Press, Cambridge, 1908. Reprinted Dover, New York, 1956.

Heath, T.L., *A History of Greek Mathematics*, vol. 1 (from Thales to Euclid), vol. 2 (from Aristarchus to Diophantus), Clarendon Press, Oxford, 1921.
[For many years the standard on Greek Mathematics.]

Heath, T.L., *Diophantus of Alexandria*, 2nd edn., Cambridge University Press, Cambridge, 1910, reprinted by Dover, 1964.

Hill, R., *A First Course in Coding Theory*, Clarendon Press, Oxford, 1986.
[Further reading on (error-correcting) codes.]

Hodges, W., *Logic*, Penguin, Harmondsworth, 1977.

Kalmanson, K., *An Introduction to Discrete Mathematics and its Applications*, Addison-Wesley, Reading, MA, 1986.

Kline, M., *Mathematical Thought from Ancient to Modern Times*, Oxford University Press, New York, 1972.
[Readable and comprehensive: controversial views on the direction of twentieth century mathematics.]

Landau, S., Zero knowledge and the Department of Defense, *Notices Amer. Math. Soc.*, **35** (1988), 5–12.

Ledermann, W., *Introduction to Group Theory*, Oliver and Boyd, Edinburgh, 1973 (reprinted, Longman, 1976).
[Clearly written.]

Li Yan and Du Shiran, *Chinese Mathematics*, translated by Crossley, J.N. and Lun, W.-C., Clarendon Press, Oxford, 1987.
[Up to date and detailed.]

Lyndon, R.C., *Groups and Geometry*, LMS Lecture Note Series vol. 101, Cambridge University Press, Cambridge, 1985.
[Readable.]

MacLane, S. and Birkhoff, G., *Algebra*, Macmillan, New York, 1967.
[A classic text, relatively advanced.]

Manheim, J.H., *The Genesis of Point Set Topology*, Pergamon, Oxford, 1964.
[Especially Chapters I–III for the development of the notion of function.]

Marcus, M., *A Survey of Finite Mathematics*, Houghton Mifflin, Boston, MA, 1969.
[A relatively advanced text on the topic.]

Needham, J., in collaboration with Wang Ling, *Science and Civilisation in China*, vol. 3 (Mathematics and the Sciences of the Heavens and the Earth), Cambridge University Press, Cambridge, 1959.
[The classic text.]

Rabin, M.O., Digitalized signatures and public-key functions as intractable as factorization, Technical Report, MIT/LCS/TR-212, MIT, 1979.

Rivest, R., Shamir, A. and Adleman, L., A method for obtaining digital signatures and public-key cryptosystems, *ACM Communications*, **21** (Feb. 1978), 120–6.

Salomaa, A., *Computation and Automata*, Cambridge University Press, Cambridge, 1985.
[Further reading on finite state machines and related topics. Quite advanced.]

Shamir, A., A polynomial time algorithm for breaking the basic Merkle–Hellman cryptosystem, *IEEE Transactions on Information Theory*, **30** (1984), 699–704.

Shurkin, J., *Engines of the Mind*, W.W. Norton and Co., New York, 1984.
[A lively account of the development of computers.]

van der Waerden, B.L., *A History of Algebra*, Springer-Verlag, Berlin, 1985.
[From the ninth century onwards.]

Venn, J., On the diagrammatic and mechanical representation of propositions and reasonings, *Philos. Mag.*, July 1880.

Weil, A., *Number Theory (An approach through history. From Hammurapi to Legendre)*, Birkhäuser, Boston, MA, 1984.
[Traces the development of number-theoretic concepts.]

Wussing, H., *The Genesis of the Abstract Group Concept*, MIT Press, Cambridge, MA, 1984, translation by A. Shenitzer of *Die Genesis des abstrakten Gruppenbegriffes*, VEB Deutscher Verlag Wiss., Berlin, 1969.

Biography

The following biographical data have been culled mainly from Gillispie, C.C., *et al.,*
Dictionary of Scientific Biography, Charles Scribner's & Sons, New York, 1970, to which
you are referred for (much) more detail. A great deal of information on the history of
mathematics, including biographies and contemporary developments, may be found at
www-gap.dcs.st-and.ac.uk/history/index.html.

Abel, Niels Henrik: b. Finnöy Island near Stavanger, Norway, 1802; d. Frøland, Norway,
    1829. Main work on elliptic integrals and the unsolvability by radicals of the general
    quintic.

Alembert, Jean le Rond d': b. Paris, France, 1717; d. Paris, France, 1783. Main work in
    mechanics; an Encyclopédiste.

Argand, Jean Robert: b. Geneva, Switzerland, 1768; d. Paris France, 1822. One of those
    who found a geometric representation of complex numbers. Also work on the
    Fundamental Theorem of Algebra.

Babbage, Charles: b. Teignmouth, Devon, England, 1792; d. London, England, 1871.
    Extremely diverse interests. Designed and partially built mechanical 'computers'.

Bachet de Meziriac, Claude-Gaspar: b. Bourg-en-Bresse, France, 1581; d. Bourg-en-
    Bresse, France, 1638. Best known for his edition of Diophantus' *Arithmetica* and his
    book of mathematical recreations and problems, *Problèmes plaisants et délectables*
    qui se font par les nombres.

Bernoulli, Daniel: b. Groningen, Netherlands, 1700; d. Basel, Switzerland, 1782. Work
    in mathematics and physics as well as medicine.

Bernoulli, Johann (Jean): b. Basel, Switzerland, 1667; d. Basel, Switzerland, 1748. Work
    in mathematics, especially the calculus.

Boole, George: b. Lincoln, England, 1815; d. Cork, Ireland, 1864. Worked on logic,
    probability and differential equations.

Brahmagupta: b. 598; d. after 665. Indian mathematician and astronomer.

Bravais, Auguste: b. Annonay, France, 1811; d. Le Chesnay, France, 1863. Main work
    on crystallography. Also made contributions in botany, astronomy and surveying.

Cantor, Georg: b. St Petersburg, Russia, 1845; d. Halle, Germany, 1918. His development
    of set theory and infinite numbers began with work on convergence of trigonometric
    series.

Cardano, Girolamo: b. Pavia, Italy, 1501; d. Rome, Italy, 1576. Practitioner of medicine. Wrote on many topics including mathematics. Was imprisoned for some months for having cast the horoscope of Christ.

Cauchy, Augustin-Louis: b. Paris, France, 1789; d. Sceaux, near Paris, France, 1857. An oustanding mathematician of the first half of the nineteenth century. Main contributions in analysis.

Cayley, Arthur: b. Richmond, Surrey, England, 1821; d. Cambridge, England, 1895. Practised as a barrister for fourteen years, during which time he wrote about 300 mathematical papers. Main contributions in invariant theory.

De Morgan, Augustus: b. Madura, India, 1806; d. London, England, 1871. Contributions in analysis and logic.

Dedekind, Richard: b. Brunswick, Germany, 1831; d. Brunswick, Germany, 1916. Work in algebra, especially number theory, and analysis.

Descartes, René du Perron: b. La Haye, Touraine, France, 1596; d. Stockholm, Sweden, 1650. Fundamental work in mathematics, physics and especially philosophy.

Diophantus (of Alexandria, Egypt): fl. AD 250. Main work is his *Arithmetica*: a collection of problems representing the high point of Greek work in number theory.

Dirichlet, Gustav Peter Lejeune: b. Düren, Germany, 1805; d. Göttingen, Germany, 1859. Important work in number theory, analysis and mechanics.

Dodgson, Charles Lutwidge: b. Daresbury, Cheshire, England, 1832; d. Guildford, Surrey, England, 1898. Better known as Lewis Carroll, author of the 'Alice' books. Some contributions to mathematics and logic.

Dyck, Walther Franz Anton von: b. Munich, Germany, 1856; d. Munich, Germany, 1934. Noteworthy contributions in various parts of mathematics.

Eratosthenes: b. Cyrene, now in Libya, c. 276 BC; d. Alexandria, Egypt, c. 195 BC. One of the foremost scholars of the time. Best known for his work on geography and mathematics.

Euclid: fl. Alexandria, Egypt (and Athens?), c. 295 BC. Author of the *Elements*, one of the most influential books on Western thought.

Euler, Leonhard: b. Basel, Switzerland, 1707; d. St Petersburg, Russia, 1783. Enormously productive mathematician (wrote and published more than any other mathematician) who also made contributions to mechanics and astronomy.

Fermat, Pierre de; b. Beaumont-de-Lomagne, France, 1601; d. Castres, France, 1665. Fundamental work in number theory.

Ferrari, Ludovico: b. Bologna, Italy, 1522; d. Bologna, Italy, 1565. Pupil of Cardano; work in algebra.

del Ferro, Scipione: b. Bologna, Italy, 1465; d. Bologna, Italy, 1526. An algebraist, first to find solution of (a particular form of) the cubic equation.

Fibonacci, Leonardo (or Leonardo of Pisa): b. Pisa, Italy, 1170; d. Pisa, Italy after 1240. Author of a number of works on computation, measurement and geometry and number theory.

Fourier, Jean Baptiste Joseph: b. Auxerre, France, 1768; d. Paris, France, 1830. Best known for his work on the diffusion of heat and the mathematics that he introduced to deal with this. Accompanied Napoleon to Egypt, where he held various diplomatic posts.

Frénicle de Bessy, Bernard: b. Paris, France, 1605; d. Paris, France, 1675. Accomplished amateur mathematician. Corresponded with other mathematicians, especially on number theory.

Galois, Evariste: b. Bourg-la-Reine near Paris, France, 1811; d. Paris, France, 1832. Determined conditions for the solvability of equations by radicals; founder of group theory. A fervent republican, he died from a wound received in a possibly contrived duel: his funeral was the occasion of a republican demonstration in Paris.

Gauss, Carl Friedrich: b. Brunswick, Germany, 1777; d. Göttingen, Germany, 1855. One of the greatest mathematicians of all time, he made fundamental contributions to many parts of mathematics and the mathematical sciences.

Gibbs, Josiah Willard: b. New Haven, CT, USA, 1839; d. New Haven, CT, USA, 1903. Important work in thermodynamics and statistical mechanics.

Gödel, Kurt: b. Brünn, now Brno, Czech Republic, 1906; d. Princeton, NJ, USA, 1978. Outstanding mathematical logician of the twentieth century.

Goldbach, Christian: b. Königsberg, Prussia (now Kaliningrad), 1690; d. Moscow, Russia, 1764. Administrator of the Imperial Academy of Sciences in St Petersburg. Corresponded with many scientists and dabbled in mathematics.

Grassmann, Hermann Günther: b. Stettin (now Szczecin, Poland), 1809; d. Stettin, Germany, 1877. Work in geometry and algebra, as well as comparative linguistics and Sanskrit.

Gregory, Duncan Farquharson: b. Edinburgh, Scotland, 1813; d. Edinburgh, Scotland, 1844. Work on laws of algebra.

Hamilton, (Sir) William Rowan: b. Dublin, Ireland, 1805; d. Dunsink Observatory near Dublin, Ireland, 1865. An accomplished linguist by the age of nine, Hamilton made important contributions to mathematics, mechanics and optics.

Hamming, Richard Wesley: b. Chicago, IL, USA, 1915; d. Monterey, CA, USA, 1998. Best known for fundamental work on codes.

Hasse, Helmut: b. Kassel, Germany 1898; d. Ahrensburg, nr. Hamburg, Germany, 1979. Work in number theory.

Hensel, Kurt: b. Königsberg, Germany (now Kaliningrad), 1861; d. Marburg, Germany, 1941. Main work in number theory and related topics.

Hollerith, Herman: b. Buffalo, NY, USA, 1860; d. Washington DC, USA, 1929. His work on the USA census led him to the use of punched card machines for processing data. Founded a company which was later to develop into IBM.

I-Hsing: flourished in China in the early part of the eighth century.

Jordan, Camille: b. Lyons, France, 1838; d. Paris, France 1921. Published in most areas of mathematics: outstanding figure in group theory.

al-Khwarizmi, Abu Ja'far Muhammad ibn Musa: b. before 800; d. after 847. Author of influential treatises on algebra, astronomy and geography.

Klein, Christian Felix: b. Düsseldorf, Germany, 1849; d. Göttingen, Germany, 1925. Contributions in most areas of mathematics, especially geometry and function theory.

Kronecker, Leopold: b. Liegnitz, Germany (now Legnica, Poland), 1823; d. Berlin, Germany, 1891. Work in a number of areas of mathematics, especially elliptic functions.

Lagrange, Joseph Louis: b. Turin, Italy, 1736; d. Paris, France, 1813. Worked in analysis and mechanics as well as algebra.

Leibniz, Gottfried Wilhelm: b. Leipzig, Germany, 1646; d. Hannover, Germany, 1716. One of the inventors of the calculus. Many contributions to mathematics and philosophy.

Liouville, Joseph: b. St-Omer, Pas-de-Calais, France, 1839; d. Paris, France, 1882. Main work in analysis.

Mathieu, Emile Léonard: b. Metz, France, 1835; d. Nancy, France, 1890. Contributions to mathematics and mathematical physics.

Mersenne, Marin: b. Oizé, Maine, France, 1588; d. Paris, France, 1648. Contributions in acoustics and optics and other areas of natural philosophy. Actively aided the development of a European scientific community by his correspondence and drawing many visitors to his convent in Paris.

Newton, Isaac: b. Woolsthorpe, Lincolnshire, England, 1642; d. London, England, 1727. Often classed with Archimedes as the greatest of scientists, his contributions in mathematics were many and he was, with Leibniz, independent co-founder of the calculus.

Pascal, Blaise: b. Clermont-Ferrand, Puy-de-Dôme, France, 1623; d. Paris, France, 1662. Work in mathematics and physics as well as writings in other areas.

Peacock, George: b. Denton, near Darlington, county Durham, England, 1791; d. Ely, England, 1858. Work important in the development of the concept of abstract algebra.

Peirce, Benjamin: b. Salem, MA, USA, 1809; d. Cambridge, MA, USA, 1880. Leading American mathematician of his time.

Peirce, Charles Sanders: b. Cambridge, MA, USA, 1839; d. 1914. Son of Benjamin Peirce, who took great care over his son's mathematical education. His main work was in logic and philosophy.

Philolaus of Crotona (now in Italy): flourished in the second half of the fifth century BC. Proposed a heliocentric astronomical system.

Qín Jiǔsháo: b. Sichuan, China, c.1202; d. Guangdong, China, c.1261. Author of the *Mathematical Treatise in Nine Sections* which includes the 'Chinese Remainder Theorem' and variants of it. A civil servant, accomplished in many areas, notorious for his inclination to poison those he found disagreeable.

Ruffini, Paolo: b. Valentano, Italy, 1765; d. Modena, Italy, 1822. Practised medicine as well as being active in mathematics including work on algebraic equations and probability.

Serret, Joseph Alfred: b. Paris, France, 1819; d. Versailles, France, 1885. Work in various mathematical areas and author of a number of popular textbooks.

Steinitz, Ernst: b. Laurahütte, Silesia, Germany (now Huta Laura, Poland), 1871; d. Kiel, Germany, 1928. Main work on the general algebraic notion of a field.

Sylow, Peter Ludvig Mejdell: b. Christiania (now Oslo), Norway, 1832; d. Christiania, Norway, 1918. Established fundamental results on the structure of finite groups.

Tartaglia (real name Fontana), Niccolò: b. Brescia, Italy, 1499 or 1500; d. Venice, 1557. Contributions to mathematics, mechanics and military science.

Taylor, Brook: b. Edmonton, Middlesex, England 1685; d. London, England 1731. Made contributions to the theory of functions, including infinite series, and physics.

Turing, Alan Mathison: b. London, England, 1912; d. Wilmslow, Cheshire, England, 1954. Known best for 'Turing machines' and his code-breaking work.

Venn, John: b. Hull, Yorkshire, England, 1834; d. Cambridge, England, 1923. Work on probability and logic.

Viète, François: b. Fontenay-le-Comte, Poitou, France, 1540; d. Paris, France, 1603. Work in trigonometry, algebra and geometry. Important innovations in use of symbolism in mathematics.

Wallis, John: b. Ashford, Kent, England, 1616; d. Oxford, England, 1703. Work on algebra and functions.

Weber, Heinrich: b. Heidelberg, Germany, 1842; d. Strasbourg, Germany (now in France), 1913. Work in analysis, mathematical physics and especially algebra.

Zermelo, Ernst Friedrich Ferdinand: b. Berlin, Germany, 1871; d. Freiburg im Breisgau, Germany, 1953. Main work in set theory.

Name index

Subject index

Boldface indicates a page on which a term is defined.

333